Lecture Notes in Control and Information Sciences

Edited by M. Thoma

Vol. 43: Stochastic Differential Systems
Proceedings of the 2nd Bad Honnef Conference
of the SFB 72 of the DFG at the University of Bonn
June 28 – July 2, 1982
Edited by M. Kohlmann and N. Christopeit
XII, 377 pages. 1982.

Vol. 44: Analysis and Optimization of Systems
Proceedings of the Fifth International
Conference on Analysis and Optimization of Systems
Versailles. December 14–17, 1982
Edited by A. Bensoussan and J. L. Lions
XV, 987 pages, 1982

Vol. 45: M. Arató
Linear Stochastic Systems
with Constant Coefficients
A Statistical Approach
IX, 309 pages. 1982

Vol. 46: Time-Scale Modeling of Dynamic Networks
with Applications to Power Systems
Edited by J. H. Chow
X, 218 pages. 1982

Vol. 47: P. A. Ioannou, P. V. Kokotovic
Adaptive Systems with Reduced Models
V, 162 pages. 1983

Vol. 48: Yaakov Yavin
Feedback Strategies for Partially
Observable Stochastic Systems
VI, 233 pages, 1983

Vol. 49: Theory and Application of Random Fields
Proceedings of the IFIP-WG 7/1
Working Conference
held under the joint auspices of the
Indian Statistical Institute
Bangalore, India, January 1982
Edited by G. Kallianpur
VI. 290 pages. 1983

Vol. 50: M. Papageorgiou
Applications of Automatic Control Concepts
to Traffic Flow Modeling and Control
IX, 186 pages. 1983

Vol. 51: Z. Nahorski, H.F. Ravn, R.V.V. Vidal
Optimization of Discrete Time Systems
The Upper Boundary Approach
V, 137 pages 1983

Vol. 52: A. L. Dontchev
Perturbations, Approximations and Sensitivity Analysis
of Optimal Control Systems
IV, 158 pages. 1983

Vol. 53: Liu Chen Hui
General Decoupling Theory of Multivariable
Process Control Systems
XI, 474 pages. 1983

Vol. 54: Control Theory for Distributed
Parameter Systems and Applications
Edited by F. Kappel, K. Kunisch,
W. Schappacher
VII, 245 pages. 1983.

Vol. 55: Ganti Prasada Rao
Piecewise Constant Orthogonal Functions
and Their Application to Systems and Control
VII, 254 pages. 1983.

Vol. 56: Dines Chandra Saha, Ganti Prasada Rao
Identification of Continuous
Dynamical Systems
The Poisson Moment Functional
(PMF) Approach
IX, 158 pages. 1983.

Vol. 57: T. Söderström, P. G. Stoica
Instrumental Variable Methods
for System Identification
VII, 243 pages. 1983.

Vol. 58: Mathematical Theory of
Networks and Systems
Proceedings of the MTNS-83 International
Symposium
Beer Sheva, Israel, June 20–24, 1983
Edited by P. A. Fuhrmann
X, 906 pages. 1984

Vol. 59: System Modelling and Optimization
Proceedings of the 11th IFIP Conference
Copenhagen, Denmark, July 25-29, 1983
Edited by P. Thoft-Christensen
IX, 892 pages. 1984

Vol. 60: Modelling and Performance
Evaluation Methodology
Proceedings of the International Seminar
Paris, France, January 24–26, 1983
Edited by F. Bacelli and G. Fayolle
VII, 655 pages. 1984

Vol. 61: Filtering and Control of Random
Processes
Proceedings of the E.N.S.T.-C.N.E.T. Colloquium
Paris, France, February 23–24, 1983
Edited by H. Korezlioglu, G. Mazziotto, and
J. Szpirglas
V, 325 pages. 1984

For information about Vols. 1–42 please contact your bookseller or Springer-Verlag.

Lecture Notes in Control and Information Sciences

Edited by M. Thoma and A. Wyner

IIASA 103

P. Varaiya
A. B. Kurzhanski (Eds.)

Discrete Event Systems: Models and Applications

IIASA Conference
Sopron, Hungary, August 3-7, 1987

Springer-Verlag
Berlin Heidelberg GmbH

Editors
Pravin Varaiya
University of California
Berkeley, CA 94708
USA

Alexander B. Kurzhanski
System and Decision Sciences Program
IIASA
2361 Laxenburg
Austria

ISBN 978-3-540-18666-3 ISBN 978-3-540-48045-7 (eBook)
DOI 10.1007/978-3-540-48045-7

Originally published by Springer-Verlag Berlin Heidelberg New York in 1988

FOREWORD

A major objective of the Systems and Decision Sciences (SDS) Program of IIASA is to further the study of new mathematical approaches to the modeling, analysis and control of real systems. SDS sponsored the conference on Discrete Event Systems in the belief that these systems will demand increasing attention. These proceedings should therefore prove to be extremely timely.

Alexander B. Kurzhanksi
Chairman
System and Decision Sciences Program

PREFACE

The purpose of these remarks is to introduce discrete event systems and to provide a framework to the contributions to this volume.

WHAT IS A DISCRETE EVENT MODEL ?

Mathematical systems theory traditionally has been concerned with systems of continuous variables modeled by difference or differential equations, possibly including random elements. However, there is a growing need for dynamical models of systems whose states have logical or symbolic, rather than numerical, values which change with the occurrence of events which may also be described in non-numerical terms. We call such models **discrete event models** or **DEMs**.

The need for DEMs stems from the effort to extend the scope of mathematical systems theory to include manufacturing systems, communication networks and other systems whose behavior is naturally described by a record or trace of the occurrence of certain discrete, qualitative changes in the system and by ignoring continuously occurring micro-changes.

For example, the behavior of a data communications network may adequately be described by sentences such as "At time t_a user A sent a packet of data to user B who received it at time t_b." The building blocks of such sentences are phrases — *packet_sent, packet_received* — that indicate the occurrence of certain discrete events. These events mark various qualitative changes in network operations and the sentence ignores micro-changes such as the actual propagation of signals over the network.

Similarly, it is easy to imagine that the flow and processing of parts in a manufacturing plant with several machines may usefully be described by sentences built from phrases such as *start_processing_part, finish_processing_part, machine_failed, buffer_empty,* etc. while the sentence ignores continuous changes in the 'amount of metal cut' or in the 'fraction of part processing task completed'.

These examples suggest that as a first approximation we may think of DEMs as follows. The "real" system behavior involves changes in variables that occur continuously over time; DEMs abstract from this behavior by recording only the occurrence of certain discrete events. We will call such a record a *trace*. We can then say that a discrete event model is a mathematical model or procedure for describing the set of traces that a system can generate. In most systems the set of traces is infinite (because of different initial conditions, control policies or random effects), whereas a model must be finite. Thus a DEM is a *finite* mathematical model representing the *infinite* set of traces of a discrete event system.

TRACE ABSTRACTIONS AND DEM FORMALISMS

Traces can be mathematically represented at three levels of abstraction which we call the *logical, temporal*, and *stochastic* levels. At the most abstract or logical level, a trace is simply a sequence of events listed in their order of occurrence. For example, the trace

$$s = \sigma_1 \, \sigma_2 \, \sigma_3 \cdots$$

might signify the behavior

$$packet\_sent \; packet\_lost \; timeout\_expired \cdots$$

At the temporal level, a trace is a sequence of pairs

$$s = (\sigma_1, t_1) \, (\sigma_2, t_2) \, (\sigma_3, t_3) \cdots$$

where $\sigma_1 \, \sigma_2 \, \sigma_3 \cdots$ is a logical trace and t_i is the time at which event σ_i is executed. Finally, at the stochastic level, the system behavior is modeled by a sequence of pairs of random variables

$$(\sigma_1, t_1) \, (\sigma_2, t_2) \, (\sigma_3, t_3) \cdots$$

where the σ_i are event-valued and the t_i are real-valued random variables such that for each sample ω

$$s(\omega) = (\sigma_1(\omega), t_1(\omega)) \, (\sigma_2(\omega), t_2(\omega)) \, (\sigma_3(\omega), t_3(\omega)) \cdots$$

is a temporal trace.

Thus a DEM gives a finite mathematical description for an infinite set of traces representing system behavior at a particular level of abstraction.[1] A DEM *formalism* is a body of mathematical and computational methods for formulating and answering questions about a family of related discrete event models. There are several formalisms at the logical level including those based on *finite state machines, Petri nets, calculus of communicating systems,* and *communicating sequential processes.*

Despite considerable practical interest, there is relatively little progress in research on DEM formalisms at the temporal level. There is work on *timed Petri nets*, and on several related formalisms such as *data flow graphs* motivated by problems in synchronous digital signal processing.

On the other hand, DEM formalisms at the stochastic level have been built on the strong foundations of the theory of stochastic processes and statistical decision theory. Of particular importance here are models of *networks of queues* and the family of mathematical and computational methods invented for the analysis and optimization of these models.

COMPARISONS OF DEMS AND THEIR USES

As noted, formalisms can work at different levels of abstraction and at each level there are several competing formalisms. Are there ways of comparing different formalisms either intrinsically or in terms of their use?

By definition trace descriptions given by logical level models are less complex than temporal level model descriptions which, in turn, are less complex than stochastic descriptions. It is therefore tempting to say that stochastic models are intrinsically more powerful (i.e. they can describe more complex phenomena) than temporal models. However, in trying to represent more aspects of trace behavior, formalisms simplify the description of each individual aspect. Thus, the logical aspect of traces of temporal models has a simpler structure than that permitted by logical level models;

---

[1] We distinguish among DEMs by how abstract their trace descriptions are. This is a focus on modeling. One could propose more technical distinctions among DEMs such as finite vs infinite state, deterministic vs non-deterministic, etc.

similarly, the temporal aspect of traces of stochastic models has a simpler structure than that permitted by temporal models. For these reasons, one usually cannot compare formalisms at different levels in terms of their descriptive power.

Such comparisons may be possible between formalisms at the same level. Let us say that two models are ***equivalent*** if they describe the same set of traces. We can then borrow an idea from the 'formal theory of languages' and say that a DEM formalism A has more ***descriptive power*** than a formalism B if every B-model has an equivalent A-model. In this sense one can assert for example that the Petri net formalism has more descriptive power than the finite state machine formalism.

Descriptive power provides an intrinsic means of comparing formalisms. It may, however, be more interesting to base comparisons on how much useful work they can do. We briefly discuss one basis — ***algebraic complexity.***

When we try to analyze or to build a system, we tend to think of it as a collection of subsystems whose operations (events) are coordinated in several ways. Similarly, every DEM formalism is an 'algebra' in the sense that it contains several 'operators' which can be used to combine one or more models in the formalism to obtain a more complex model.[2] We can then say that a DEM formalism has greater algebraic complexity, relative to a particular application domain, if its algebraic operators correspond more naturally to the ways in which real systems are coordinated. The notion of algebraic complexity is not formal like the notion of descriptive power, but it has heuristic value since it points to an important aspect of modeling.

How are discrete event models used to formulate and address interesting questions about manufacturing systems, communication networks, etc? In the first place, by abstracting from the real system behavior in terms of traces, models can exhibit the structural similarity between very different systems (e.g. a manufacturing system and a digital signal processing system). Analysis of the model then permits prediction of certain properties of system behavior. For instance, a logical level model may reveal that the system is free of 'deadlock', or it may show the possibility of incorrect transmission of data. A temporal level model may be analyzed to determine the 'throughput' of a manufacturing system, i.e. the maximum rate of production. Finally, a stochastic model may show the distribution of the build-up of inventories of work-in-progress in a manufacturing system.

The preceding examples illustrate the use of DEMs for purposes of analysis. DEMs may also serve the need for synthesis. For example, logical models have been used to propose, synthesize, and verify control algorithms (protocols) that guarantee error-free transmission of data packets in communication networks; temporal models can suggest how to speed up computation in digital signal processing algorithms; stochastic models can be used to determine optimal buffer sizes in manufacturing systems.

CONTRIBUTIONS TO THIS VOLUME

Work in discrete event systems has just begun. There is a great deal of activity now, and much enthusiasm. There is considerable diversity reflecting differences in the intellectual formation of workers in the field and in the applications that guide their effort. This diversity is manifested in a proliferation of DEM formalisms. Some of the formalisms are essentially different. Some of the 'new' formalisms are

[2] For example, linear time-invariant systems may be modeled by transfer functions. The operations of sum and product of two transfer functions then correspond respectively to the parallel and cascade connection of two systems.

reinventions of existing formalisms presented in new terms. These 'duplications' reveal both the new domains of intended application as well as the difficulty in keeping up with work that is published in journals on computer science, communications, signal processing, automatic control, and mathematical systems theory — to name the main disciplines with active research programs in discrete event systems.

The first eight papers deal with models at the logical level, the next four are at the temporal level and the last six are at the stochastic level. Of these eighteen papers, three focus on manufacturing, four on communication networks, one on digital signal processing, the remaining ten papers address methodological issues ranging from simulation to computational complexity of some synthesis problems. The authors have made good efforts to make their contributions self-contained and to provide a representative bibliography. The volume should therefore be both accessible and useful to those who are just getting interested in discrete event systems.

ACKNOWLEDGEMENTS

These papers were presented at a conference on "Discrete Event Systems" held in Sopron, Hungary, from August 3 to August 7, 1987. The conference was sponsored by the Systems and Decision Sciences (SDS) Program of the International Institute for Applied Systems Analysis (IIASA), Laxenburg, Austria. It is a pleasure to acknowledge Professor Alexander Kurzhanksi, Program Leader of SDS, for his enthusiastic support of the conference. The local arrangements in Hungary were handled by the Hungarian NMO of IIASA; I am very grateful to Ms. Zsófia Zámori for providing everything needed for the conference. Ms. Erica Mayhew of SDS saw to it that everything ran smoothly starting with preparations months before the conference and ending with the publication of this volume. Without her sure competence, humor and discipline, this enterprise may have foundered many times. I owe her many thanks.

The conference attracted 30 participants from 11 countries. Twenty-one papers were presented over five days; 18 of these are included here. The participants were deliberately chosen to reflect diverse intellectual backgrounds and application interests. Indeed one organizational criterion was that no participant should have known more than one-third of the others. Another criterion was that each participant would be able to present work at a level that was sufficiently abstract so that one could readily detect both common features and distinguishing characteristics. Every participant made the extra effort needed to establish communication across disciplinary biases. I think the conference was much more successful than my original expectations. It brought together researchers in discrete event systems who would ordinarily not have become aware of each others' work. The resulting interaction will undoubtedly enrich their research.

Pravin Varaiya
Department of Electrical Engineering & Computer Science
University of California
Berkeley, CA 94720, USA

and

System and Decision Sciences Program
IIASA

CONTENTS

FINITELY RECURSIVE PROCESSES

Kemal Inan and Pravin Varaiya
Department of Electrical Engineering & Computer Sciences
University of California
Berkeley, CA 94720, USA

ABSTRACT

We present a new class of discrete event models called Finitely Recursive Processes (FRP). These models are proposed to help in the specification, implementation and simulation of communication protocols and supervisory control strategies. We believe that the 'algebra' of FRPs offers certain advantages over models based on state machines. The formal structure of FRPs builds on Hoare's Communicating Sequential Processes (CRP). The main differences with CRP are: (1) a FRP is specified via recursion equations which clearly bring out the dynamic evolution of a process, (2) some additional operators for combining FRPs are introduced, (3) the structure of the 'algebra' of FRPs is exploited to suggest methods for simulating them.

INTRODUCTION

The behavior of a discrete event system is described by the sequences of events or traces that it generates. Let A be the finite set of events and A^* the set of all finite sequences of events in A, including the empty trace $<>$. A^* represents the universe of all possible behaviors, whereas the behavior of a particular system is given by a subset $L \subset A^*$. The set L will usually be infinite. A discrete event model (DEM) is a finite mathematical description of this infinite set L.

To say that $s = a_1 a_2 \cdots$ is a trace of the system means only that in this behavior the event a_{i+1} occurs some time after event a_i, but the trace gives us no information about the real time at which the event occurs. We may say that the trace only reflects the *precedence* constraints of the system but not its real time constraints. Second, an event is *atomic*. In the 'real' system each atomic event may be subdivided into several operations; however, this finer granularity is not reflected in the model.

Our aim is to present a family of DEMs which we call Finitely Recursive Processes (FRP). For purposes of exposition it is convenient to contrast FRPs with Finite State Machines (FSM), in part because FSMs are more familiar. Formally, a FSM M is a 4-tuple

$$M = (Q, A, f, q_0)$$

where Q is the finite set of states, $q_0 \in Q$ is the initial state, A is the finite set of events, and f: $Q \times A \to Q$ is a partial function called the state transition function. f is partial means that $f(q, a)$ is only defined for a subset of pairs (q, a). f can be extended to a partial function on $Q \times A^*$ by induction:

$$f(q, <>) := q$$

$$f(q, s\hat{}a) := \begin{cases} \text{undefined if either } f(q, s) \text{ or } f(f(q, s), a) \text{ is undefined} \\ f(f(q, s), a), \text{ otherwise} \end{cases}$$

Here $r\hat{}t$ is the concatenation of two strings r and t in A^*. The traces of M is the set

$$trM := \{s \in A^* \mid f(q_0, s) \text{ is defined}\}.$$

It is known that trM is a regular subset of A^*. Conversely, if $L \subset A^*$ is closed[1] and regular, then there is an FSM M such that $trM = L$; see, e.g. Hopcroft and Ullman (1979). Thus, for example, the closure of the set

$$L := \{a^n b^n \mid n \geq 0\}$$

is not regular and so it cannot be described by any FSM.

In a state machine (whether finite or infinite) the notions of state and state transition are fundamental while the notion of trace is derivative. By contrast, in a process (defined below) the notion of trace is primary while state is a derived notion.

PROCESSES AND RECURSIONS

A **process** P is a triple $(trP, \alpha P, \tau P)$, where $trP \subset A^*$ is the set of **traces** of P, αP: $trP \to 2^A$ is the **event function**, and τP: $trP \to \{0,1\}$ is the **termination function**. trP is the set of traces that P can *execute*, $\alpha P(s)$ is the set of next events that P can *engage* in, i.e. execute or block, after it executes s, and $\tau P(s)$ specifies whether P terminates or continues after executing s. This interpretation explains why a process P must satisfy the following conditions:

(1) $<> \in trP$

(2) $s\hat{}t \in trP \Longrightarrow s \in trP$

(3) $s\hat{}a \in trP \Longrightarrow a \in \alpha P(s)$

(4) $\tau P(s) = 1 \Longrightarrow s\hat{}t \notin trP$ unless $t = <>$

Two trivial but useful processes are:

$$STOP_B := (\{<>\}, \alpha STOP_B(<>) = B, \tau STOP_B(<>) = 0) ,$$

$$SKIP_B := (\{<>\}, \alpha SKIP_B(<>) = B, \tau SKIP_B(<>) = 1) .$$

The only difference between them is that $STOP$ never terminates, while $SKIP$ terminates immediately.

A process $Q = (trQ, \alpha Q, \tau Q)$ is a **subprocess** of P, denoted $Q \subset P$, if $trQ \subset trP$, and if αQ, τQ agree with the functions αP, τP restricted to the domain trQ. Thus, for example, $STOP_B$ is a subprocess of every process P with $\tau P(<>) = 0$ and $\alpha P(<>) = B$. For each integer n let $P{\uparrow}n$ be the subprocess of P which contains all traces of P with length at most n.

A fundamental notion is that of the **post-process** P/s defined for $s \in trP$ by

$$tr(P/s) := \{t \mid s\hat{}t \in trP\} ,$$

$$\alpha(P/s)(t) := \alpha P(s\hat{}t) ,$$

$$\tau(P/s)(t) := \tau P(s\hat{}t) .$$

[1] L is closed means that if $r\hat{}t \in L$ then $r \in L$.

Thus P/s is the process that follows P after it has executed s. Since $(P/s)/a = P/(s\hat{}a)$, we may associate with P a (possibly infinite) state machine M_P,

$$M_P = (\{P/s \mid s \in trP\}, A, f, P/<> = P),$$

where the transition function is defined by

$$f(P/s, a) := \begin{cases} P/(s\hat{}a) \text{ if } s\hat{}a \in trP \\ \text{undefined if } s\hat{}a \notin trP \end{cases}$$

M_P is 'equivalent' to P in the sense that $trM_P = trP$.

Let Π be the set of all processes (with the same set A of events). A function f: $\Pi \to \Pi$ is **continuous** if for every increasing sequence of processes $P_1 \subset P_2 \subset \cdots$

$$f(\bigcup_i P_i) = \bigcup_i f(P_i) .$$

A function f: $\Pi^n \to \Pi$ is continuous if it is continuous in each argument with the others held fixed, and $f = (f_1, \cdots, f_m)$: $\Pi^n \to \Pi^m$ is continuous if each f_i is continuous.

In order to propose a recursive construction of a process whose traces unfold step by step in time, Hoare (1985) introduced the following notions of non-anticipative functions.

Definition

f: $\Pi \to \Pi$ is **constructive** or *con* if for $X \in \Pi$ and $n > 0$

$$f(X \uparrow n) \uparrow n = f(X) \uparrow n+1 ;$$

f is **non-destructive** or *ndes* if

$$f(X \uparrow n) \uparrow n = f(X) \uparrow n ;$$

f is **strictly non-destructive** or *sndes* if it is *ndes* but not *con.*

Thus f is *con* if the $(n+1)$st event executed by the process $f(X)$ is determined by the first n events executed by X. If f is a function of several arguments these definitions apply if they apply to each argument with the others held fixed.

Theorem 1

Consider the equation

$$X = f(X) \tag{1}$$

where f: $\Pi^n \to \Pi^n$. For any set of initial conditions

$$X_i \uparrow 0 = Z_{0i} , \quad i = 1, \cdots, n \tag{2}$$

that is consistent, i.e.

$$Z_{0i} = f_i(Z_{01}, \cdots, Z_{0n}) \uparrow 0 , \quad i = 1, \cdots, n$$

equation (1) has a unique solution $X \in \Pi^n$ that satisfies the initial conditions (2), provided f satisfies the following two conditions:

(C1) Each f_i is continuous and *ndes*;

(C2) f contains no *sndes* loop, i.e. there is no sequence of indexes $\{i_1, \cdots, i_m = i_1\}$ such that f_{i_k} is *sndes* in $X_{i_{k+1}}$.

These basic definitions and Theorem 1 (for the one-dimensional case) are from Hoare (1985). The multi-dimensional extension and other results given below are from Inan and Varaiya (1987).

Theorem 1 allows a recursive definition of processes in a way that is analogous to defining trajectories in R^n via the difference equation

$$x(t) = f(x(t-1)) = \cdots = f^{(t)}(x(0)) \ , \ t = 1, 2, \cdots$$

This analogy becomes clearer in the one-dimensional version of (1), $n = 1$. In this case (C2) implies that f is *con* and then the unique solution is given by

$$X = \bigcup_{t>0} f^{(t)}(Z_0)$$

where $f^{(t)} = f \circ f \circ \cdots \circ f$ (t times).

THE SPACE Ω^n

Our strategy is to use Theorem 1 to propose a class of DEMs. To do this we must provide a finite procedure for constructing a class of functions f that satisfy conditions (C1) and (C2). To be useful for the intended application this class of functions should have good 'algebraic' properties in the sense that it must be possible to combine functions in ways that reflect coordination of real systems. This class of functions Ω^n is constructed out of five operators that serve as building blocks.

Deterministic choice

Given $A_0 \subset A$, distinct elements $a_1, \cdots, a_n$ from A_0, and $\tau_0 \in \{0,1\}$, this operator maps $(P_1, \cdots, P_n) \in \Pi^n$ into $Q \in \Pi$ denoted by

$$Q = \left[a_1 \to P_1 \mid \cdots \mid a_n \to P_n\right]_{A_0, \tau_0} .$$

(By convention, if A_0, τ_0 is omitted, then $A_0 = \{a_1, \cdots, a_n\}$ and $\tau_0 = 0$.) Q is defined by:

If $\tau_0 = 1$, then $Q = SKIP_{A_0}$.

If $\tau_0 = 0$, then

$$trQ := \{<>\} \cup \bigcup_i \{a_i\hat{\ }s \mid s \in trP_i\} ,$$

$$\alpha Q(<>) := A_0 \ , \ \text{and } \alpha Q(a_i\hat{\ }s) := \alpha P_i(s) \ \text{ for } s \in trP_i \ ,$$

$$\tau Q(<>) := 0 \ , \ \text{and } \tau Q(a_i\hat{\ }s) := \tau P_i(s) \ \text{ for } s \in trP_i \ .$$

Thus if $\tau_0 \neq 1$ so that Q does not terminate immediately, then Q executes an event a_i and then follows the corresponding process P_i.

Example

Deterministic choice and recursion can be combined to obtain a process 'equivalent' to any finite state machine M. We associate with each state $i = 1, \cdots, n$ of M the process X_i and the equation

$$X_i = \left[a_{i_1} \to X_{i_1} \mid \cdots \mid a_{i_k} \to X_{i_k}\right] ,$$

which includes the 'choice' $a \to X_j$ if and only if in M the event a leads to a transition from state i to state j. By Theorem 1 these n equations uniquely define processes $X_1, \cdots, X_n$. It is easy to see that if the initial state of M is i, then $trM = trX_i$.

Synchronous composition

First define by induction the projection of $s \in A^*$ on a process P, denoted $s\uparrow P$: $<>\uparrow P := <>$, and

$$s\hat{\ }a\uparrow P := \begin{cases} \text{undefined if } s\uparrow P \notin trP \\ (s\uparrow P)\hat{\ }a \text{ if } a \in \alpha P(s\uparrow P) \\ (s\uparrow P) \text{ if } a \notin \alpha P(s\uparrow P) \end{cases}$$

$P\,||\,Q$ denotes the synchronous composition of P and Q. Its traces are defined inductively by: $<> \in tr(P\,||\,Q)$, and

If $s \in tr(P\,||\,Q)$, then $s\hat{\ }a \in tr(P\,||\,Q)$ if and only if
$s\hat{\ }a\uparrow P \in trP$, $s\hat{\ }a\uparrow Q \in trQ$, and $a \in \alpha P(s\uparrow P) \cup \alpha Q(s\uparrow Q)$.

Next,

$$\alpha(P\,||\,Q)(s) := \alpha P(s\uparrow P) \cup \alpha Q(s\uparrow Q) ,$$

and,

$$\tau(P\,||\,Q)(s) = 1 \iff \begin{cases} \tau P(s\uparrow P) = 1, \text{ and } \tau Q(s\uparrow Q) = 1 \text{ ; or} \\ \tau P(s\uparrow P) = 1, \text{ and } \alpha Q(s\uparrow Q) \subset \alpha P(s\uparrow P) \text{ ; or} \\ \tau Q(s\uparrow Q) = 1, \text{ and } \alpha P(s\uparrow P) \subset \alpha Q(s\uparrow Q) \end{cases}$$

$$\tau(P\,||\,Q)(s) = 0 \text{ , otherwise.}$$

In essence, $P\,||\,Q$ can execute event a if P and Q simultaneously execute a, or if one of these processes, say P, executes a and it is not blocked by the other (i.e. it is not in the event function of Q). Similarly $P\,||\,Q$ terminates after executing s if both P and Q terminate, or if one process, say P, terminates and subordinates the other, i.e. $\alpha Q(s) \subset \alpha P(s)$.

It is not difficult to see that the binary operator ' $||$ ' is associative and commutative.

Sequential composition

The sequential composition of P and Q is the process $P;Q$ which first follows P and, once P terminates, it follows Q. Formally,

$$tr(P;Q) := trP \cup \{s\hat{\ }trQ \mid s \in trP \text{ and } \tau P(s) = 1\}$$

$$\alpha(P;Q)(s) := \begin{cases} \alpha P(s) \text{ if } s \in trP \text{ and } \tau P(s) = 0 \\ \alpha Q(t) \text{ if } s = r\hat{\ }t,\ r \in trP \text{ and } \tau P(r) = 1 \end{cases}$$

$$\tau(P;Q)(s) := \begin{cases} 1 \text{ if } s = r\hat{\ }t,\ r \in trP,\ \tau P(r) = \tau Q(t) = 1 \\ 0 \text{ otherwise} \end{cases}$$

It is trivial that ' ; ' is associative.

Example

The two-dimensional recursion

$$Y = \left(a \rightarrow Y;X \mid d \rightarrow SKIP_{\{\}}\right)_{A,0}$$

$$X = \left(b \rightarrow SKIP_{\{\}}\right)_{A,0}$$

gives

$$trY = \{<>\} \cup \{a^n d b^n \mid n \geq 0\}$$

which is not regular and hence Y is not 'equivalent' to any finite state machine.

Example

Take $A = \{a, b, c\}$ and

$$X = \left[a \to X; (a \to SKIP_{\{\}}) \mid b \to X; (b \to SKIP_{\{\}}) \mid c \to SKIP_{\{\}}\right]$$

Then

$$trX = \{s\hat{\ }c\hat{\ }s^T \mid s \in \{a, b\}^*\},$$

where s^T is the same as s written in reverse order. This example is interesting because trX cannot be generated by a Petri net; see Theorem 6-9 in Peterson (1981).

Local and global change of event set

The previous four operators were introduced by Hoare. The following unary operators were introduced by Inan and Varaiya; they change the event function of a process and eliminate traces if necessary to conform to the definition of a process. Let B, C be subsets of A. The *local* change operator corresponding to B, C maps P into the process $P^{[-B+C]}$ where

$$trP^{[-B+C]} := \{s \in trP \mid b \in B \Longrightarrow b \text{ is not the first entry of } s\}$$

$$\alpha P^{[-B+C]}(<>) := \{\alpha P(<>) \setminus B\} \cup C \text{ , and } \alpha P^{[-B+C]}(s) := \alpha P(s), \text{ for } s \neq <>$$

$$\tau P^{[-B+C]}(s) := \tau P(s), \text{ for } s \in trP^{[-B+C]}$$

The *global* change operator corresponding to B, C maps P into $P^{[[-B+C]]}$ where

$$trP^{[[-B+C]]} := \{s \in trP \mid b \in B \Longrightarrow b \text{ is not an entry of } s\}$$

$$\alpha P^{[[-B+C]]}(s) := \{\alpha P(s) \setminus B\} \cup C$$

$$\tau P^{[[-B+C]]}(s) := \tau P(s), \text{ for } s \in trP^{[[-B+C]]}$$

Example

In combination with synchronous composition, the event change operators can be used to block the occurrence of certain events. For example, in the recursion

$$Y = \left[a \to X \,||\, Y\right]$$

$$X = \left[b \to SKIP_{\{\}}\right]$$

Y can execute event a arbitrarily often before X can execute event b and generate a trace of the form

$$a^{i_1}\, b\, a^{i_2}\, b\, a^{i_3}\, b \cdots$$

However, if the equation for Y is changed to

$$Y = \left[a \to X^{[+a]} \,||\, Y\right],$$

(where $X^{[+a]} := X^{[-B+C]}$ with $B = \emptyset$ and $C = \{a\}$), then $X^{[+a]}$ blocks repeated executions of a by Y and so the traces of Y are restricted to be of the form

$$a\, b\, a\, b\, a\, b \cdots$$

The introduction of these event change operators was partly motivated by the 'supervisory control' problem formulation of Ramadge and Wonham (1987). On this problem, also see Bochmann and Merlin (1978), Ramadge and Wonham (1986), Lin and Wohman (1986), Cieslak *et al* (1986) and Cho and Marcus (1987).

The space Ω^n

Let Ω^n be the smallest class of formulas f of n variables $x_1, \cdots, x_n$ that satisfy the rules (1)-(4):

(1) For every $B \in 2^A$, the 'constants' $STOP_B$ and $SKIP_B$ are formulas.

(2) For each i, x_i is a formula.

(3) If f is a formula, $f^{[-B+C]}$ and $f^{[[-B+C]]}$ are formulas for all B, C in 2^A.

(4) If f, g are formulas, $(f \,||\, g)$ and $(f;g)$ are formulas.

To each formula $f \in \Omega^n$ we associate a function (also denoted by f) which maps Π^n into Π: its value at the vector process $(X_1, \cdots, X_n) \in \Pi^n$ is obtained by 'substituting' these for the variables $(x_1, \cdots, x_n)$ and then evaluating each of the rules (1)-(4) used in constructing the formula f as the corresponding operator described earlier. Thus, for example, if the formula is $f = ((x_1;x_2) \,||\, x_2) \,||\, x_3$, and P, Q, R are processes, then $f(P, Q, R)$ is the process $((P;Q) \,||\, Q) \,||\, R$ which is the same as the process $(P;Q) \,||\, Q \,||\, R$ since ' $||$ ' is associative. It is not difficult to show that all these functions are *ndes*.

Two different formulas may yield the same function. For example, $x_1 \,||\, SKIP_A$ and $SKIP_A$ yield the same constant function $f \equiv SKIP_A$; similarly, the formulas $x_1 \,||\, x_1$ and x_1 give the function $f(X_1) \equiv X_1$.

FINITELY RECURSIVE PROCESSES

The next definition is fundamental to the construction of FRPs. $(X_1, \cdots, X_n) \in \Pi^n$ is **mutually recursive** if for every i and trace $s \in trX_i$, the post-process X_i/s has a representation

$$X_i/s = f(X_1, \cdots, X_n)$$

for some $f \in \Omega^n$.

Example

Reconsider the recursion

$$Y = \left[a \rightarrow Y;X \mid d \rightarrow SKIP_{\{\}}\right]_{A,0}$$

$$X = \left[b \rightarrow SKIP_{\{\}}\right]_{A,0}$$

for which trY is the closure of $\{a^n db^n \mid n \geq 0\}$, and $trX = \{<>, b\}$. Since

$$Y/a^n = Y;X;\cdots;X \ (n \text{ times}),$$

$$Y/a^n d = X;\cdots;X \ (n \text{ times}),$$

$$Y/a^n db^m = X;\cdots;X \ (n-m \text{ times}),$$

it follows that $(Y, X, SKIP_{\{\}})$ is mutually recursive.

The notion of mutually recursive processes brings us closer to our aim of finding an 'algebraically rich' class of processes having a finite representation. In the last example the state machine 'equivalent' to Y has infinitely many states (the post-processes of Y) because X, $X;X$, $X;X;X$, $\cdots$ are all different processes. Nevertheless, since each state is a process of the form $f(X, Y)$ for some $f \in \Omega^2$, one can say that we have a finite representation, especially in light of Theorems 2 and 3 which show that the representation can be calculated recursively as the process unfolds over time.

Theorem 2

$X = (X_1, \cdots, X_n)$ is mutually recursive if and only if X is the unique solution of the recursion equation

$$Y = f(Y), \quad Y_i \uparrow 0 = X_i \uparrow 0, \quad i = 1, \cdots, n\ ,$$

where each component f_i of f has the form

$$f_i(X) = \left(a_{i1} \rightarrow f_{i1}(X) \mid \cdots \mid a_{ik_i} \rightarrow f_{ik_i}(X)\right)_{A_i, \tau_i}\ ,$$

and each $f_{ij} \in \Omega^n$.

Definition

$Y \in \Pi$ is a **finitely recursive process** (FRP) if it can be represented as

$$X = f(X)$$

$$Y = g(X)$$

where f is in the form of Theorem 2 and $g \in \Omega^n$.

Theorem 3

Fix f as in Theorem 2. There is an effectively calculable partial function Φ_f: $A \times \Omega^n \rightarrow \Omega^n$ such that (a, g) is in the domain of Φ_f if and only if

$$Y := g(X) \Longrightarrow a \in tr\,Y\ .$$

Furthermore, if $g_a := \Phi_f(a, g)$, then we have the representation

$$Y/a = g_a(X)\ .$$

FRP simulator

Theorem 3 suggests how to simulate any FRP Y. The simulator is an interactive program that takes as initial data the pair f, g. The user enters a sequence of events $a_1\, a_2 \cdots$ one at a time. After each event is entered, the simulator either returns *reject* to indicate that Y cannot execute that event, or it returns *accept* and changes its internal state accordingly. More precisely, suppose the user has successfully entered trace s. The state of the simulator is then given by (the formula) g_s such that $g_s(X) = Y/s$. If the user now enters event a, the simulator will return

reject, if $\Phi_f(a, g_s)$ is undefined; or return

accept and replace g_s by $g_{s\hat{}a} := \Phi_f(a, g_s)$, if $\Phi_f(a, g_s)$ is defined.

USES OF FRP FORMALISM

From a formal viewpoint FRPs have strictly greater descriptive power than Petri nets. This follows from an example given above and a result in Inan and Varaiya (1987) showing how to construct an FRP with the same traces as a given Petri net.

However, the real test of a DEM formalism must come from its usefulness in modeling, simulation and performance evaluation of discrete event systems.[2] The preceding section outlined one approach to simulation of FRPs. In this section and the next we offer some observations on modeling and performance evaluation.

Modeling

The FRP operators can be used in a flexible way to model or design systems in a 'top down' manner following the precepts of structured programming. As a simple exercise let us model a job shop with two machines. There are K types of jobs, each job must be processed by both machines in either order. A machine can process only one job at a time, and once it starts work on a job it cannot be interrupted until that job is finished. The shop has a capacity of one job of each type, counting the jobs being processed. To create a FRP model of shop operations from this informal description we define the following events:

$a_k :=$ admission of new type k job

$b_i^k :=$ beginning of job k on machine i

$f_k :=$ finishing of job on machine i

Then the following $K+3$ recursion equations form a possible FRP model.

$$Y = \left[a_1 \to (X_1^{[+a_1]}; SKIP_{\{\}}) \,||\, Y \mid \cdots \mid a_K \to (X_K^{[+a_K]}; SKIP_{\{\}}) \,||\, Y\right]$$

$$X_k = \left[b_k^1 \to (S^1; b_k^2 \to S^2) \mid b_k^2 \to (S^2; b_k^1 \to S^1)\right], \; k = 1, \cdots, K$$

$$S^i = \left[f^i \to SKIP_{\{\}}\right]_{\{B^i, f^i\}, 0}, \; i = 1, 2$$

where $B^i := \{b_1^i, \cdots, b_K^i\}$.

To understand these equations think of Y as the 'master' process which never terminates. Its task is to admit new jobs (execute events a_k) while maintaining the capacity constraint. Thus, after Y executes a_k, we get the post-process

$$Y/a_k = (X_k^{[+a_k]}; SKIP_{\{\}}) \,||\, Y .$$

Y can no longer admit another job of type k until $X_k^{[+a_k]}$ terminates. However, Y can execute any event $a_j \neq a_k$ since a_j is not blocked by $X_k^{[+a_k]}$. If it does this we get the post-process

$$Y/a_k a_j = (X_k^{[+a_k]}; SKIP_{\{\}}) \,||\, (X_j^{[+a_j]}; SKIP_{\{\}}) \,||\, Y .$$

The process $X_k^{[+a_k]}$ guarantees the processing of job k by both machines in either order. Finally, once S^i starts, it executes f^i (i.e. machine i finishes its job) before another event b_k^i is executed, thereby ensuring uninterrupted processing of a job.

[2] Other aspects that should be considered are verification and testing.

One can give an 'equivalent' finite state machine description. However, the FRP description has several advantages. First, it is more compact. The state machine description can be given in a 'modular' fashion using $K+2$ sub-machines, one for each job type and each processing machine. The job sub-machine will have eight states, and each processor sub-machine will have $K+1$ states, giving a total of $8^K \times (K+1)^2$ states.

Second, each of the $K+3$ processes introduced in the FRP description can be interpreted as a 'task': Y admits jobs without exceeding capacity, X_k guarantees correct processing of type k job, etc. Since one often thinks of a system in terms of the tasks that must be performed, the FRP formalism offers a natural language for describing such systems. The deterministic choice, event set change, synchronous and sequential composition operators provide flexible ways of describing task coordination, including precedence constraints, mutual exclusion, etc. Moreover, this task decomposition orientation makes it quite easy to alter or to add new tasks. For example, suppose we want to admit the possibility that when machine 1 is processing a job, it may have a failure (denoted by the event e^1), and then processing on this machine should stop, processing on the other machine should continue, no more jobs should be admitted, and the master process Y should 'switch' to an emergency process Z not yet described. These changes can easily be accommodated by changing the description of Y and S^1 as follows.

$$Y = \left[a_1 \to (X_1^{[+a_1]}; SKIP_{\{\}}) \,||\, Y \mid \cdots \mid a_K \to (X_K^{[+a_K]}; SKIP_{\{\}}) \,||\, Y \mid e^1 \to Z\right]$$

$$S^1 = \left[f^1 \to SKIP_{\{\}} \mid e^1 \to SKIP_{\{\}}\right]_{\{B^1, f^1 \cdot e^1\},0}$$

Coordination through shared memory

Following Hoare (1985) one can use the synchronous composition operator to construct a 'communication channel' as a device for coordinating two FRPs. Here we introduce the notion of a 'shared memory' for coordination.

In addition to the event set A we assume given a finite set V and a mapping $\gamma: A \to F$, where F is the set of all partial functions from V into V. Let $\delta(a) \subset V$ denote the domain of the partial function $\gamma(a)$. V will serve as the possible values (assignments) of the shared memory. It is used in the following way. Suppose the current value of the memory is u, and suppose a FRP can execute an event a. Then

(1) The event a can be executed only if $u \in \delta(a)$; and

(2) After a is executed the value of the memory changes to $\gamma(a)(u)$.

Thus the memory serves only to *restrict* the possible traces of a process. In particular, if P is a process without memory, then the process with memory whose initial value is v is the subprocess $P_{\gamma,v}$ consisting of those traces $s \in trP$ for which $\gamma(s)(v)$ is defined.

Example

We illustrate the use of shared memory to describe a buffer of size K. New arrivals can occur only if there is space in the buffer; if the buffer is full, arrivals will be blocked. Similarly, a departure can occur only if the buffer is not empty. Take $A = \{a, d\}$ representing arrivals and departures. Let $V = \{0, \cdots, K\}$ and

$$\delta(a) = \{v \mid v < K\}\ ,\ \gamma(a)(v) = v+1\ ,$$

$$\delta(d) = \{v \mid v > 0\}\ ,\ \gamma(d)(v) = v-1\ .$$

Thus v represents number in the buffer. Now consider the recursion

$$X = Y \,||\, Z \,, \quad Y = \left[a \rightarrow Y\right], \quad Z = \left[d \rightarrow Z\right]. \tag{3}$$

Take initial value v. Then

$$trX = \{a, d\}^* \,,$$

but $a_1 \cdots a_m \in trX_{\gamma,v}$ if and only if

$$0 \leq v + \sum_1^j 1(a_i = a) - \sum_1^j 1(a_j = d) \leq K \,, \quad \text{for all } j \,,$$

as required.

Of course, we can do without the memory. For instance, take the recursion with K processes

$$P_0 = \left[a \rightarrow P_1\right], \; P_1 = \left[a \rightarrow P_2 \mid d \rightarrow P_0\right], \cdots, \; P_K = \left[d \rightarrow P_{K-1}\right]. \tag{4}$$

Then it is easy to see that $X_{\gamma,v} = P_v$. However, the description using shared memory is more compact and convenient.

In this example Y is the arrival process and Z is the server process. Suppose now we have two parallel servers Z_1 and Z_2 and the same buffer. Then in (3) we should replace the equation for Z by

$$Z = Z_1 \,||\, Z_2 \,,$$

where each Z_i is a server, i.e.

$$Z_i = \left[d_i \rightarrow Z_i\right], \; \delta(d_i) = \{v > 0\} \,, \; \gamma(d_i)(v) = v - 1 \,.$$

We can do a similar modification using (4) and dispense with the shared memory.

Now let us suppose we have a job shop with capacity K and two machines. Each job must be served by both machines in either order. (This is the same example as before, except that there is only one job type.) Then in (3) we should define Z by $Z = Z_{12} \,||\, Z_{21}$, and

$$Z_{12} = \left[d_1 \rightarrow S_1 ; b_2 \rightarrow S_2 ; Z_{12}\right]_{d_1,0}$$

$$S^1 = \left[f_1 \rightarrow SKIP_{\{\}}\right]_{\{d_1,b_1,f_1\},0}$$

$$Z_{21} = \left[d_2 \rightarrow S_2 ; b_1 \rightarrow S_1 ; Z_{21}\right]_{d_2,0}$$

$$S^2 = \left[f_2 \rightarrow SKIP_{\{\}}\right]_{\{d_2,b_2,f_2\},0}$$

Moreover, $\gamma(d_i)(v) = v - 1$, whereas $\gamma(f_i) = \gamma(b_i)$ are identity functions on V. The memory serves to enforce the following coordination constraint between Y and Z

$0 \leq v$ + No. of Y cycles (arrivals) - No. of Z cycles (departures) $\leq K$.

It is more difficult now to represent the process $X_{\gamma,v}$ without using shared memory.

The introduction of this shared memory is simply a modeling convenience, since it does not increase the descriptive power of FRPs as shown by the next result.

Theorem 4

If X is a FRP, so is $X_{\gamma,v}$.

INCORPORATING REAL TIME IN FRP

We briefly discuss one way of enriching the concept of an FRP X so that to each $s \in trX$ we can associate a number $tX(s)$ which represents the 'real time' needed by X to execute s.

Let us start with the simplest example,

$$X = \left[a \to X\right],$$

whose traces are a^n, $n \geq 0$. Suppose X models Hoare's candy machine which responds to a coin inserted in it by releasing a piece of candy. Suppose that if the coin is inserted at time t, the candy is released at time $t + t(a)$, so that we can say that $t(a)$ is the time needed by the machine to respond to the command a. This suggests the following *command-response* model: If coins are inserted at times

$$t_1, t_2, \cdots,$$

then the machine responds at times

$$t_1 + t(a),\ t_2 + t(a),\ \cdots . \tag{5}$$

But this is not quite right. Suppose $t_2 < t_1 + t(a)$, i.e. the second command is issued before the first one is executed. Assume, moreover, that the machine is insensitive or blinded to a new command issued while it is executing the previous command. Then we must impose the restriction

$$t_2 \geq t_1 + t(a),\ \cdots,\ t_{i+1} \geq t_i + t(a),\ \cdots$$

on the times at which commands can be issued so that (5) gives the correct machine response times. It seems reasonable to summarize this by defining $tX(s)$ to be the minimum time needed by the candy machine to execute s, i.e.

$$tX(a^n) = n\, t(a) .$$

Suppose now that the machine issues candy or gum in response to the command a or b respectively,

$$X = \left[a \to X \mid b \to X\right], \tag{6}$$

and the corresponding single event response times are $t(a)$, $t(b)$. If the sequence of commands is $a_1, a_2, \cdots$ issued at times $t_1, t_2, \cdots$, then the constraint on these times should be

$$t_2 \geq t_1 + t(a_1), \cdots, t_{i+1} \geq t_i + t(a_i), \cdots$$

and the corresponding response times are

$$t_1 + t(a_1),\ t_2 + t(a_2),\ \cdots ;$$

and the minimum time needed by the machine to execute a trace s is

$$tX(s) = n(a, s)\, t(a) + n(b, s)\, t(b) ,$$

where $n(a_i, s)$ is the number of times a_i occurs in s.

Suppose now that the machine has two independent mechanisms — one dispensing candy, the other dispensing gum — that can operate concurrently. This machine can be represented by the process Y,

$$Y = P \,||\, Q,\quad P = \left[a \to P\right],\quad Q = \left[b \to Q\right], \tag{7}$$

and suppose $t(a)$, $t(b)$ are as before. Let $s = a_1, a_2, \cdots$ be the sequence of commands issued at times $t_1, t_2, \cdots$. To obtain the constraint imposed on these times by the machines, let

$$s\uparrow P = a_{i_1}a_{i_2}\cdots, \quad s\uparrow Q = a_{j_1}a_{j_2}\cdots.$$

Then the constraint is

$$t_{i_2} \geq t_{i_1} + t(a)\,,\; t_{i_3} \geq t_{i_2} + t(a)\,,\; \cdots \quad t_{j_2} \geq t_{j_1} + t(b)\,,\; t_{j_3} \geq t_{j_2} + t(b)\,,\; \cdots$$

and when this constraint is met, the Y-machine response times are

$$t_1 + t(a),\; t_2 + t(a_2), \cdots$$

As before, let $tY(s)$ be the minimum time needed to respond to $s \in trY$. For example, if $t(a) = t(b) = 1$, then

$$tX(aaabbb) = 6,\; tY(aaabbb) = 5,\;\; tX(ababab) = 6,\; tY(ababab) = 3.$$

From this simple example we see that although $trX = trY$ (X given by (6) and Y given by (7)), the response times of the two machines, tX and tY, are quite different. Thus the response time of a process X is *not* reducible to its logical behavior trX. This raises the question:

Q1 Suppose we are given the 'elementary' execution times $t(a)$, $a \in A$; how do we model the response time of a process X?

From the example it seems that the response time tX should depend upon the actual *implementation* of X, i.e. the recursion equations used to implement X. Thus (6) and (7) are different implementations of the same process. In the example the implementation (7) has a greater degree of concurrency than (6), and it is sensible to say that (7) is a *faster* implementation than (6) since $tY(s) \geq tX(s)$ for all s. If we can answer Q1 satisfactorily then one can ask a follow-up question:

Q2 Consider all possible implementations of a process X. Does there always exist a fastest one?

In the remainder of this section we outline one possible approach to an answer to Q1.

Implementing FRP

The approach rests on an 'execution model' of FRPs. We assume that a user issues a sequence of commands, one at a time, to a machine that implements a FRP Y. Assume that the sequence is a trace of Y. Each command is first received by a *scheduler* which can either *accept* or *block* the command. If the command is accepted, the user may issue another command; if the command is blocked, another command may not be issued until it is unblocked (accepted). Thus the scheduler may buffer (block) at most one command. (This is simply the generalization of the constraints on the command times we had in the preceding example.)

To discuss the scheduler further we need to propose the execution model of an implementation. We assume given formulas f_{ij} and g in Ω^n. Interpreted as functions these are used to specify Y and the MRP X:

$$X_i = \left(a_{i1} \to f_{i1} \mid \cdots \mid a_{in_i} \to f_{in_i}\right)_{A_i, r_i}\,,\quad 1 \leq i \leq n\,, \tag{8}$$

$$Y = g(X_1, \cdots, X_n)\,. \tag{9}$$

Thus an implementation of a process is the set of formulas $\{f_{ij}, g\}$. Recall that the same process can have several implementations.

We assume that in all the formulas of an interpretation there is only one occurrence of each variable. Repeated occurrence of the same variable is accommodated simply by adding another subscript. Thus $X_{i1}, X_{i2}, \cdots, X_{ij}$ will all be considered as instances of the same variable X_i. For example,

$$X_1 = \left[a \to X_1 \mid b \to X_1\right], \quad Y = X_1 ; X_1$$

would be replaced by

$$X_1 = \left[a \to X_{11} \mid b \to X_{12}\right], \quad Y = X_{13} ; X_{14} .$$

Because $X_{ij} = X_i$, the process Y given by (9) is unchanged by this procedure.

Now suppose (8), (9) is an implementation of Y (with the understanding that each variable occurs only once in the f_{ij} and g). To simplify the discussion below, it is assumed that the event change operators do not occur in the implementation.

We will associate with g three items:

1. The set αg of *enabled* events of g;
2. The set βg of executable *next* events of g;
3. For each $a \in \beta g$, a subset $\pi g(a)$ of the set of processes that occur in the formula g; these are the processes that must execute a.

These items are constructed as follows by induction on the number of steps needed to construct g as a formula in Ω^n.

1. If g is $STOP_B$ or $SKIP_B$, then $\alpha g = B$, $\beta g = \emptyset$.
2. If $g(X) = X_{ij}$, then (see (8))
   $\alpha g = A_i$, $\beta g = \{a_{i1}, \cdots, a_{in_i}\}$, and $\pi g(a) = \{X_{ij}\}$, $a \in \beta g$.

Now suppose we know how to define αf, βf, πf for formulas $f = g$ and $f = h$.

3. Suppose $f(X) = g(X) \,||\, h(X)$. Then

   $$\alpha f = \alpha g \cup \alpha h ,$$

   and

   $$\begin{aligned} a \in \beta f \text{ if and only if either } \; & a \in \beta g \text{ and } a \notin \alpha h \quad (1) \\ \text{or } \; & a \notin \beta g \text{ and } a \in \alpha h \quad (2) \\ \text{or } \; & a \in \beta g \text{ and } a \in \beta h \quad (3) \end{aligned}$$

   To define πf in case (1) take $\pi f(a) = \pi g(a)$; in case (2) take $\pi f(a) = \pi h(a)$; in case (3) take $\pi f(a) = \pi g(a) \cup \pi h(a)$.

4. Suppose $f(X) = g(X) ; h(X)$. We assume the formula reduction

   $$SKIP_B ; f \equiv f ,$$

   so that we may assume $g \neq SKIP_B$. Then, take

   $$\alpha f = \alpha g, \quad \beta f = \beta g, \quad \pi f = \pi g .$$

Lemma 1

(1) $\alpha g = \alpha Y(<>)$, and

(2) $\beta g = \{a \mid <a> \in tr Y\}$; moreover,

(3) if $X_{ij} \in \pi g(a)$, then $a \in trX_i$.

Updating procedure

Now let $a \in \beta g$. Let g/a be the formula obtained from g by the following two-step procedure which we call $Proc(a)$.

Step 1. In g replace each variable $X_{ij} \in \pi g(a)$ by the variable x_i.

Step 2. By Lemma 1(3), $a = a_{ik_i}$ which appears on the right hand side of (8). Now replace each occurrence of x_i by the formula f_{ik_i}. After all occurrences of x_i are replaced, make sure that in the formula g/a each variable has a single occurrence by replacing repeated occurrences with new instances.

Lemma 2

$Proc(a)$ computes the correct formula in the sense that

$$Y/a = g/a(X).$$

Lemma 3

Suppose $a, b, c, \cdots$ are in βg and $\pi g(a), \pi g(b), \pi g(c), \cdots$ are all disjoint. Then applying $Proc(a)$, $Proc(b)$, $Proc(c), \cdots$ at most once to g leads to the same formula independent of the order in which these procedures are applied.[3] As a consequence, any sequence s with at most one occurrence of $a, b, c, \cdots$ is in $tr Y$; moreover, if s and r are two such sequences that consist of the same event (but in different order), then

$$Y/s = Y/r$$

We say that a subset $\{a, b, c, \cdots\} \subset \beta g$ is **non-interfering** (in state g) if $\pi g(a)$, $\pi g(b) \cdots$ are disjoint. Non-interference provides faster response because the implementation can execute these events concurrently.

The scheduler

We can now describe the operation of the scheduler. We start at time $t = 0$. The 'state' of the scheduler at this time is denoted $S(0)$,

$$S(0) = (g; \alpha g, \beta g, \gamma g, \delta g, \pi g), \quad \pi g\colon a \mapsto \pi g(a),\ a \in \beta g\ , \tag{10}$$

where formula g is the current *configuration* of the implementation; $\alpha g, \beta g, \pi g$ are as before; $\gamma g, \delta g$ are subsets of βg and at time $t = 0$, $\gamma g = \delta g = \emptyset$. We will see that γg is the set of commands being executed and δg (which contains at most one event) is the set of blocked commands.

Suppose the user issues command a_1 at time t_1. The scheduler responds to this request as follows:

1. If $a_1 \in \beta g$, the scheduler *accepts* the command, and asks every process in $\pi g(a_1)$ to execute event a_1 and assigns

   $$\gamma g \leftarrow \gamma g \cup \{a_1\}\,.$$

2. If $a_1 \in \alpha Y \backslash \beta Y$, then the scheduler *blocks* for ever and gives the response *STOP*.
3. If $a_1 \notin \alpha Y$, then the scheduler *rejects* the command.

More generally, suppose at time t the situation is as follows. The user has *issued* the sequence of commands $s\hat{\ }r$; the sequence s has been *executed* (by the implementation). Suppose $r = a_1 \cdots a_n$. There are two possible cases.

[3] Two formulas are considered to be the same modulo a 1-1 mapping of instances of the same variable. Thus, e.g. $X_{12} \,||\, X_{23}$ and $X_{11} \,||\, X_{22}$ are the same formulas.

Case 1

The scheduler has *accepted* all the commands in r, in which case its state is

$$S(t) = (h = g/s;\ \alpha h,\ \beta h,\ \gamma h = \{a_1, \cdots, a_n\},\ \delta h = \emptyset,\ \pi h)\,. \tag{11}$$

In this case the scheduler's state will change either because

(1) the processes executing the commands in γh are all finished at some future time t_+ before the user issues another command and then the state changes to

$$S(t_+) = (f = [g/s]/r;\ \alpha f,\ \beta f,\ \gamma f = \emptyset,\ \delta f = \emptyset,\ \pi f)\ ,$$

and the scheduler reports *executed r* to the user; or because

(2) the user issues another command, say a, at time t_+ before the commands in γh are all executed; the scheduler then tests whether a interferes with any command being executed, i.e. whether

$$\pi h(a) \cap \{\bigcup_{b \in \gamma h} \pi h(b)\} = \emptyset\ ?$$

If it is empty, the scheduler *accepts* the user's command a, and its state changes to (compare (11))

$$S(t_+) = (h = g/s;\ \alpha h,\ \beta h,\ \gamma h \leftarrow \gamma h \cup \{a\},\ \delta h = \emptyset,\ \pi h),$$

which has the same form as (11). If it is not empty, the scheduler *blocks* the user's command a, and its state changes to

$$S(t_+) = (h = g/s;\ \alpha h,\ \beta h,\ \gamma h,\ \delta h = \{a\},\ \pi h)$$

which has the same form as in *Case 2*.

Case 2

The scheduler has *accepted* commands $a_1, \cdots, a_{n-1}$ (the first $n-1$ commands in r) and it has *blocked* the last command a_n. In this case its state is (compare (11))

$$S(t) = (h = g/s;\ \alpha h,\ \beta h,\ \gamma h = \{a_1, \cdots, a_{n-1}\},\ \delta h = \{a_n\},\ \pi h). \tag{12}$$

Since the scheduler has blocked the user, it will not accept any more commands until all the commands in γh have been executed. Suppose this happens at time t_+. The scheduler's state then becomes

$$S(t_+) = (f = h/a_1 \cdots a_{n-1};\ \alpha f,\ \beta f,\ \gamma f = \{a_n\},\ \delta f = \emptyset,\ \pi f)\,,$$

and the scheduler returns a_n *accepted* to the user who may then issue another command.

Real time performance

With this detour we can define as follows the real time needed to execute a sequence s of commands by a particular *implementation*, i.e. a particular set of formulas $I = \{f_{ij},\ g\}$ that realizes a process Y (see (8), (9)). Let $s \in tr Y$,

$$s = a_1 \cdots a_n\,.$$

Find integers $k_1, k_2, \cdots$ such that

$$s = a_1 \cdots a_{k_1} a_{k_1+1} \cdots a_{k_2} \cdots a_{k_m} \cdots a_{k_{m+1}} (= a_n)$$

satisfies the following property:

For each $i = 0, \cdots, m$, the set of events $A(i) := \{a_{k_i+1}, \cdots, a_{k_{i+1}}\}$ is a non-interfering set for the state $g/a_1 \cdots a_{k_i}$; however, $a_{k_{i+1}+1}$ interferes with $A(i)$.

Thus each of the sets $A(0), \cdots, A(m)$ can be executed concurrently. Moreover, in the implementation I this is the maximum amount of concurrency. Hence we propose to define the time taken by the implementation I to execute s as

$$tI(s) := \sum_{i=0}^{m} \max \{t(a) \mid a \in A(i)\} .$$

Example

Within the framework provided by this definition the fastest implementation of a process with traces $\{a, b\}^*$ is given by

$$X = P \,||\, Q, \; P = \left[a \rightarrow P\right], \; Q = \left[b \rightarrow Q\right].$$

To see this we simply note that the largest possible non-interfering set is $\{a, b\}$ and the implementation always achieves this maximum. Hence it has maximum concurrency.

CONCLUSION

We have introduced a new class of discrete event models called finitely recursive processes (FRP). Most of the discussion concerned properties of FRP at the logical level. In our opinion, FRPs have certain advantages over state machines and Petri nets in terms of (1) compactness of descriptions, (2) descriptive power, and (3) the kinds of system coordination that can modeled by the operators of the FRP 'algebra'. We briefly discussed how the FRP formalism lends itself to simulation, and we indicated one approach to enriching this formalism to model real-time aspects of system behavior.

ACKNOWLEDGEMENT

The research reported here was supported by Pacific Bell, Bell Communication Research, the State of California MICRO Program, the National Science Foundation, and the International Institute for Applied Systems Analysis (IIASA).

REFERENCES

Cho, H. and Marcus, S.I. (1987). On supremal languages of classes of sublanguages that arise in supervisor synthesis problems with partial observations. Preprint. Department of Electrical and Computer Engineering, University of Texas, Austin, TX.

Cieslak, R., Desclaux, C., Fawaz, A. and Varaiya, P. (1986). Supervisory control of discrete event systems with partial observations. Preprint. Electronics Research Laboratory, University of California, Berkeley, CA.

Hoare, C.A.R. (1985). Communicating Sequential Processes. Prentice-Hall International, U.K. Ltd.

Hopcroft, J.E. and Ullman, J.D. (1979). Introduction to Automata theory, Languages, and Computation. Addison-Wesley.

Inan, K. and Varaiya, P. (1987). Finitely recursive process models for discrete event systems. Preprint. Electronics Research Laboratory, University of California, Berkeley, CA.

Lin, F. and Wonham, W.M. (1986). Decentralized supervisory control of discrete event systems. University of Toronto, Systems Control Group Report No. 8612.

Merlin, P. and Bochmann, G.v. (1983). On the construction of submodule specifications and communication protocols. ACM Trans. Prog. Lang. and Syst. Vol 5(1).

Peterson, J.L. (1981). Petri Net Theory and the Modelling of Systems. Prentice-Hall.

Ramadge, P.J. and Wonham, W.M. (1986). Modular supervisory control of discrete event systems. INRIA Conference. Paris.

Ramadge, P.J. and Wonham, W.M. (1987). On the supremal controllable sublanguage of a given language. SIAM J. Control and Optimization, Vol 25(3).

REDUCIBILITY IN ANALYSIS OF COORDINATION

R. P. Kurshan

AT&T Bell Laboratories
Murray Hill, New Jersey 07974

ABSTRACT

The use of automata to model non-terminating processes such as communication protocols and complex integrated hardware systems is conceptually attractive because it affords a well-understood mathematical model with an established literature. However, it has long been recognized that a serious limitation of the automaton model in this context is the size of the automaton state-space, which grows exponentially with the number of coordinating components in the protocol or system. Since most protocols or hardware systems of interest have many coordinating components, the pure automaton model has been all but dismissed from serious consideration in this context; the enormous size of the ensuing state-space has been thought to render its analysis intractable.

The purpose of this paper is to show that this is not necessarily so. It is shown that through exploitation of symmetries and modularity commonly designed into large coordinating systems, an apparently intractable state space may be tested for a regular-language property or "task" through examination of a smaller associated state space. The smaller state space is a "reduction" relative to the given task, with the property that the original system performs the given task if and only if the reduced system performs a reduced task.

Checking the task-performance of the reduced system amounts to testing whether the ω-regular language associated with the reduced system is contained in the language defining the reduced task. For a new class of automata defined here, such testing can be performed in time linear in the number of edges of the automaton defining each reduced language. (For Büchi automata, testing language containment is P-SPACE complete.) All ω-regular languages may be expressed by this new class of automata.

1. Introduction

Finite state automata which accept sequences (rather than strings) define the ω-regular languages. This class of automata is established as a model in logic, topology, game theory and computer science [Bu62, Ra69, Ra72, Ch74, Ku85a, SVW85, etc.]. In computer science, such automata are used to model non-terminating processes such as communication protocols and complex integrated hardware systems [CE81, MP81, AKS83, MW84, Ku85, etc.].

In the context of modelling non-terminating processes, one is given two automata, Λ and Γ. The automaton Λ models the process under study (the protocol or hardware system, for example), whereas the automaton Γ models the task which the process is intended to perform; although the state-space of Λ may be too large to permit it to be constructed explicitly, it may be defined implicitly in terms of a tensor product of components [AKS83, Ku85]. Likewise, the task-defining automaton Γ (defining a property for which Λ is to be tested) also may be defined implicitly, in terms of components. Determination of whether or not the process performs the specified task is ascertained by checking whether or not the language $\tau(\Lambda)$ defined by Λ is contained in the language $\tau(\Gamma)$ defined by Γ.

In this paper conditions are given under which the test: $\tau(\Lambda) \subset \tau(\Gamma)$, may be replaced by the test: $\tau(\Lambda') \subset \tau(\Gamma')$, for smaller automata Λ' and Γ', which smaller test may then be conducted in time linear in the number of edges of Λ' and of Γ'. The automata Λ' and Γ' are derived from Λ and Γ through *co-linear automaton homomorphisms*, maps which are graph homomorphisms, "preserve" the transition structure of the respective automaton, and agree on a Boolean algebra associated with the underlying alphabet. These homomorphisms may be constructed implicitly from (explicit) homomorphisms on components defining each of Λ and Γ, thus avoiding construction of the product spaces Λ and Γ themselves. Component homomorphisms may be generated through an $O(n \log n)$ algorithm for each n-state component, together with a user-designated Boolean-algebra homomorphism. While there is no guarantee that a particular intractably large problem can be thus rendered tractable, experience [HK86] has demonstrated the utility of this approach.

This program is conducted in the context of a newly defined class of automata, termed *L-automata* (where L is the Boolean algebra associated with the underlying alphabet). Three issues germane to these automata are complexity, expressiveness and syntax. Classically, given automata Λ and Γ, in order to test whether $\tau(\Lambda) \subset \tau(\Gamma)$, one first constructs an automaton $\bar{\Gamma}$ which defines the complementary language $\tau(\Gamma)'$, then one constructs an automaton $\Lambda * \bar{\Gamma}$ satisfying $\tau(\Lambda * \bar{\Gamma}) = \tau(\Lambda) \cap \tau(\bar{\Gamma})$ and finally one tests whether $\tau(\Lambda * \bar{\Gamma}) = \emptyset$. This entire procedure is at least as complicated as constructing $\bar{\Gamma}$, and since Λ may be taken to define all sequences (over the given alphabet), testing language containment is at least as hard as testing whether $\tau(\bar{\Gamma}) = \emptyset$, the so-called "emptiness of complement" problem. This problem is PSPACE-complete in the number of states for Büchi automata [SVW85]. For L-automata, as already mentioned, the language-containment test is linear in the size of Λ and in the size of Γ.

In the context of testing task-performance, it is often natural to take the task-defining automaton Γ to be deterministic. The reason for this is that properties of physical systems are often portrayed in deterministic terms, with conditional branches described causally. While it is logically possible to define a task requirement nondeterministically so that each nondeterministic branch corresponds implicitly to some behavior, it is more customary to condition each branch on some causal event. For example, if a task for a communication protocol has a conditional branch associated with whether or not a message is lost, it is customary to define the "event" *message-lost* and condition the branch upon its truth-value, thereby rendering the branch deterministic. Alternative acceptable behaviors are expressible in a *deterministic* automaton through alternative acceptance structures.

The process under study, on the other hand, represented by the automaton Λ, is often nondeterministic, represented in terms of incompletely defined information (*e.g.*, whether the channel loses or passes the message may be represented as a nondeterministic choice).

Given an ω-regular language τ, a nondeterministic L-automaton Λ may be found such that $\tau = \tau(\Lambda)$, while a finite number of deterministic L-automata $\Gamma_1, \ldots, \Gamma_n$ may be found such that $\tau = \bigcap_{i=1}^{n} \tau(\Gamma_i)$. In order to test $\tau(\Lambda) \subset \bigcap_i \tau(\Gamma_i)$, one tests $\tau(\Lambda) \subset \tau(\Gamma_i)$ for $i = 1, \ldots, n$. Each test $\tau(\Lambda) \subset \tau(\Gamma_i)$ may be completed in time linear in the number of edges of Λ and linear in the number of edges of Γ_i. The several individual tests $\tau(\Lambda) \subset \tau(\Gamma_i)$, $i = 1, \ldots, n$ defined by the task decomposition $\tau = \bigcap_i \tau(\Gamma_i)$, provide a greater potential for reduction than an undecomposed representation $\tau = \tau(\Gamma)$; each test may be separately reducible to a test $\tau(\Lambda') \subset \tau(\Gamma_i')$, with each Λ' different for different i.

Every homomorphism Φ can be decomposed as $\Phi = \Phi_1 \circ \Phi_2$ where Φ_1 is trivial on the underlying Boolean algebra and Φ_2 is trivial on the underlying graph. Hence, in order to explicitly construct homomorphisms, it is enough to construct separately Φ_1 and Φ_2. It turns out that for deterministic automata there exist unique maximally reducing Φ_1 and Φ_2. The maximally reducing Φ_1 corresponds to minimal reduction of string-accepting automata (in the Huffman-Moore sense), and Hopcroft's $O(n \log n)$ algorithm [Ho71] for state minimization (n states) serves to construct Φ_1. Construction of the optimal Φ_2 is less fortuitous, being equivalent to the NP-complete "set-basis" problem. However, as a practical matter, one will guess (suboptimal) candidates for Φ_2 through knowledge of the modularity and symmetry in a given process, and this guess is verified in time linear in the size of the graph and the underlying Boolean algebra.

In this report the central ideas are developed; a complete report, including proofs, is available from the author.

2. Preliminaries

Conventionally, an automaton is viewed as a set of states and a successor relation which takes a "current" state and "current" input and returns a "next" state (for deterministic automata) or a set of "next" states, in general. I prefer to view an automaton as a directed graph whose vertices are the automaton states, and each edge of which is labelled with the set of inputs which enables that state transition (*cf.* [Ts59]). The labelled graph is defined in terms of its adjacency matrix. It is convenient to give the set of all subsets of inputs the structure of a Boolean algebra. Then an automaton over a given alphabet of inputs may be described as a matrix over the Boolean algebra whose atoms are those inputs. (A definition of automaton initial states and acceptance conditions must also be given.)

2.1 Boolean Algebra

Let L be an atomic Boolean algebra, the set of whose atoms is denoted by $S(L)$, with meet (product) $*$, join (sum) $+$, complement $\sim$ (placed to the left of or over an element), multiplicative identity 1 and additive identity 0 [Ha74]. For the purposes of this paper, little is lost if one thinks of L as 2^S, the *power field* over S, which is the set of all subsets of a finite set S (the "alphabet") where $1=S$, $0=\varnothing$, $*$ is set intersection, $+$ is set union and $\sim$ is set complementation in S; the atoms of L in this case are the (singleton sets comprised of) the elements of S. A Boolean algebra admits of a partial order $\leq$ defined by $x \leq y$ iff $x * y = x$. If $x \leq y$ and $x \neq y$, write $x < y$. Atoms are minimal elements with respect to this order. A homomorphism of Boolean algebras is a map which is linear with respect to $*$, $+$ and $\sim$ (*i.e.*, $\phi(x*y)=\phi(x)*\phi(y)$, $\phi(x+y)=\phi(x)+\phi(y)$ and $\phi(\sim x) = \sim\phi(x)$). Any homomorphism is order-preserving ($x<y \Rightarrow \phi(x)<\phi(y)$). If Boolean algebras L and M are isomorphic, this will be denoted by writing $L \cong M$. Every Boolean algebra contains as a subalgebra the trivial Boolean algebra $\mathbb{B} = \{0, 1\}$. A sum or product indexed over the empty set is 0, 1 respectively.

2.2 L-Matrix; Graph

Let L be a Boolean algebra, let V be a nonempty set and let M be a map

$$M: V^2 \to L$$

(where $V^2 = V \times V$ is the cartesian product). Then M is said to be an *L-matrix* with *state space* $V(M) = V$. The elements of $V(M)$ are said to be *states* or *vertices* of M. An *edge* of an L-matrix M is an element $e \in V(M)^2$ for which $M(e) \neq 0$. ($M(e)$ is the "label" on the edge e.) The set of edges of M is denoted by $E(M)$. If $e=(v, w) \in E(M)$, let $e^- = v$ and $e^+ = w$. If M is an L-matrix and $L \subset L'$ then surely M is an L'-matrix as well.

If G is an L-matrix and $W \subset V(G)$, then $G|_W$, the *restriction* of G to W, is the L-matrix defined by $V(G|_W) = W$ and $G|_W(e) = G(e)$ for all $e \in W^2$.

A *graph* is a $\mathbb{B}$-matrix. The *graph* of the L-matrix M is the graph $\overline{M}$ with state space $V(\overline{M}) = V(M)$, defined by

$$\overline{M}(e) = \begin{cases} 1 & \text{if } M(e) \neq 0 \\ 0 & \text{otherwise.} \end{cases}$$

A *path* in a graph G of *length* n is an $(n+1)$-tuple $\mathbf{v} = (v_0, \ldots, v_n) \in V(G)^{n+1}$ such that $G(v_i, v_{i+1}) = 1$ for all $0 \leq i < n$; the path $\mathbf{v}$ is a *cycle* if $v_n = v_0$. The path $\mathbf{v}$ is said to be *from* v_0 *to* v_n. The path $\mathbf{v}$ *contains* the edge $(v, w) \in E(G)$ if for some i, $0 \leq i < n$, $v_i = v$ and $v_{i+1} = w$. If $C \subset V(G)$ and each $v_i \in C$, then $\mathbf{v}$ is *in* C. A cycle (v, v) of length 1 is called a *self-loop* (*at* v). A vertex $v \in V(G)$ is *reachable from* $I \subset V(G)$ if for some $v_0 \in I$, there is a path in G from v_0 to v. Any statement about a "path" in a L-matrix M is to be construed as a statement about $\overline{M}$.

Let G be a graph. A set $C \subset V(G)$ containing more than one element is said to be *strongly connected* provided for each pair of distinct elements $v, w \in C$ there is a path from v to w. A singleton set $\{v\} \subset V(G)$ is *strongly connected* if $(v, v) \in E(G)$. A maximal strongly connected set is called a strongly connected *component* (of G). Clearly, for every graph G, $V(G)$ is uniquely partitioned into strongly connected components and a non-strongly connected set, each vertex of which has no self-loop. (The requirement that a single vertex have a self-loop in order to be strongly connected, at some variance with the customary definition, is important to the theory developed here.)

Let G, H be graphs and let $\Phi: V(G) \to V(H)$ be a map which satisfies $(v, w) \in E(G) \Rightarrow (\Phi(v), \Phi(w)) \in E(H)$. Then Φ extends to a map $\Phi: E(G) \to E(H)$ and we say Φ is a *homomorphism* from G to H, and write

$$\Phi: G \to H.$$

We say Φ is 1-1 or onto according to the behavior of Φ on $V(G)$, and call Φ a *monomorphism* or *epimorphism* accordingly. If $\Phi: E(G) \to E(H)$ is onto we say Φ is *full*. The inclusion map of a restriction is a full monomorphism. A homomorphism which is 1-1, onto and full is said to be an *isomorphism*. The *image* of G under Φ is the graph $\Phi(G)$ with $V(\Phi G) = \Phi V(G)$ and $E(\Phi G) = \Phi E(G)$.

Let M and N be L-matrices. Their *direct sum* is the L-matrix $M \oplus N$ with $V(M \oplus N) = V(M) \cup V(N)$, defined by

$$(M \oplus N)(v, w) = \begin{cases} M(v, w) & \text{if } v, w \in V(M), \\ N(v, w) & \text{if } v, w \in V(N), \\ 0 & \text{otherwise;} \end{cases}$$

their *tensor product* is the L-matrix $M \otimes N$ with $V(M \otimes N) = V(M) \times V(N)$, defined by

$$(M \otimes N)((v, w), (v', w')) = M(v, v') * N(w, w').$$

The direct sum and tensor product can be extended to a commutative, associative sum and an associative product, respectively, of any finite number of L-matrices. If L is complete (*i.e.*, closed under infinite sums and products), the direct sum and tensor product can be extended to infinite sums and products as well.

Let G, H be graphs. The projection

$$\Pi_G: V(G \otimes H) \to V(G)$$

induces a (not necessarily onto) projection

$$\Pi_G: E(G \otimes H) \to E(G).$$

If G and H are matrices, Π_G will denote the projection on the underlying graph $\overline{G}$. Given G_1, $G_2, \ldots$, the projections Π_{G_i} may be written as Π_i, for convenience.

An L-matrix M is *lockup-free* if for all $v \in V(M)$ the sum $\sum_{w \in V(M)} M(v, w) = 1$. An L-matrix M is *deterministic* if for all $u, v, w \in V(M)$, $v \neq w \Rightarrow M(u, v) * M(u, w) = 0$.

Lemma 1: *The tensor product of deterministic L-matrices is deterministic. The tensor product of lockup-free L-matrices is lockup-free.*

2.3 L-Automata

An *L-automaton* is a 4-tuple

$$\Gamma = (M_\Gamma, I(\Gamma), R(\Gamma), Z(\Gamma))$$

where M_Γ, the *transition matrix* of Γ, is a lockup-free L-matrix, $\emptyset \neq I(\Gamma) \subset V(M_\Gamma)$, the *initial* states of Γ, $R(\Gamma) \subset E(M_\Gamma)$, the *recurring* edges of Γ and $Z(\Gamma) \subset 2^{V(M_\Gamma)}$, the *cycle* sets of Γ, is a (possibly empty) set of non-empty subsets of $V(M_\Gamma)$. Set $V(\Gamma) = V(M_\Gamma)$, $E(\Gamma) = E(M_\Gamma)$ and $\Gamma(v, w) = M_\Gamma(v, w)$ for all $v, w \in V(\Gamma)$. Let $R^-(\Gamma) = \{e^- \mid e \in R(\Gamma)\}$ and $R^+(\Gamma) = \{e^+ \mid e \in R(\Gamma)\}$. Define the *graph* of Γ, $\overline{\Gamma} = \overline{M}_\Gamma$.

Suppose Λ, Γ are L-automata satisfying $V(\Lambda) \subset V(\Gamma)$, $M_\Lambda = M_\Gamma|_{V(\Lambda)}$, $I(\Lambda) \subset I(\Gamma)$, $R(\Lambda) \subset R(\Gamma)$ and $Z(\Lambda) \subset Z(\Gamma)$. Then Λ is said to be a *subautomaton* of Γ, denoted $\Lambda \subset \Gamma$.

Given an L-automaton Γ, the *reachable* subautomaton of Γ is the L-automaton Γ^* defined as follows: $V(\Gamma^*)$ is the set of states reachable from $I(\Gamma)$, $M_{\Gamma^*} = M_\Gamma|_{V(\Gamma^*)}$, $I(\Gamma^*) = I(\Gamma)$, $R(\Gamma^*) = R(\Gamma) \cap E(\Gamma^*)$ and $Z(\Gamma^*) = \{C \cap V(\Gamma^*) \mid C \in Z(\Gamma)\}$.

An L-automaton Γ is *finite-state* if $V(\Gamma)$ is finite. (No general assumption is made on the finiteness of L, except that L is assumed atomic — guaranteed, if finite —, and in discussions of complexity, it is assumed that for any $x \in L$, the question of whether $x = 0$, can be settled in constant time.) Let $|\Gamma| = card V(\Gamma)$.

A sequence of states $\mathbf{v} = (v_0, v_1, \cdots) \in V(\Gamma)^\omega$ is *Γ-cyclic* if for some integer N and some $C \in Z(\Gamma)$, $v_i \in C$ for all $i > N$, while $\mathbf{v}$ is *Γ-recurring* if $\{i \mid (v_i, v_{i+1}) \in R(\Gamma)\}$ is unbounded.

A sequence $\mathbf{c} = (c_0, c_1, \cdots) \in (V(\Gamma) \times S(L))^\omega$ with elements $c_i = (c_i^{(1)}, c_i^{(2)})$ (where $c_i^{(1)} \in V(\Gamma)$ and $c_i^{(2)} \in S(L)$) is a *chain* of Γ if $(c_i^{(1)})$ is either Γ-cyclic or Γ-recurring and

$$(1) \quad c_0^{(1)} \in I(\Gamma),$$

$$(2) \quad c_i^{(2)} * \Gamma(c_i^{(1)}, c_{i+1}^{(1)}) \neq 0 \text{ for all } i \geq 0;$$

$\mathbf{c}$ is said to be *Γ-cyclic* or *Γ-recurring* according to the behavior of $(c_i^{(1)})$. The set of chains of Γ is denoted by $\mathscr{C}(\Gamma)$. A sequence $\mathbf{x} \in S(L)^\omega$ is an *L-tape* (or *tape* when L is understood). Given a chain $\mathbf{c} \in \mathscr{C}(\Gamma)$, the tape $\tau(\mathbf{c}) = (c_i^{(2)})$ is the tape *represented* by $\mathbf{c}$. (Sometimes it is said, conversely, that $(c_i^{(1)})$ is a "run" in Γ of the tape $\mathbf{x}$.) The set $\tau\mathscr{C}(\Gamma)$ of tapes represented by chains of Γ is denoted by $\tau(\Gamma)$ and called the set of tapes *accepted* by Γ, or the *language defined* by Γ. Clearly, $\mathscr{C}(\Gamma^*) = \mathscr{C}(\Gamma)$, $\tau(\Gamma^*) = \tau(\Gamma)$ and $\Lambda \subset \Gamma \Rightarrow \tau(\Lambda) \subset \tau(\Gamma)$. If Λ and Γ are L-automata satisfying $\tau(\Lambda) = \tau(\Gamma)$, then they are said to be *equivalent*.

The L-automaton Γ is said to be *deterministic* if M_{Γ^*} is deterministic; if Γ is deterministic and $card I(\Gamma) = 1$ then Γ is said to be *strongly* deterministic. (Customarily, "deterministic" has been used in the literature to mean what is here called "strongly deterministic"; however, this leads to unnecessary restriction, for example, in automata complementation and minimization, where strong determinism is not required.)

Lemma 2: *A tape accepted by a strongly deterministic L-automaton is represented by a unique chain.*

For any Boolean algebra L and any $n > 0$, a *string* (*in* L, of *length* n) is an n-tuple of atoms $\mathbf{x} = (x_0, \ldots, x_{n-1}) \in S(L)^n$. Given an L-automaton Γ, a string $\mathbf{x}$ in L of length n and a path (cycle) $\mathbf{v}$ in $\overline{M}_\Gamma$, also of length n, then $\mathbf{v}$ is said to be a path (cycle) *of* $\mathbf{x}$ provided for all $i = 0, \ldots, n-1$

$$x_i * \Gamma(v_i, v_{i+1}) \neq 0 .$$

For simplicity, we will say that $\mathbf{v}$ is a path of $\mathbf{x}$ "in Γ". Since the transition matrix of an L-automaton is lock-up free, for any $v \in V(\Gamma)$ and any string $\mathbf{x}$ in L, there is at least one path of $\mathbf{x}$ in Γ, from v. When Γ is deterministic, this path is unique. In this case, if the path is to $w \in V(\Gamma)$, we denote w as $v^{\mathbf{x}} \equiv w$. In this context, it is notationally convenient to admit among the strings in L the string $\mathbf{x}$ "of length 0" with the property that $v^{\mathbf{x}} = v$. (This simply provides a short-cut wherein any assertion about $v^{\mathbf{x}}$ for an arbitrary string $\mathbf{x}$ includes the case in which $v^{\mathbf{x}}$ is replaced by v. It has nothing to do with "silent" or "$\epsilon -$" transitions considered by some authors, and except for the convenience it affords, it could be dropped with no effect to the theory.)

If L is a Boolean algebra and $x \in L$, let x^n denote the set of strings of atoms in x of length n, let y^ω denote the set of sequences of atoms in y, let $x^n y^\omega$ denote the set of sequences formed by concatenation of each string of the first set by each sequence of the second, and let $x^+ = \bigcup_{n=1}^{\infty} x^n$. (There should be no confusion here, since there is no call for writing the n-fold product of x as an element of L, inasmuch as that product is just x.)

3. Expressiveness of L-automata

Let S be a nonempty set. An *ω-regular language τ over S* is an element of the Boolean algebra 2^{S^ω} of the form $\tau = \sum_{i=1}^{n} A_i B_i^\omega$ where A_i and B_i are regular sets over S [HU79] and juxtaposition denotes concatination.

An L-automaton Γ is said to be *pseudo-Büchi* (*pseudo-Muller*) provided $Z(\Gamma) = \emptyset$ (respectively, $R(\Gamma) = \emptyset$). If B is an n-state Büchi automaton [Ch74] (respectively, deterministic Büchi automaton {which by definition has a unique initial state}) which defines the language τ over S, then there exists a pseudo-Büchi (respectively, strongly deterministic pseudo-Büchi) 2^S-automaton Γ satisfying $\tau(\Gamma) = \tau$ and $|\Gamma| = n$; if Γ is a (strongly deterministic) pseudo-Büchi L-automaton, there exists a (deterministic {with unique initial state}) Büchi automaton B with n states which defines the language $\tau(\Gamma)$, satisfying $n \le 2|\Gamma|$ ([Ku85a; (2.3)]). Since every ω-regular language is defined by some (nondeterministic) Büchi automaton [Ch74; thm. 6.16], the following is immediate.

Theorem 1: *The ω-regular languages over a nonempty finite set S are exactly the respective sets of languages defined by the finite-state pseudo-Büchi 2^S-automata.*

Let $\Gamma_1, \Gamma_2, \ldots$ be L-automata. Define their *direct sum* to be the L-automaton $\oplus\Gamma_i$ defined by

$$\oplus\Gamma_i = (\oplus M_{\Gamma_i}, \cup I(\Gamma_i), \cup R(\Gamma_i), \cup Z(\Gamma_i)).$$

Proposition 1: *Let $\Gamma_1, \Gamma_2, \ldots$ be L-automata. Then*

1. $\tau(\oplus\Gamma_i) = \cup\,\tau(\Gamma_i)$;
2. $|\oplus\Gamma_i| = \sum |\Gamma_i|$;
3. $(\oplus\Gamma_i)^* = \oplus\Gamma_i^*$;
4. $\oplus\Gamma_i$ *is deterministic if each* Γ_i *is.*

Notice that the direct sum of two or more strongly deterministic L-automata fails to be strongly deterministic. This can be rectified, at the cost of a state space which grows exponentially with the number of summands. Define the *weak product* of L-automata Γ_1, $\Gamma_2, \ldots$ to be the L-automaton

$$\vee\Gamma_i = (\otimes M_{\Gamma_i}, \mathbf{X}\, I(\Gamma_i), \cup\Pi_i^{-1}R(\Gamma_i), \cup\Pi_i^{-1}Z(\Gamma_i)).$$

Proposition 2: *Let $\Gamma_1, \Gamma_2, \ldots$ be L-automata. Then*

1. $\tau(\vee\Gamma_i) = \cup\,\tau(\Gamma_i)$;
2. $|\vee\Gamma_i| = \Pi|\Gamma_i|$;
3. $(\vee\Gamma_i)^* \subset \vee\Gamma_i^*$;
4. $\vee\Gamma_i$ *is (strongly) deterministic if each* Γ_i *is.*

Corollary: *Let Γ be a deterministic L-automaton. Then there exists an equivalent strongly* *deterministic L-automaton Γ' satisfying*

$$|\Gamma'| = |\Gamma|^{card\, I(\Gamma)},$$

$$card\, R(\Gamma') \leq card\, R(\Gamma) \cdot card\, I(\Gamma),$$

$$card\, Z(\Gamma') \leq card\, Z(\Gamma) \cdot card\, I(\Gamma).$$

Things become more complicated with "product". There can be no general determinism-preserving definition of a "product" of L-automata with the property that the set of tapes of the "product" is the intersection of the respective sets of tapes of the factors. Nonetheless, one may define such a product in two special cases.

Suppose $\Gamma_1, \Gamma_2, \ldots$ are L-automata. Define their *tensor product* to be the pseudo-Muller L-automaton

$$\otimes \Gamma_i = (\otimes M_{\Gamma_i},\ \mathbf{X}\, I(\Gamma_i),\ \varnothing,\ \mathbf{X}\, Z(\Gamma_i)).$$

"Tensor product" properly applies only to pseudo-Muller automata; it is defined for general L-automata in order to provide a simple notational device to describe, for L-automata Λ and Γ, the "subgraph" of $M_\Lambda \otimes M_\Gamma$ reachable from $I(\Lambda) \times I(\Gamma)$, namely $(\Lambda \otimes \Gamma)^*$ (which has nothing to do with acceptance of tapes; see Theorem 4 below). Define the *projection* $\Pi_{\Gamma_i} : V(\otimes \Gamma_j) \to V(\Gamma_i)$ to be the canonical map; as usual, this extends to a map $\Pi_{\Gamma_i} : E(\otimes \Gamma_j) \to E(\Gamma_i)$.

Proposition 3: *Let $\Gamma_1, \Gamma_2, \ldots$ be pseudo-Muller L-automata. Then*

1. $\tau(\otimes \Gamma_i) = \cap \tau(\Gamma_i)$;
2. $|\otimes \Gamma_i| = \Pi |\Gamma_i|$;
3. $(\otimes \Gamma_i)^* \subset \otimes \Gamma_i^*$;
4. $\otimes \Gamma_i$ *is (strongly) deterministic if each* Γ_i *is.*

Let L' be a Boolean algebra containing independent subalgebras $L, M \subset L'$ with $L' = L \cdot M$. For any L'-matrix H, define the L-matrix $\Pi_L H$ by

$$V(\Pi_L H) = V(H),$$

$$(\Pi_L H)(e) = \Pi_L(H(e))$$

for all $e \in E(H)$.

Note: Clearly, $\Pi_L H$ deterministic $\Rightarrow$ H deterministic. (One may readily see that the converse of this fails.)

Lemma 1: *Suppose H and K are L'-matrices and $L' = L \cdot M$ for independent subalgebras L, M. Then*

1. *K lockup-free $\Rightarrow$ $\Pi_L K$ lockup-free;*
2. *$\Pi_L H$, K lockup-free $\Rightarrow$ $\Pi_L(H \otimes K)$ lockup-free;*
3. *$\Pi_L H$, K deterministic, $\Pi_M H(e) \in S(M)$ for all $e \in E(H)$ $\Rightarrow$ $\Pi_L(H \otimes K)$ deterministic.*

Proposition 4: *Let $\Gamma_1, \ldots, \Gamma_k$ be pseudo-Büchi L-automata. Then there exists a pseudo-Büchi L-automaton Γ satisfying*

1. $\tau(\Gamma) = \cap \tau(\Gamma_i)$;

2. $|\Gamma| \le k \prod_{i=1}^{k} |\Gamma_i|$;

3. *Γ is (strongly) deterministic if each Γ_i is.*

Theorem 2: *Let $\Gamma_1, \ldots, \Gamma_k$ be L-automata. Then there exists an L-automaton Γ satisfying*

1. $\tau(\Gamma) = \bigcap \tau(\Gamma_i)$;

2. $|\Gamma| \le (k+2)\left(\prod_{i=1}^{k} |\Gamma_i|\right)\left(\prod_{i=1}^{k} (card\, Z(\Gamma_i) + 1)\right)$.

Given an L-automaton Γ, define $\tau(\Gamma)' = \underline{S}(L)^{\omega} \setminus \tau(\Gamma)$. The problem of "complementing" Γ, that is, finding an L-automaton $\tilde{\Gamma}$ satisfying $\tau(\tilde{\Gamma}) = \tau(\Gamma)'$, is rather different than the same problem for automata accepting strings, for which one usually produces a complement by first determinizing the given automaton via the Rabin-Scott "subset" construction [RS59]. This approach does not work for L-automata, since it does not work for pseudo-Büchi L-automata [Ku85a; (3.12.2)]. Nonetheless, the following proposition, together with Theorem's 2 and 3, does give a decomposition construction for complement.

Say that an L-automaton Γ is *node-recurring* if $(v, w) \in R(\Gamma)$ and $(v', w) \in E(\Gamma)$ imply that $(v', w) \in R(\Gamma)$. From [Ku85a; (2.2), (2.3)] it easily follows that for any (deterministic) L-automaton Γ one may construct an equivalent (deterministic) node-recurring L-automaton Γ' satisfying $card\, I(\Gamma') = card\, I(\Gamma)$, $card\, R(\Gamma') = card\, R(\Gamma)$, $card\, Z(\Gamma') = card\, Z(\Gamma)$ and $|\Gamma'| \le |\Gamma| + card\, R^{+}(\Gamma)$. Obviously, in a node-recurring L-automaton Γ, a chain $\mathbf{c} \in \mathscr{C}(\Gamma)$ is Γ-recurring iff $\{i \mid c_i^{(1)} \in R^{+}(\Gamma)\}$ is unbounded.

Lemma 2: *If Γ is a strongly deterministic pseudo-Büchi (respectively, pseudo-Muller) L-automaton, then there exists a strongly deterministic pseudo-Muller (respectively, pseudo-Büchi) L-automaton $\tilde{\Gamma}$ such that*

1. $\tau(\tilde{\Gamma}) = \tau(\Gamma)'$;

2. $|\tilde{\Gamma}| \le |\Gamma| + card\, R^{+}(\Gamma)$ *(respectively,* $|\tilde{\Gamma}| \le |\Gamma| \max\{2, card\, Z(\Gamma)\}$*).*

Proposition 5: *Given a deterministic L-automaton Γ, there exist strongly deterministic L-automata Γ_B and Γ_M, pseudo-Büchi and pseudo-Muller respectively, such that for $r = card\, I(\Gamma)$,*

1. $\tau(\Gamma_B) \cap \tau(\Gamma_M) = \tau(\Gamma)'$;

2. $|\Gamma_M| \le |\Gamma|^r + r \cdot card\, R^{+}(\Gamma)$,
   $|\Gamma_B| \le |\Gamma|^r \max\{2, r \cdot card\, Z(\Gamma)\}$.

Let Γ be an L-automaton. For each chain $\mathbf{c} \in \mathscr{C}(\Gamma)$, define $\mu(\mathbf{c}) = \{v \in V(\Gamma) \mid card\{i \mid c_i^{(1)} = v\} = \aleph_0\}$, the *Muller* set of $\mathbf{c}$ (*cf.* [Ch74]). Obviously, if Γ is finite-state, $\mu(\mathbf{c}) \ne \varnothing$. For any tape $\mathbf{x}$, let $\mu_\Gamma(\mathbf{x}) = \{\mu(\mathbf{c}) \mid \mathbf{c} \in \mathscr{C}(\Gamma),\ \tau(\mathbf{c}) = \mathbf{x}\}$. Clearly, $\mu_\Gamma(\mathbf{x}) \ne \varnothing$ iff $\mathbf{x} \in \tau(\Gamma)$. If Γ is strongly deterministic then $\mu_\Gamma(\mathbf{x}) = \varnothing$ or $\mu_\Gamma(\mathbf{x}) = \{C\}$ for some $C \subset V(\Gamma)$. In this case, when Γ is finite-state, write $\mu_\Gamma(\mathbf{x}) = C$; then $\mu_\Gamma(\mathbf{x}) \ne \varnothing$ iff $\mathbf{x} \in \tau(\Gamma)$.

The *Muller* tapes accepted by an L-automaton Γ are the tapes $\tau_\mu(\Gamma) = \{\mathbf{x} \in \tau(\Gamma) \mid \mu_\Gamma(\mathbf{x}) \cap Z(\Gamma) \ne \varnothing\}$. For any ω-regular language τ over a nonempty finite set S, there exists a finite-state strongly deterministic 2^S-automaton Γ such that $\tau_\mu(\Gamma) = \tau$ [Ch74].

Theorem 3: *For every ω-regular language τ over a nonempty finite set S, there exist deterministic finite-state 2^S-automata $\Gamma_1, \ldots, \Gamma_n$ such that* $\bigcap_{i=1}^{n} \tau(\Gamma_i) = \tau$.

4. Testing Containment

For a graph G and a set $R \subset E(G)$, let G/R denote the graph with $V(G/R) = V(G)$,

$$(G/R)(e) = \begin{cases} G(e) & \text{if } e \in V(G)^2 \backslash R \\ 0 & \text{otherwise}. \end{cases}$$

(G/R is the result of removing the edges R from G.) It is convenient to extend the definition to an arbitrary set R, defining $G/R = G/(R \cap E(G))$.

Theorem 4: *Let Λ, Γ be finite-state L-automata and let $\mathcal{K}$ be the set of strongly connected components of the graph*

$$\overline{(\Lambda \otimes \Gamma)^*} \,/\, \Pi_\Gamma^{-1} R(\Gamma).$$

If for each $K \in \mathcal{K}$, either

1. *for some $C \in Z(\Gamma)$, $\Pi_\Gamma(K) \subset C$,*

or

2. *for each $e \in K^2$, $\Pi_\Lambda(e) \not\in R(\Lambda)$, and for each $C \in Z(\Lambda)$ and each strongly connected component k of*

$$\overline{(\Lambda \otimes \Gamma)} \,|_{\Pi_\Lambda^{-1}(C) \cap K}$$

there exists a $D \in Z(\Gamma)$ such that $\Pi_\Gamma(k) \subset D$,

then $\tau(\Lambda) \subset \tau(\Gamma)$. Furthermore, if Γ is strongly deterministic, then the converse holds as well.

5. Complexity

Theorem 4 may be translated into an algorithm for testing the containment $\tau(\Lambda) \subset \tau(\Gamma)$, based upon a procedure NEXTS() which takes a state v and generates all the "next states": NEXTS(v) = $\{w \in V(\Lambda \otimes \Gamma) \mid (v, w) \in E(\Lambda \otimes \Gamma)\}$. The procedure NEXTS() makes its computation as follows. If $v = (v_\Lambda, v_\Gamma)$, NEXTS($v$) computes NEXTS($v_\Lambda$), NEXTS($v_\Gamma$) through local manipulations in the respective definitions of Λ and Γ. For each $w = (w_\Lambda, w_\Gamma) \in$ NEXTS(v_Λ) $\times$ NEXTS(v_Γ), NEXTS(v) tests

$$(1) \qquad (\Lambda \otimes \Gamma)(v, w) \neq 0$$

and includes $w \in$ NEXTS(v) when (1) is *true*. There are a variety of means to test (1). One is to test

$$(2) \qquad \Lambda(v_\Lambda, w_\Lambda) * \Gamma(v_\Gamma, w_\Gamma) \neq 0$$

for each pair (w_Λ, w_Γ) for which $\Lambda(v_\Lambda, w_\Lambda) \neq 0$ and $\Gamma(v_\Gamma, w_\Gamma) \neq 0$. Testing (2) could be done symbolically, or exhaustively through expansion to a sum of atoms (disjunctive normal form). Another way to test (1) is to test, for each atom $t \in S(L)$,

$$t * \Lambda(v_\Lambda, w_\Lambda) \neq 0$$

for all $w_\Lambda \in V(\Lambda)$ such that $\Lambda(v_\Lambda, w_\Lambda) \neq 0$ and then for the same atom t, make the analogous test in Γ. Since t is an atom,

$$(3) \qquad t * \Lambda(v_\Lambda, w_\Lambda) \neq 0 \text{ and } t * \Gamma(v_\Gamma, w_\Gamma) \neq 0$$

implies (2) and hence (1). Conversely, if (3) fails for every atom t, then (1) fails. Hence, to compute NEXTS(v), one may find

(4) $$R_\Lambda(v_\Lambda, t) = \{w_\Lambda \in V(\Lambda) \mid t * \Lambda(v_\Lambda, w_\Lambda) \neq 0\}$$

and $R_\Gamma(v_\Gamma, t)$ defined analogously, for each t, and then set

(5) $$\mathrm{NEXTS}(v_\Lambda, v_\Gamma) = \bigcup_{t \in S(L)} (R_\Lambda(v_\Lambda, t) \times R_\Gamma(v_\Gamma, t)).$$

In the general worst general case, to test (2), one must parse each of $\Lambda(v_\Lambda, w_\Lambda)$ and $\Gamma(v_\Gamma, w_\Gamma)$ into a disjunctive normal form and compare the two forms. If the two forms contain n atoms and n atoms respectively, then testing (2), as a practical matter (*cf.* [Kn73; p. 391]), consists of nm comparisons. If there are r edges in Λ outgoing from v_Λ whose edge labels contain, respectively, $a_1, \ldots, a_r$ atoms, s edges in Γ from v_Γ whose edge labels contain, respectively, $b_1, \ldots, b_s$ atoms, and n atoms in S, then the total cost of computing NEXTS(v) via (2) can be as much as

$$\Sigma\, a_i b_j = (\Sigma\, a_i)(\Sigma\, b_j) \geq n^2$$

comparisons, with equality when both Λ and Γ are deterministic (Σa_i, $\Sigma b_i \geq n$ since an automaton by definition is lockup-free). In addition to that is the cost of transforming each edge label into normal form. On the other hand, in order to compute (5), one computes $R_\Lambda(v_\Lambda, t)$ and $R_\Gamma(v_\Gamma, t)$ and then forms the product. This appears a bit more costly, as to compute (4) for each atom t involves $\Sigma a_i \geq n$ comparisons, and this is repeated n times, once for each atom, resulting in at least $2n^2$ comparisons for (5). However, using an efficient internal computer representation for an edge label can reduce the cost of each test of the form (3) to that of one comparison. In this case, the cost of (5) reduces to $n(r+s)$ which could be substantially less than n^2.

This procedure is incorporated into a depth-first traversal of $\overline{(\Lambda \otimes \Gamma)^* / \Pi_\Gamma^{-1} R(\Gamma)}$, which is used in conjunction with Tarjan's algorithm [Ta72] to compute the set $\mathcal{H}$ of strongly connected components of this graph, by traversing each edge exactly once. In the course of this traversal, prior to its completion, strongly connected components may be found. As each component is found, first property 1. of Theorem 4 is tested, and failing that, the same algorithm is applied to find the strongly connected components of the graph defined in property 2.

In summary, the cost of checking $\tau(\Lambda) \subset \tau(\Gamma)$ using the algorithm outlined above is linear in each of the size of $E(\Lambda)$, $E(\Gamma)$, $Z(\Gamma)$ and $Z(\Lambda)$. It is assumed that the size of Z is constant. The next section shows how the basis of this cost may be reduced through exploitation of "regularities" and symmetries in the definitions of Λ and Γ.

6. Homomorphism

Let Γ be an L-automaton and let Γ' be an L'-automaton. A *homomorphism*

$$\Phi\colon\ \Gamma \to \Gamma'$$

is a pair of maps $\Phi = (\phi, \phi')$ where

$$\phi\colon\ \overline{\Gamma} \to \overline{\Gamma'}$$

is a graph homomorphism satisfying

$$(1) \qquad \phi I(\Gamma) = I(\Gamma'),$$

$$(2) \qquad \phi R(\Gamma) = R(\Gamma'),$$

$$(3) \qquad Z(\Gamma') = \{\phi(C) \mid C \in Z(\Gamma)\}$$

and

$$\phi' : L' \to L$$

is a Boolean algebra homomorphism such that for each $(v, w) \in V(\Gamma)^2$, ϕ and ϕ' jointly satisfy

$$(4) \qquad \Gamma(v, w) \leq \phi' \Gamma'(\phi(v), \phi(w)).$$

We say Φ is a *monomorphism* or an *epimorphism* or *full* according to the behavior of ϕ. The homomorphism Φ is *level* if ϕ' is an isomorphism and is *flat* if ϕ' is the identity map; if $\phi: V(\Gamma) \to V(\Gamma')$ is onto and

$$(2°) \qquad R(\Gamma) = \{e \in E(\Gamma) \mid \phi(e) \in R(\Gamma')\}$$

and

$$(3°) \qquad Z(\Gamma) = \{\phi^{-1}(C) \mid C \in Z(\Gamma')\}$$

then Φ is said to be *exact*. We say Φ is an *isomorphism* if ϕ is an isomorphism, Φ is level and exact, and (4) is an equality. If $\Phi: \Gamma \to \Gamma'$ is an isomorphism, write $\Gamma \cong \Gamma'$. If Φ is an isomorphism and $\Gamma' = \Gamma$, we say Φ is an *automorphism*. We may denote both ϕ and ϕ' by Φ.

Note that in view of (4), it was not necessary to require that ϕ be a homomorphism (it is a consequence of (4)).

Theorem 5: *If* $\Phi: \Gamma \to \Gamma'$ *is a homomorphism then* $\sum_{\substack{\Phi(r) = \Phi(v) \\ \Phi(s) = \Phi(w)}} \Gamma(r, s) \leq \Phi\Gamma'(\Phi(v), \Phi(w))$, *with equality holding when* Γ' *is deterministic.*

Lemma 1: *Let* $\Phi: \Gamma \to \Gamma'$ *be a homomorphism. If* Γ' *is deterministic then* Φ *is a full epimorphism.*

Let Γ be an L-automaton, let Γ' be an L'-automaton and let $\Phi: \Gamma \to \Gamma'$ be a homomorphism. An L' automaton $\Phi(\Gamma)$, the *image* of Γ under Φ, is defined in terms of $S(L')$ as follows:

$$V(\Phi(\Gamma)) = \Phi(V(\Gamma)),$$

$$(\Phi(\Gamma))(\Phi v, \Phi w) = \sum_{t \in Q(v, w)} t$$

where for $v, w \in V(\Gamma)$, $Q(v, w) = \{t \in S(L') \mid$ for some $r, s \in V(\Gamma)$, $\Phi(r) = \Phi(v)$, $\Phi(s) = \Phi(w)$ and $\Gamma(r, s) * \Phi(t) \neq 0\}$;

$$I(\Phi(\Gamma)) = \Phi I(\Gamma),$$

$$R(\Phi(\Gamma)) = \Phi R(\Gamma),$$

$$Z(\Phi(\Gamma)) = \Phi Z(\Gamma)$$

(which are, respectively, $I(\Gamma')$, $R(\Gamma')$ and $Z(\Gamma')$, by definition).

Lemma 2: $\Phi(\Gamma)$ *is an* L'*-automaton, and* $\Phi: \Gamma \to \Phi(\Gamma)$ *is onto and full.*

Note: Simple examples show that it is possible for Γ to be deterministic while $\Phi\Gamma$ is not deterministic, as well as for $\Phi\Gamma$ to be deterministic while Γ is not deterministic.

Corollary: *If* $\Phi: \Gamma \to \Gamma'$ *and* Γ' *is deterministic, then* $\Phi(\Gamma) = \Gamma'$.

Let $\phi': L' \to L$ *be a homomorphism of Boolean algebras and let* $\mathbf{x}'$ *be a tape in* L'. *Define* $\phi'(\mathbf{x}') \equiv \{\mathbf{x} \in S(L)^\omega \mid x_i * \phi'(x'_i) \neq 0\}$. *Thus, for any tape* $\mathbf{x}'$ *in* L', $\phi'(\mathbf{x}')$ *is a set of tapes in* L.

Theorem 6: *If* Γ *is an* L*-automaton,* Γ' *is an* L'*-automaton and* $\Phi: \Gamma \to \Gamma'$ *is a homomorphism, then* $\tau(\Gamma) \subset \cup \Phi\tau(\Gamma')$, *with equality holding when* Γ' *is deterministic and* Φ *is exact.*

Note: Containment in Theorem 6 may be proper even when $\Gamma' = \Phi(\Gamma)$ and Φ is exact (but Γ' is not deterministic).

Corollary: *If* Γ, Γ' *are* L*-automata,* Γ' *deterministic and if* $\Phi: \Gamma \to \Gamma'$ *is exact and flat, then* $\tau(\Gamma) = \tau(\Gamma')$.

Lemma 3: *If* $\Phi = (\phi, \phi')$ *and* $\Psi = (\psi, \psi')$ *are homomorphisms,* $\Phi: \Gamma \to \Gamma'$ *and* $\Psi: \Gamma' \to \Gamma''$, *then* $\Psi \circ \Phi \equiv (\psi \circ \phi, \phi' \circ \psi')$ *is a homomorphism* $\Gamma \to \Gamma''$; *if* Φ *and* Ψ *are exact then so is* $\Psi \circ \Phi$.

6.1 Lifting

Let $\Gamma_1, \Gamma_2, \ldots$ be a family of L-automata, let $\Gamma'_1, \Gamma'_2, \ldots$ be a family of L'-automata and for $i = 1, 2, \ldots$ let

$$\Phi_i: \Gamma_i \to \Gamma'_i$$

be a family of homomorphisms. Then $\Phi_1, \Phi_2, \ldots$ are said to be *co-linear* provided they agree on L' and either the family is finite, or they are complete on L'. In this case, for $\Gamma = \otimes\Gamma_i$ or $\vee\Gamma_i$, and $\Gamma' = \otimes\Gamma'_i$ or $\vee\Gamma'_i$ respectively, define the *lifting*

$$\Phi \equiv \otimes \Phi_i: \Gamma \to \Gamma'$$

to be the common homomorphism on $L' \to L$ and for all $\mathbf{v} \in V(\Gamma)$,

$$\Phi(\mathbf{v}) = (\Phi_i(v_i)).$$

Lemma 4: *A lifting* $\Phi = \otimes\Phi_i: \Gamma = \otimes\Gamma_i \to \Gamma' = \otimes\Gamma'_i$ *is an automata homomorphism.*

Lemma 5: *A lifting of exact homomorphisms is exact.*

Lemma 6: *Suppose* $L_1, L_2, \ldots, L_k$ *are independent subalgebras of a Boolean algebra* $L = \prod_i L_i$, $L'_1, L'_2, \ldots, L'_k$ *are independent subalgebras of a Boolean algebra* $L' = \prod_i L'_i$ *and* $\phi_i: L'_i \to L_i$ *is a homomorphism, for all* $i = 1, 2, \ldots, k$. *Then* $\phi = \prod_i \phi_i$ *defined by* $\phi(x) = \phi_1(x) * \cdots * \phi_k(x)$ *is a homomorphism,* $\phi: L' \to L$.

6.2 Reduction

Theorem 7: *Let* $\Phi: \Lambda \to \Lambda'$, $\Psi: \Gamma \to \Gamma'$ *be co-linear homomorphisms. If* $\tau(\Lambda) = \cup\Phi\tau(\Lambda')$ *then* $\tau(\Lambda) \subset \tau(\Gamma) \Rightarrow \tau(\Lambda') \subset \tau(\Gamma')$; *if* $\tau(\Gamma) = \cup\Psi\tau(\Gamma')$ *then* $\tau(\Lambda') \subset \tau(\Gamma') \Rightarrow \tau(\Lambda) \subset \tau(\Gamma)$.

Corollary: *Let* $\Phi: \Lambda \to \Lambda'$, $\Psi: \Gamma \to \Gamma'$ *be exact, co-linear homomorphisms and suppose* Λ' *and* Γ' *are deterministic. Then*

$$\tau(\Lambda) \subset \tau(\Gamma) \Leftrightarrow \tau(\Lambda') \subset \tau(\Gamma').$$

Note: If in Theorem 7 Φ and Ψ are flat, then the theorem follows trivially from the corollary to Theorem 6.

Let Λ, Λ', Γ and Γ' be as in Theorem 7, with $\tau(\Lambda) = \cup \Phi\tau(\Lambda')$ and $\tau(\Gamma) = \cup \Psi\tau(\Gamma')$. Then the pair (Λ', Γ') is called a *homomorphic co-reduction* of the pair (Λ, Γ). By Theorem 7, if (Λ', Γ') is a homomorphic co-reduction of (Λ, Γ), then $\tau(\Lambda) \subset \tau(\Gamma) \iff \tau(\Lambda') \subset \tau(\Gamma')$. The homomorphic co-reduction (Λ', Γ') is *exact* if Φ and Ψ are exact, is *deterministic* if Λ' and Γ' are both deterministic, and is *minimal state* if $card\, V(\Lambda' \otimes \Gamma')$ $(= card\, V(\Lambda') \cdot card\, V(\Gamma'))$ is minimal among all homomorphic co-reductions. (The benefit of a co-reduction, as measured by the saving associated with using Theorem 4, was shown in section 5 to be a function not only of the size of the reduced state space, but also the sizes of R, Z and, perhaps most importantly, $S(L')$. However, as discussed below, many facets of co-reduction, of which state space reduction is just one, remain unresolved.)

For a single L-automaton Γ, a *homomorphic reduction (exact homomorphic reduction)* of Γ is an equivalent L-automaton $\Gamma' = \Phi(\Gamma)$, for some flat (exact) epimorphism $\Phi: \Gamma \to \Gamma'$. (Recall, Γ' is equivalent to Γ if $\tau(\Gamma') = \tau(\Gamma)$.) A *homomorphic minimization (exact homomorphic minimization)* of Γ is a homomorphic reduction (exact homomorphic reduction) of Γ with fewest states.

Little is known in general about exact homomorphic (co-)reduction of a pair (Λ, Γ) or a single automaton Γ. In the class of deterministic L-automata, life is easier: the Corollary to Theorem 7 proves that *any* pair of exact co-linear homomorphisms Φ, Ψ give rise to a homomorphic co-reduction of (Λ, Γ); likewise, by the Corollary to Theorem 6, *any* exact homomorphism of Γ produces a homomorphic reduction of Γ. In Theorem 8 below it is shown that the problem of homomorphic co-reduction of a pair (Λ, Γ) can be reduced to the problem of homomorphic reduction of Λ and Γ separately. In section 6.3 an algorithm will be presented for finding the (unique up to isomorphism) exact homomorphic minimization of a deterministic L-automaton.

Theorem 8: *Let Γ be an L-automaton, Γ' be an L'-automaton and $\Phi: \Gamma \to \Gamma'$ a homomorphism. Then $\Phi = \Phi_1 \circ \Phi_2$ where $\Phi_2: \Gamma \to \Gamma''$ is flat and $\Phi_1: \Gamma'' \to \Gamma'$ is the identity on $V(\Gamma'')$.*

Corollary: *Let Λ, Γ be L-automata. Then (Λ', Γ') is a minimal-state deterministic homomorphic co-reduction of (Λ, Γ) iff for homomorphic minimizations Λ'' of Λ and Γ'' of Γ, $card\, V(\Lambda') = card\, V(\Lambda'')$ and $card\, V(\Gamma') = card\, V(\Gamma'')$.*

Let Λ, Γ be L-automata and $\phi': L' \to L$ a homomorphism of Boolean algebras. If (id_Λ, ϕ'), (id_Γ, ϕ') are homomorphisms on Λ and Γ respectively, then ϕ' is said to be *co-linear for* Λ and Γ.

6.3 Reducing

Let Λ and Γ be L-automata for which we wish to prove or disprove that $\tau(\Lambda) \subset \tau(\Gamma)$. The theme of this paper is to "reduce" that problem to a simpler equivalent problem of showing $\tau(\Lambda') \subset \tau(\Gamma')$ for some respectively smaller automata Λ' and Γ'. If the relationship between the pair (Λ, Γ) and its "co-reduction" (Λ', Γ') is undefined, then the problem of finding a "minimal state co-reduction" of (Λ, Γ) is meaningless. (Let $\tau(\Lambda') = S(L)^\omega$ (for which one state suffices) and let $\Gamma' = \Lambda'$ if $\tau(\Lambda) \subset \tau(\Gamma)$, while otherwise $\tau(\Gamma') = \emptyset$ (also realizable with one state).) One may meaningfully define a *co-reduction* of (Λ, Γ) to be a pair of L'-automata (Λ', Γ') derived from (Λ, Γ) through replacement of Λ and Γ by any other L-automata Λ'', Γ'' which accept the same respective sets of tapes and for which (Λ', Γ') is a homomorphic reduction of (Λ'', Γ'') having the property that the complexity, as described in section 5, of proving or disproving $\tau(\Lambda') \subset \tau(\Gamma')$ using Theorem 4, is less than the associated complexity for (Λ, Γ). With this definition one may seek a co-reduction of (Λ, Γ), minimal with respect to that complexity measure. Almost nothing is known about this problem in such generality. For example, it is only conjectured that such a minimal co-reduction (Λ', Γ') of (Λ, Γ) must satisfy

$$card\,V(\Lambda' \otimes \Gamma')^* \leq card\,V(\Lambda \otimes \Gamma)$$

as, conceivably, the size of R, Z, $S(L')$ or even $E(\Lambda' \otimes \Gamma')$ may be considerably reduced by slightly increasing the size of the state space. Furthermore, even if simply minimizing the size of the state space were found to be desirable, it is not known how to do this, short of an exhaustive search.

Things do not improve even if one restricts to the class of deterministic finite-state L-automata. Unlike deterministic finite-state automata which accept strings, for arbitrary ω-regular languages there is no analogue to the Myhill-Nerode right-invariant equivalence relation [RS59], and a minimal-state deterministic finite-state L-automaton which accepts a given ω-regular language need not be unique up to isomorphism [St83].

On the other hand, finding a minimal state exact homomorphic reduction in the class of deterministic L-processes, will be shown to be tractable.

It should be emphasized that, in view of the forgoing, there exist equivalent deterministic Γ, Γ' such that the homomorphic minimization of Γ has fewer states than the homomorphic minimization of Γ' (see the examples below). It is also true that there are reductions which are better than any homomorphic reduction.

In the case of general co-reductions, it was already observed that whether or not (Λ', Γ') is a co-reduction of (Λ, Γ) may depend upon the complexity measure associated with testing containment. In the case of homomorphic co-reduction this problem does not arise, as a homomorphic co-reduction reduces (or leaves unchanged) the size of each parameter which enters into the complexity measure (see section 5). Hence (assuming the complexity measure is monotone), every homomorphic co-reduction is a co-reduction.

The temptation to reduce the size of the state space of Γ through an equivalence relation on $V(\Gamma)$ which is analogous with the equivalence relation defined in association with the Huffman-Moore minimization algorithm for automata accepting strings [HU79], does not appear to work. Let Γ_1 be Γ except that $I(\Gamma_1) = V(\Gamma)$. An equivalence relation for an L-automaton Γ analogous to the Huffman-Moore equivalence relation is: $v \approx w$ if for every tape $\mathbf{x}$ in L, $\mathbf{x}$ is represented by a chain $\mathbf{c} \in \mathscr{C}(\Gamma_1)$ with $c_0^{(1)} = v$ iff $\mathbf{x}$ is represented by a chain $\mathbf{d} \in \mathscr{C}(\Gamma_1)$ with $d_0^{(1)} = w$. Clearly, $\approx$ is an equivalence relation on $V(\Gamma)$. Letting $[v]$ denote the equivalence class of $v \in V(\Gamma)$, we would be tempted to define the "minimization" of Γ to be the L-automaton $[\Gamma]$ defined by

$$V([\Gamma]) = \{[v] \mid v \in V(\Gamma)\},$$

$$[\Gamma]([v], [w]) = \sum_{\substack{r \approx v \\ s \approx w}} \Gamma(r, s),$$

$$I([\Gamma]) = \{[v] \mid [v] \cap I(\Gamma) \neq \emptyset\},$$

$$R([\Gamma]) = \{([v], [w]) \mid \text{for some } (v', w') \in R(\Gamma),\ v' \approx v \text{ and } w' \approx w\},$$

$$Z([\Gamma]) = \{\{[v] \mid [v] \cap C \neq \emptyset\} \mid C \in Z(\Gamma)\}.$$

(It should be clear that $[\Gamma]$ is indeed an L-automaton.)

Lemma 7: *The map* $\Phi: V(\Gamma) \to V([\Gamma])$ *defined by* $\Phi(v) = [v]$ *defines a flat epimorphism* $\Phi: \Gamma \to [\Gamma]$ *and thus* $\tau(\Gamma) \subset \tau([\Gamma])$.

The reason that $[\Gamma]$ may not be a homomorphic reduction of Γ is that Φ may not be exact. This may cause what appears to be an irreconcilable problem (that $\tau(\Gamma) \subsetneq \tau([\Gamma])$), as the next example shows.

Examples: 1. Let Γ be defined by

$$M_\Gamma = \begin{pmatrix} x & \bar{x} & 0 \\ 0 & x & \bar{x} \\ \bar{x} & 0 & x \end{pmatrix}$$

where the states of Γ are the indices of the rows of M_Γ, with $I(\Gamma)$ any nonempty subset of $V(\Gamma)$, $R(\Gamma) = \emptyset$ and $Z(\Gamma) = \{\{1\}, \{2\}, \{3\}\}$. It is easy to show that $\tau(\Gamma) = 1^* x^\omega$. There are several pairwise non-isomorphic minimal state automata defining the same ω-language (see [St83; ex. 2]) of which one is Λ with

$$M_\Lambda = \begin{pmatrix} x & \bar{x} \\ \bar{x} & x \end{pmatrix}$$

and $Z(\Lambda) = \{\{1\}, \{2\}\}$. While it is easily shown that there are flat homomorphisms $\Gamma \to \Lambda$, none can be exact. In fact, Γ is homomorphically minimal. It follows from this example that for any monotone complexity measure, there are minimal reductions less complex than any minimal homomorphic reduction. On the other hand, it is easily shown that the respective states of Γ and of Λ are equivalent, and hence $[\Gamma]$ and $[\Lambda]$ are both isomorphic to the one-state automaton which defines 1^ω, properly containing $\tau(\Gamma) = \tau(\Lambda)$.

2. There seems little hope, even for special cases, that $\tau(\Gamma) = \tau([\Gamma])$, as the next example shows. If Λ is as in the first example, except with $Z(\Lambda) = \emptyset$ and $R(\Lambda) = \{(1,2)\}$, then $\tau(\Lambda) = (x^*\bar{x}x^*\bar{x})^\omega$. Clearly, $[\Lambda]$ has a single state and $\tau([\Lambda]) = 1^\omega$. Things are much the same if the two examples are combined, with $R(\Lambda) = \{(1,2)\}$ and $Z(\Lambda) = \{\{1\},\{2\}\}$. Then $\tau(\Lambda) = (x^*\bar{x}x^*\bar{x})^\omega + 1^*x^\omega$ while $[\Lambda]$ again has one state and $\tau([\Lambda]) = 1^\omega$.

In the case of a finite-state deterministic L-automaton Γ for which $R(\Gamma) = \emptyset$, and $Z(\Gamma)$ has, "essentially", a single element (*i.e.*, $C, D \in Z(\Gamma) \Rightarrow$ any cycle in $C \cup D$ is in C or D) then $\tau([\Gamma]) = \tau(\Gamma)$. However, this is an uninteresting generalization of a homomorphic minimization: if each $C \in Z(\Gamma)$ is a union of strongly connected components (which one may as well assume, by discarding the complement of the union of the strongly connected components of C, an operation which leaves $\tau(\Gamma)$ unaltered), then $\Gamma \to [\Gamma]$ is exact, so $\tau(\Gamma) = \tau([\Gamma])$ by the corollary to Theorem 6.

Likewise, an analogous uninteresting generalization of homomorphic minimization is to add to $R(\Gamma)$ any set of edges contained in no cycle of Γ. Clearly, this has no effect upon $\tau(\Gamma)$.

While the Huffman-Moore minimization algorithm does not generalize directly to L-automata (as just shown), it can be adapted to homomorphic minimization of L-automata, which is unique up to isomorphism. Because of this, one may consider that minimization of deterministic automata which accept strings corresponds to *homomorphic minimization* of automata which define ω-regular languages, rather than the more general unqualified minimization. (Homomorphic minimization of string acceptors can be defined analogously to the definition given here, in which case for deterministic string acceptors, "minimization" and "homomorphic minimization" coincide.)

For a node-recurring deterministic L-automaton Γ with $v, w \in V(\Gamma)$, write $v \equiv w$ provided for every string $\mathbf{x}$ in L,

1. $v^{\mathbf{x}} \in R^+(\Gamma) \Longleftrightarrow w^{\mathbf{x}} \in R^+(\Gamma)$, and

2. for every $C \in Z(\Gamma)$, $v^{\mathbf{x}} \in C \Longleftrightarrow w^{\mathbf{x}} \in C$.

Clearly, $\equiv$ is an equivalence relation on $V(\Gamma)$. Let the equivalence class of v be denoted by $[v]$, and let $[\Gamma]$ be defined as before but relative to $\equiv$ rather than $\approx$ (from here on, there is no further mention of $\approx$, and "[ ]" refers only to $\equiv$).

Lemma 7°: *For any node-recurring deterministic L-automaton* Γ, *the map* $\Phi: V(\Gamma) \to V([\Gamma])$

defined by $\Phi(v) = [v]$ *defines a flat exact epimorphism* $\Phi: \Gamma \to [\Gamma]$.

Example 3. One could be tempted to try to weaken the above definition of $\equiv$ by allowing general (non-node-recurring) automata and replacing condition 1. with the condition

$$(v^{\mathbf{x}}, v^{(\mathbf{x},t)}) \in R(\Gamma) \text{ iff } (w^{\mathbf{x}}, w^{(\mathbf{x},t)}) \in R(\Gamma)$$

for all atoms $t \in S(L)$. However, with this definition Φ is not necessarily exact, as the following example shows. Let Γ be defined by

$$M_\Gamma = \begin{pmatrix} 0 & x & 0 & \tilde{x} \\ 0 & 0 & \tilde{x} & 0 \\ 0 & \tilde{x} & 0 & x \\ \tilde{x} & 0 & 0 & 0 \end{pmatrix}$$

where the states of Γ are the indices of the rows of M_Γ, with $I(\Gamma) = \{1\}$, $R(\Gamma) = \{(1,4), (3,2)\}$ and $Z(\Gamma) = \emptyset$. It is easily seen that $\tau(\Gamma) = ((1\tilde{x})^* \tilde{x}^2)^\omega$. However, with this suggested weaker definition of $\equiv$, $1 \equiv 3$ and $2 \equiv 4$, as may be easily checked. Thus,

$$M_{[\Gamma]} = \begin{pmatrix} 0 & 1 \\ \tilde{x} & 0 \end{pmatrix}$$

with $R([\Gamma]) = \{(1,2)\}$, so $\tau(\Gamma) \subsetneq \tau([\Gamma]) = (1\tilde{x})^\omega$ and by the corollary to Theorem 6, the flat epimorphism $\Phi: \Gamma \to [\Gamma]$ (weaker $\equiv$) cannot be exact.

Lemma 8: $[\Gamma]$ *is deterministic.*

Theorem 9: *Let* Γ *be a deterministic node-recurring L-automaton. Then* $\tau([\Gamma]) = \tau(\Gamma)$.

Lemma 9: *If* Γ *is node-recurring and* Φ *is exact then* $\Phi(\Gamma)$ *is node-recurring.*

Theorem 10: *Let* $\Psi: \Gamma \to \Gamma'$ *be a flat, exact epimorphism of deterministic L-automata,* Γ *node-recurring. Then*

a) $\Psi(v) = \Psi(w) \iff v \equiv w$, for all $v, w \in V(\Gamma^*)$;

b) $[\Gamma] \cong [\Gamma']$.

Corollary: *The exact homomorphic minimization of a deterministic node-recurring L-automaton is unique up to isomorphism.*

If $card\, V(\Gamma^*) = n$, an $O(n \log n)$ algorithm for computing the exact homomorphic minimization $[\Gamma]$ of a deterministic node-recurring L-automaton Γ is given by the Hopcroft algorithm [Ho71]. The Hopcroft algorithm applies by partitioning into separate blocks states $v, w \in V(\Gamma^*)$ which, for some string $\mathbf{x}$, give rise to respective values $v^{\mathbf{x}}$, $w^{\mathbf{x}}$ which violate either condition 1. or 2. above for $v \equiv w$.

Returning to Theorem 8, in the context of a homomorphic minimization, it remains to minimize the Boolean algebra which underlies the image of a homomorphism. Theorem 8 tells us that we may assume for this purpose that the graph of the image is fixed. Suppose that the L-automaton Γ is that image. The problem then can be posed thus: find a smallest subalgebra $L' \subset L$ such that Γ is an L'-automaton. Clearly, L' is the (unique) subalgebra generated by

$$\lambda \equiv \{\Gamma(e) \mid e \in E(\Gamma^*)\}.$$

Since L' is determined by its atoms $S(L')$ and each atom in $S(L')$ is a union of atoms of L, the problem of finding L' reduces to the problem of finding the coarsest partition of $S(L)$ such that every element of λ is a sum of partition elements. This problem is equivalent to the following problem, for $n = card\, S(L)$ and $m = card\, \lambda$. Given sets $\lambda_1, \ldots, \lambda_m \subset \{1, \ldots, n\}$, find a minimal partition $b_1 \cup \cdots \cup b_r = \{1, \ldots, n\}$ (b_i's pairwise disjoint) such that each λ_i is a union of b_j's. Unfortunately, this problem, known as the "set basis" problem, is NP-complete [St75]. A solution is given by the following proposition. Denote $\lambda_i^{-1} \equiv \{1, \ldots, n\} \backslash \lambda_i$, $\lambda_i^1 \equiv \lambda_i$ and for $\lambda \in (2^{\{1, \ldots, n\}})^m$, $\alpha \in \{-1, 1\}^m$, define $\lambda^\alpha \equiv \bigcap_{i=1}^{m} \lambda_i^{\alpha_i}$. Say that $\{b_1, \ldots, b_r\}$ (above) is a *basis* for $\lambda_1, \ldots, \lambda_m$.

Proposition 6: *Distinct sets* $\lambda_1, \ldots, \lambda_m \subset \{1, \ldots, n\}$ *admit of a unique basis, namely* $\{\lambda^\alpha \mid \alpha \in \{\pm 1\}^m\} \backslash \{\emptyset\}$.

Corollary: *Let* Γ *be an L-automaton. To find the smallest* L' *such that* Γ *is an* L'*-automaton is NP-complete in* $card\, E(\Gamma)$.

7. Analysis of Coordination

It is now described how the results of the previous sections may be applied to reduce the complexity of analysis of a system of coordination components. (For example, computer software and hardware often can be represented as such a system [AKS83, Ku85, KK86, GK87].) The system is defined in terms of its coordinating components $A_1, \ldots, A_k$ as a *product* $A = A_1 \otimes \cdots \otimes A_k$ (all of this is defined below). It is required to know whether the system A possesses a certain given behavioral trait. This is determined through formal analysis, formulated in terms of testing A for its "performance" of a *task* defined by an L-automaton Γ, where L is a Boolean algebra associated with A.

The components $A_1, \ldots, A_k$ as well as A are Moore-like [HU79] state machines called *L-processes*. An L-process A induces an L-automaton $A^\#$ which characterizes the behavior of A. Testing A for performance of the task defined by Γ amounts to testing whether or not $\tau(A^\#) \subset \tau(\Gamma)$.

The cardinality of the state space of A is the product of that of each component $A_1, \ldots, A_k$. Thus, simply the size of A may preclude testing language containment directly as above (it may be impossible to construct A explicitly). In this case, one seeks to find co-linear homomorphisms $\Phi: A \to B$ and $\Psi: \Gamma \to \Gamma'$, thereby reducing the complexity of testing task-performance to that of the test $\tau(B^\#) \subset \tau(\Gamma')$. The homomorphism Φ may be defined as a lifting (*cf.* §6.1) $\Phi = \otimes \Phi_i$, where $\Phi_i: A_i \to B_i$ and $B = \otimes B_i$. In this way, one avoids constructing A (or Φ) explicitly. Determination of whether or not $\tau(B^\#) \subset \tau(\Gamma')$ may be accomplished using the test of Theorem 4.

7.1 L-Process

Let L be a Boolean algebra. Although the theory holds in greater generality, assume now that L is finite. An *L-process* is a triple $A \equiv (M_A, S_A, I(A))$, where M_A is an L-matrix (the *transition matrix* of A, with the associated notation $V(A) \equiv V(M_A)$, $E(A) \equiv E(M_A)$ and for all $v, w \in V(A)$, $A(v, w) \equiv M_A(v, w)$), $I(A) \subset V(A)$ (the set of *initial states* of A) satisfies $I(A) = \emptyset$, and $S_A: V(A) \to 2^L$ (the *selector* of A) satisfies $S_A(v) \neq \emptyset$ for each $v \in V(A)$; for $S(A) \equiv \bigcup_{v \in V(A)} S_A(v)$, the set of *selections* of A, it is required that each selection $x \in S(A)$ is an atom in the Boolean algebra generated by $S(A)$; finally, it is required that for all $v, w \in V(A)$,

$$A(v, w) \leq \sum_{x \in S_A(v)} x .$$

If in fact equality holds for all v, w then A is said to be *lockup-free*.

The interpretation of L-process is similar to that of L-automaton. Given an L-process A, for each $v \in V(A)$ the selections $x \in S_A(v)$ are the "outputs" possible from v. If A is interpreted as a Moore-like state machine, then while A is "in" state v, the selection of A is a nondeterministically chosen element of $S_A(v)$; as long as A "stays" in v, this selection may repeatedly change to any other element of $S_A(v)$. For any $e \in E(A)$, the "edge label" $A(e)$ is an enabling predicate, expressed in terms of selections, for the transition along the edge e. The selection $x \in S_A(v)$ "enables" the transition (v, w) provided $x * A(v, w) \neq 0$. A is lockup-free if and only if every selection $x \in S_A(v)$ enables some transition (v, w). Nondeterminism in selection facilitates modelling incompletely specified actions (*e.g.*, from one state a process may select to "send message m_i", for $i = 1, 2, \ldots$, the procedure for such selection remaining unspecified). In keeping with the classical theory of finite state machines, process *state* models unobservable (private) memory, and a processes "behavior" is defined in terms of its *selections* (output) alone.

Given L-processes $A_1, \ldots, A_k$ define their (*tensor*) *product* to be

$$\bigotimes_{i=1}^{k} A_i = \left(\bigotimes_{i=1}^{k} M_{A_i}, \ \prod_{i=1}^{k} S_{A_i}, \ \mathop{\mathsf{X}}_{i=1}^{k} I(A_i) \right)$$

where $\left(\prod_i S_{A_i} \right) (v_1, \ldots, v_k) = \{ x_1 * \cdots * x_n \mid x_i \in S_{A_i}(v_i), \ i = 1, \ldots, k \}$.

Lemma 1: *If $A_1, \ldots, A_k$ are L-processes, then their product $\otimes A_i$ is an L-process.*

The L-processes $A_1, \ldots, A_k$ are said to be *independent* provided $x_i \in S(A_i)$, $i = 1, \ldots, k \Rightarrow x_1 * \cdots * x_k \neq 0$.

Proposition 7: *If $A_1, \ldots, A_k$ are independent, lockup-free L-processes, then their product $\otimes A_i$ is lockup-free.*

The discrete-event behavior of a system modelled by an L-process $A = A_1 \otimes \cdots \otimes A_k$ is interpreted in terms of the coordination among its independent, lockup-free components $A_1, \ldots, A_k$ as follows. At each time, in each process A_i, a selection x_i possible at the "current" state v_i (*i.e.*, $x_i \in S_{A_i}(v_i)$) is chosen non-deterministically. The product $x = x_1 * \cdots * x_k$ defines a "current global selection", *i.e.*, a selection of the product A at the state $(v_1, \ldots, v_k)$. At each time, in each process, the current global selection determines a set of possible "next" states, namely those states to which the transition from the current state is enabled by the current global selection. In A_i the transition from state v_i to state w_i is enabled by x iff $x * A_i(v_i, w_i) \neq 0$. Each process *resolves* the current global selection by choosing (non-deterministically) one of these possible next states.

A system progresses in time by repeatedly "selecting" and "resolving". This interpretation correctly describes the behavior of the product since by the atomicity assumption on each $S(A_i)$ it follows that for $v = (v_1, \ldots, v_k)$ and $w = (w_1, \ldots, w_k)$,

$$x * A_i(v_i, w_i) \neq 0 \ \text{ for } \ 1 \leq i \leq k \Leftrightarrow x * A(v, w) \neq 0 .$$

This same interpretation may be used to model systems of processes coordinating asynchronously in continuous time [Ku86, GKR86].

7.2 Task Performance

Note that if $A_1, \ldots, A_k$ are independent L-processes and $k > 1$ then the selections of each A_i are *not* atoms of L, *i.e.*, $S(A_i) \cap S(L) = \varnothing$. If, on the other hand, A is an L-process and $S(A) = S(L)$ then A is said to be *autonomous* (its behavior depends upon no other independent L-process).

Now, suppose A is an autonomous, lockup-free L-process. Let $A^{\#}$ denote the pseudo-Muller L-automaton defined as follows:

$$V(A^{\#}) = V(A) \cup \{\#\}$$

where $\#$ is a symbol not in $V(A)$; for $v, w \in V(A)$, $A^{\#}(v, w) = A(v, w)$ while $A^{\#}(v, \#) = \sum_{x \in S(A) \setminus S_A(v)} x$, $A^{\#}(\#, \#) = 1$ and $A^{\#}(\#, v) = 0$; $I(A^{\#}) = I(A)$ and

$Z(A^{\#}) = \{V(A)\}$. Thus, $\mathcal{C}(A^{\#})$ is the set of sequences $\mathbf{c} = (c_0, c_1, \ldots)$ where $c_i = (v_i, x_i)$ with $v_i \in V(A)$, $x_i \in S_{A_i}(v_i)$, $v_0 \in I(A)$ and $x_i * A(v_i, v_{i+1}) \neq 0$ for $i = 0, 1, \ldots$. A *task* for A is a set

$$\mathcal{T} \subset 2^{(V(A) \times S(A))^{\omega}}$$

i.e., a set of sets of sequences of (state, selection) pairs of A. It is said that A *performs* the task $\mathcal{T}$ if $\mathcal{C}(A^{\#}) \in \mathcal{T}$. ($\mathcal{T}$ defines "acceptable behavior" of A by defining all "acceptable versions" of A.)

This paper focuses upon testing task performance for a particular class of tasks: the *regular language* tasks relative to L, those being the tasks of the form $\mathcal{T} = 2^{\tau}$ for some ω-regular language τ over $S(L)$. Thus, if $\mathcal{T}$ is such as task and $\tau = \tau(\Gamma)$ for some L-automaton Γ, then A performs $\mathcal{T}$ iff $\tau(A^{\#}) \subset \tau(\Gamma)$.

7.3 Reduction

Let A be an L-process and B be an L'-process. In a manner strictly analogous to the development of §6, a *homomorphism* $\Phi: A \to B$ is defined. All the details of this definition are as in §6, with the exception of (2) and (3) which pertain to automaton acceptance structures, and the following treatment of selections, which is added: for each $v \in V(A)$ and $x \in S_A(v)$, it is required that there be some $y \in S_B(\Phi v)$ such that $x \leq \Phi(y)$.

Proposition 8: *Let A be a lockup-free L-process, let B be an L'-process and let $\Phi: A \to B$ be a epimorphism. Then B is lockup-free. Furthermore, for each $v \in V(A)$ and $y \in S_B(\Phi v)$, setting $Y = \{x \in S_A(v) \mid x \leq \Phi(y)\}$,*

$$\sum_{x \in Y} x = \Phi(y).$$

Thus, one effect of a process homomorphism is to associate to each selection of B, a nonempty set of selections of A.

If A and B as above are both autonomous and lockup-free, then the process homomorphism $\Phi: A \to B$ extends naturally to an automaton homomorphism $\Phi: A^{\#} \to B^{\#}$ with $\Phi(\#) = \#$. Furthermore, if $A_1, \ldots, A_k$ are independent, lockup-free L-processes and $\Phi_i: A_i \to B_i$ are co-linear process homomorphisms then the "lifting" results of §6.1 for automata carry over to here as well, and

$$\Phi = \otimes \Phi_i : \otimes A_i \to \otimes B_i$$

is a process homomorphism. If $A = \otimes A_i$ and $B = \otimes B_i$ are autonomous, one thus may reduce the test of whether for some L-automaton Γ, $\tau(A^{\#}) \subset \tau(\Gamma)$ to a test of whether $\tau(B^{\#}) \subset \tau(\Gamma)$. If furthermore each Φ_i is co-linear with an automaton homomorphism $\Psi: \Gamma \to \Gamma'$, the latter test may be reduced further to the test $\tau(B^{\#}) \subset \tau(\Gamma')$. Specifically, the results of Theorems 7, 8, 10 and their respective corollaries pertain.

These results suggest several techniques for reduction. The most obvious (and easiest to automate) is state reduction via an exact homomorphic minimization applied to several A_i at once, replacing each remaining A_j with a *free* version: a single-state process A_j' such that $S(A_j') = S(A_j)$. (The freeing of the several A_j's renders the exact homomorphic minimization algorithm tractable.) This procedure can be continued, applying exact homomorphic minimization to the reduced A_i's together with several of the (unreduced) A_j's, freeing the remaining A_j's. If this procedure is continued until it has been applied to all of $A_1, \ldots, A_k$, then the result is the exact homomorphic minimization of $\otimes A_i$ (this having been obtained without computing $\otimes A_i$ explicitly).

Another reduction technique, which may be used in conjunction with the above, is to guess a Boolean algebra homomorphism which induces full co-linear maps $\Phi_i: A_i \to B_i$. The correctness of such a guess may be checked using Theorem 4. Such "guessing" is often facilitated by inherent symmetry or modularity among the A_i's; for example, see [GK87].

Finally, if the structure of the A_i's is sufficiently simple and regular, one may apply a mathematical argument to demonstrate the existence of a homomorphism $\otimes A_i \to B$, as in [Ku85].

8. Conclusions

The overriding motive behind this paper has been to provide a theoretical basis for machinery with which to perform formal analysis of large systems of coordinating processes. A measure of the success of this approach can be taken from the derivative software system [HK86] which performs formal analysis of systems with trillions or more states through applications of Theorems 4, 7, 8 and 10 above.

The machinery developed here might be considered in the context of automated theorem-proving [BL84]. Classically, theorem-proving algorithms consist of inference rules applied to axiomatic systems. Because of the typically very general nature of these rules, it may be hard to classify the tractably provable theorems. However, since the tractable theorems are all characterized as provable in a (relatively) small number of steps from a fixed reportoire of instructions, they may be considered to be those theorems reducible to small regular-language assertions. The approach presented in this paper proceeds in the opposite direction: the class of small regular-language assertions is extended to the class of homomorphically reducible assertions.

REFERENCES

[RS59] M. O. Rabin, D. Scott, "Finite Automata and their Decisions Problems", IBM J. Res. and Dev. **3** (1959) 114-125. (Reprinted in [Mo64] 63-91.)

[Ts59] M. L. Tsetlin, "Non-primitive Circuits" (in Russian) Problemy Kibernetiki **2** (1959).

[Bu62] J. R. Büchi, "On a Decision Method in Restricted Second-Order Arithmetic", Proc. Internat. Cong. on Logic, Methodol. and Philos. of Sci., 1960, 1-11 (Stanford Univ. Press, 1962).

[Ra69] M. O. Rabin, "Decidability of Second-Order Theories and Automata on Infinite Trees", Trans. Amer. Math. Soc. **141** (1969) 1-35.

[Ho71] J. E. Hopcroft, "An $n \log n$ Algorithm for Minimizing the States in a Finite Automaton" in *Theory of Machines and Computations* (Kohavi, Paz, eds.) Academic Press, 189-196.

[Ra72] M. O. Rabin, *Automata on Infinite Objects and Church's Problem*. Amer. Math. Soc., 1972.

[Ta72] R. Tarjan, "Depth-First Search and Linear Graph Algorithms", SIAM J. Comput. **1** (1972), 146-160.

[Kn73] D. E. Knuth, *Sorting and Searching, (The Art of Computer Programming, v. 3)* Addison-Wesley, 1973.

[Ch74] Y. Choueka, "Theories of Automata on ω-Tapes: A Simplified Approach", J. Comput. Syst. Sci. **8** (1974), 117-141.

[Ha74] P. Halmos, *Lectures on Boolean Algebras*, Springer-Verlag, N.Y., 1974.

[St75] L. Stockmeyer, "The Set Basis Problem is NP-Complete", unpublished.

[HU79] J. E. Hopcroft, J. D. Ullman, *Intro. to Automata Theory, Languages and Computation*, Addison-Wesley, N.Y. 1979.

[CE81] E. M. Clarke, E. A. Emerson, "Synthesis of Synchronization Skeletons from Branching Time Temporal Logic", Proc. Logic of Programs Workshop, 1981, Lect. Notes in Comput. Sci. **131**, Springer-Verlag, 1982, 52-71.

[MP81] Z. Manna, A. Pnueli, "Verification of Concurrent Programs: The Temporal Framework", Stanford Univ. Tech. Report CS-81-836.

[AKS83] S. Aggarwal, R. P. Kurshan, K. K. Sabnani, "A Calculus for Protocol Specification and Validation" in *Protocol Specification, Testing and Verification, III*, North-Holland, 1983, 19-34.

[St83] L. Staiger, "Finite-State ω-Languages", J. Comput. Syst. Sci. **27** (1983) 434-448.

[BL84] W. W. Bledsoe, D. W. Loveland (eds.), *Automated Theorem Proving: After 25 Years*, Amer. Math. Soc. (*Contemp. Math.* v. **29**), 1984.

[MW84] Z. Manna, P. Wolper, "Synthesis of Communicating Processes from Temporal Logic Specifications", ACM Trans. on Programming Languages and Systems **6** (1984) 68-93.

[Ku85] R. P. Kurshan, "Modelling Concurrent Processes", Proc. Symp. Applied Math. **3** (1985) 45-57.

[Ku85a] R. P. Kurshan, "Complementing Deterministic Büchi Automata in Polynomial Time", J. Comput. Syst. Sci. (to appear).

[SVW85] A. P. Sistla, M. Y. Vardi, P. Wolper, "The Complementation Problem for Büchi Automata, with Applications to Temporal Logic", in Proc. 12th Internat. Coll. on Automata, Languages and Programming, *Lect. Notes Comp. Sci.*, 1985, Springer-Verlag.

[HK86] Z. Har'El, R. P. Kurshan, "COSPAN User's Guide", in preparation.

[Ku86] R. P. Kurshan, "Modelling Coordination in a Continuous-Time Asynchronous System", preprint.

[KK86] J. Katzenelson and R. P. Kurshan, "S/R: A Language For Specifying Protocols and Other Coordinating Processes", Proc. 5th Ann. Int'l Phoenix Conf. Comput. Commun., IEEE, 1986, 286-292.

[GK87] I. Gertner, R. P. Kurshan, "Logical Analysis of Digital Circuits", Proc. 8th Intn'l. Conf. Comput. Hardware Description Languages, 1987, 47-67.

[GKR87] I. Gertner, R. P. Kurshan, M. I. Reiman, "Stochastic Analysis of Coordinating Systems", preprint.

Distributed Reachability Analysis for Protocol Verification Environments

Sudhir Aggarwal

Bell Communications Research
435 South Street
Morristown, NJ 07960

Rafael Alonso

Department of Computer Science
Princeton University
Princeton, NJ 08544

Costas Courcoubetis

AT&T Bell Laboratories
600 Mountain Av.
Murray Hill, NJ 07974

ABSTRACT

A topic of importance in the area of distributed algorithms is the efficient implementation of formal verification techniques. Many such techniques are based on coupled finite state machine models, and reachability analysis is central to their implementation. SPANNER is an environment developed at AT&T Bell Laboratories, and is based on the selection/resolution model (S/R) of coupled finite state machines. It can be used for the formal specification and verification of computer communication protocols. In SPANNER, protocols are specified as coupled finite state machines, and analyzed by proving properties of the joint behavior of these machines. In this last step, reachability analysis is used in order to generate the "product" machine from its components, and constitutes the most time consuming part of the verification process. In this paper we investigate aspects of distributing reachability over a local area network of workstations, in order to reduce the time needed to complete the calculation. A key property which we exploit in our proposed design is that the two basic operations performed during reachability, the new state generation, and the state tabulation, can be performed asynchronously, and to some degree independently. Furthermore, each of these operations can be decomposed into concurrent subtasks. We provide a description of the distributed reachability algorithm we are currently in the process of implementing in SPANNER, and an investigation of the scheduling problems we face.

1. Introduction

Designing reliable distributed software such as computer communication protocols is extremely difficult and challenging. Informal specifications of such software are often imprecise and incomplete, and are not sufficient to ensure correctness of many simple distributed algorithms. One reason is that the concurrent execution of components typically results in an exploding number of execution histories. This makes the prediction of all possible erroneous behavior of such systems prohibitively complex for the human mind, and as a result the designer *must* rely on formal methods for specifying and analyzing the software. There is an increasingly extensive literature on such formal methods and tools; for example, see [Bo87] for a survey.

Among the formal specification methods, finite state models are one of the most popular. In these models, the system is described as a set of coupled finite state machines (FSMs), each machine modeling a concurrently executing component. The reason FSMs are widely used to describe complex systems is that it is conceptually easier to describe such a system in terms of a large number of small components, and then derive the complete system by taking the "product" of these components. In addition to this, in many cases, FSM descriptions can be directly translated to implementable code or implemented in hardware, see [AC85, AK84, GK87].

There are many existing tools for the analysis of finite state models, for example see [ABM86, An86, Fl87, CG86]. The way these tools work can be summarized in the following steps. First, they provide an environment in which the designer specifies the FSMs by describing the components of the system. Usually, the designer also specifies the task that must be satisfied by the system in order to ensure correct execution. In the second step, the system uses the description of the components to construct the "product" FSM which models the complete system. This step constitutes the reachability analysis of the system, during which a database of all the reachable states and transitions of the product FSM is constructed from the specifications of the component FSMs. The last step consists of checking the validity of the task formula on the product FSM. This can be accomplished in a "partial" way by assigning probabilities to the FSM and checking correctness on some finite set of most probable histories, obtained by simulation, or in a complete way by doing model checking of the FSM and the task formula, see [CE82, QS82]. In some systems, the last step is embedded in the second step, and corresponds to doing reachability analysis on a larger number of FSMs, some of which model the task requirements, see [ACW86]. One can also use various reduction techniques depending on the underlying model, so that the product FSM is substituted by a smaller one. Further details about these techniques can be found in the literature, and are beyond the scope of this paper.

From the previous discussion, it follows that a limiting factor for the practical application of the FSM methods is the time it takes to perform the reachability analysis. With the current technology, graphs of 10^4 to 10^6 states can be analyzed in times on the order of minutes or hours, by running the tools on single dedicated workstations. In this paper we focus on how to move this limit substantially further by performing the reachability analysis in parallel. Note that an important issue in favor of distributing the reachability analysis is the size of the table of the explored states. For large graphs, this table cannot fit in the main memory of a single workstation and the tabulation becomes increasingly slower as the number of explored states grows. We describe a parallel reachability algorithm and its implementation aspects, for the already existing tool SPANNER [ABM87], in a local area network of SUN workstations. We should emphasize that the design we propose is general enough to be used for parallelizing the reachability analysis in most of the existing FSM tools.

We will elaborate now further on the ideas presented in the paper. A "centralized" reachability analysis is performed as follows. The input consists of the description of N FSMs. A "global" state is a state of the product FSM, and consists of a vector of N local states, one per component machine. The underlying model provides a way to compute for each global state, the set of all possible successor global states. The reachability analysis starts with some initial global state, and completes when all global states reachable (in any number of steps) from this initial state, are found. While doing this, there are two basic operations involved: the *state generation*, which given a global state, computes its successor global states, and the *state tabulation*, which given a global state, checks if it has been already found by keeping an updated table of the global states visited so far. The distributed reachability analysis we propose is basically performed as follows. There are *n state generators* and *m state tabulators*. Each state generator receives (global) states from the tabulators, computes their successor states, and sends them to the tabulators. A state whose successor states have been computed is considered to be "explored". The tabulators receives newly generated states from the generators, filter out the states that have already be found, and send unexplored new states to the generators. The key issues we address in our design is how to distribute the newly found state information among the tabulators, and the scheduling of the work requests among the tabulators and the generators, so that the workload of different processors remains balanced. As it turns out, the scheduling problem involved is non trivial, and has many generic aspects. This is due to the large number of messages, and the comparable magnitude of the time involved in processing the work carried by a message, with the message delay of the network.

The paper is organized as follows. In section 2, we briefly describe the SPANNER system and the model of FSMs on which SPANNER is based. In section 3 we describe the design of the distributed software, and its implementation environment. In section 4 we examine the underlying scheduling problem, and we provide a queuing model for the system. This model can be used as a basis for simulating the performance of the system, with different scheduling parameters. We also mention two open scheduling problems which abstract different parts of the original problem and seem interesting for further research. At the end of section 4, we investigate performance issues related with the gain in speed of the reachability analysis due to parallelism. In section 5 are the conclusions of this work.

2. The Selection/Resolution model and the SPANNER system

2.1. The Selection/Resolution model

For completeness, we review the selection/resolution model and the SPANNER system. For simplicity we discuss the model in terms of its operational semantics. Further details are available in [GK82, AKS83, ABM87]. The selection/resolution model is a formal method of describing a complex system as a finite set of coupled FSMs. Each component FSM (called a process) is specified as an edge labelled directed graph, see figure 2.1.

The vertices of the graph are *states* of the process, and the directed edges describe a state transition that is possible in one time step. A state encapsulates the past history of the process and is private to that process. That is, no component FSM can know

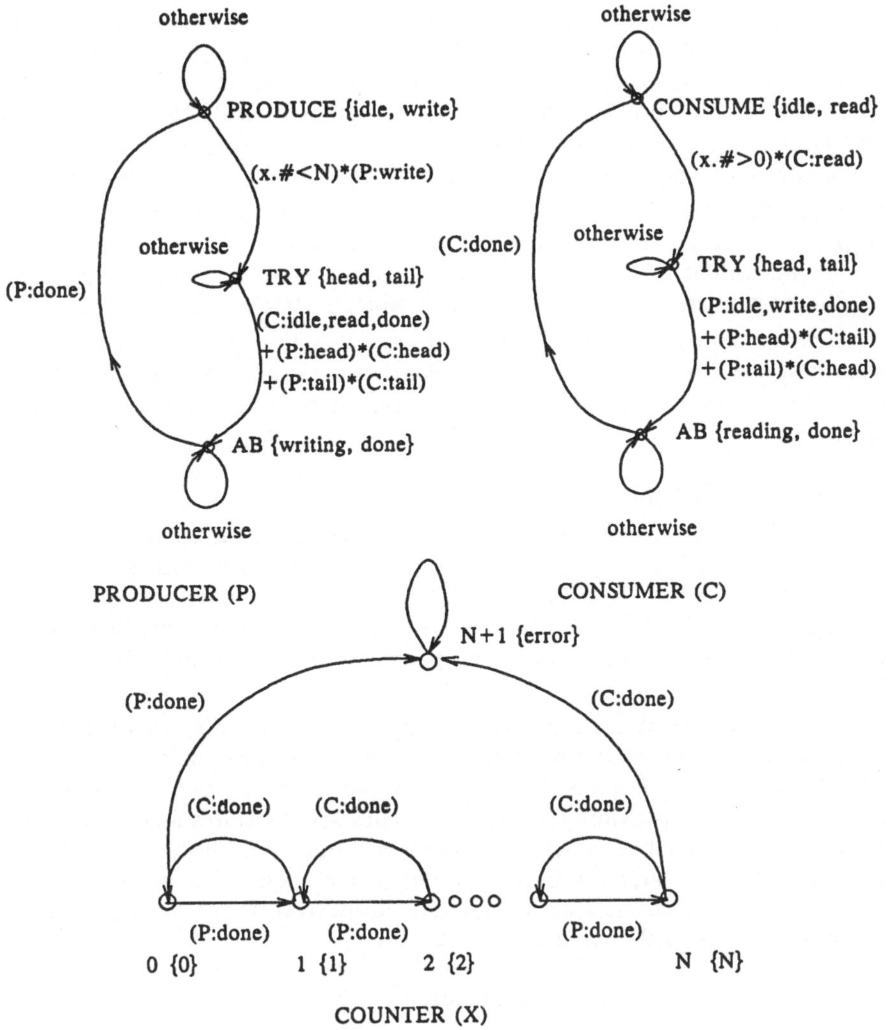

Figure 2.1

about this state directly. In each state, a process can nondeterministically choose from a set of *selections* (enclosed in braces next to the state). The selections are signals processes use to coordinate. They can be viewed as indicating the "intention" of the process. The component FSMs use these selections to determine their transitions.

The directed edges of the component FSMs are labelled by elements of a Boolean algebra generated by the selections of the processes. We use * to indicate the multiplication operator (Boolean *and*), and we use + to indicate the addition operator (Boolean *or*). We use ˜ for the Boolean *negation*.

After each process has made its selection, each process decides on a transition to a new state. This *resolution* is done as follows. First, calculate the global selection of the processes. This is done by multiplying together the current selections of all the processes. Note that, by definition, this product is the *and* of all the current selections. Next, each component FSM independently determines which transitions out of the current state are enabled. It determines if a transition is enabled by the global selection by multiplying the edge label by the global selection and checking if the result is 0 in the Boolean algebra. If the product is 0, the transition is not enabled. Otherwise, it is enabled (a valid transition). Finally, each process chooses one of the enabled transitions and transitions along that edge to its new state.

Consider the case of k processes $P_1, \ldots, P_k$ and let the selection of P_i be s_i at time step t. Then the global selection $s = s_1 s_2 \ldots s_k$ is the *AND* (in the Boolean algebra) of the individual selections. If process P_i is in state v at time step t, and the label on the edge from v to w is ℓ, then w is a possible state at time $t+1$ if $s \cdot \ell \neq 0$.

A *chain* of a process is a sequence of state-selection pairs consistent with the dynamics described above. Intuitively, a chain is a sample path of the behavior or possible history of the process, where at each time step we record the state and selection of the process.

2.2. The Spanner System

SPANNER is an environment consisting of a set of modules for specifying and analyzing protocols. The underlying formal model is the selection/resolution model discussed above. SPANNER allows the user to specify a protocol as a set of coupled FSMs using the SPANNER specification language. The parser module checks the specification for syntactic correctness, and produces an intermediate description used by other modules.

The basic construct of the specification language is a process; this corresponds to a labelled directed graph of the s/r model. The initial declaration of the process simply describe the ranges of states and selections and gives the user the option (using the keyword **valnm**) of providing descriptive names for the states and selections. The *import* declaration describes which processes' selections are visible in that process. The *init* declaration declares the initial state of that process. The *trans* section is the transitions section and consists of blocks that define transitions from sets of states. The format of a block is shown in figure 2.2.

```
current state      {selection list}

    > next state   : condition;

         .             .

         .             .

    > next state   : condition;
```

Figure 2.2

Figure 2.3 shows the processes of a simple producer-consumer problem in the specification language.

```
constants N = 3
process P        /*the producer*/
 import C
 states 0..2     valnm [PRODUCE:0, AB:1, TRY:2]
 selections 0..5 valnm [idle:0,write:1,writing:2, done:3, head:4,tail:5]
 init PRODUCE
 trans
   PRODUCE       {idle, write}
     > TRY   :write);
     > $     :otherwise;

   TRY       {head,tail}
     > AB    :(C:idle,read,done) + (P:head)*(C:head) + (P:tail)*(C:tail);
     > $     :otherwise;

   AB        {writing, done}
     > PRODUCE:(P:done);
     > $     :otherwise;
end

process C        /*the consumer*/
 import P
 states 0..2     valnm [CONSUME:0, AB:1, TRY:2]
 selections 0..5  valnm [idle:0, read:1, reading:2, done:3, head:4, tail:5]
 init CONSUME
 trans
   CONSUME       {idle, read}
     > TRY   :(X.# > 0) * (C:read);
     > $     :otherwise;

   TRY        {head,tail}
     > AB    :(P:idle,write,done) + (P:head)*(C:tail) + (P:tail)*(C:head);
     > $     :otherwise;

   AB        {reading, done}
     > CONSUME:(C:done);
     > $     :otherwise;
end

process X        /*the counter*/
 states  0..N+1     valnm [ERROR:N+1]
 selections 0..N+1  valnm [error:N+1]
 init 0
 trans
   $            {$}
     >($ + 1)%(N+2): (P:done);
     >($ - 1)%(N+2): (C:done);
     > $        : otherwise;
end
```

Figure 2.3

SPANNER provides a variety of other constructs that make it easier to specify large systems. This includes the notion of *cluster*, to facilitate hierarchical development, and the notion of *process type* as a template for the instantiation of similar processes. These constructs are discussed in the references.

The SPANNER system allows the user to experiment with and study the system of coupled FSMs in two ways. First, the system can be studied using the reachability graph. The reachability graph is a graph whose vertices are global states (vectors of local states), and whose directed edges are valid transitions between global states.

The latest version of SPANNER is actually based on an extension of the s/r model which allows reasoning about infinite paths (chains) [ACW86]. It turns out that many questions about protocols such as deadlocks, livelocks, and liveness can be answered solely by proper investigation of the reachability graph. The general mechanism that we use is to add monitor processes that either check for or ensure certain properties of interest. For example, to an existing protocol, we could add a process that ensures that a particular component always makes progress and does not pause forever (a liveness property). Similarly, we could add a process that checks that a particular task such as receiving messages in the order sent was met. In this approach, proving the validity of arbitrary temporal logic formulas is done by checking properties of a reachability graph.

In order to make it convenient to study the reachability graph, SPANNER produces a database that consists of three tables (relations). The *global reachable states table* (table *r*) has as attributes *index* (the number of the global state), and the *local state* for each process identified by the name of the process. In addition, each global state has the attribute *cc* that identifies to which strongly connected component that state belongs. The transitions table (table *t*) has as attributes *to-state* and *from-state* that specify the global state numbers for the one step transitions. Using a set of commands, the user can query the relations to determine those table entries that satisfy a particular condition. For example, in the producer-consumer protocol of figure 2.1, we could ask if there is a global reachable state with process *P* in state *AB* and process *C* in state *AB*, corresponding to both processes accessing a common buffer at the same time, and we would find that the answer is no. In addition, the database has a third relation called table *c* that is used in checking for liveness properties. For details, see [ABM88].

Another way of studying the system of coupled FSMs is through simulation. This is particularly useful for very complex protocols, since interesting constraints can be imposed on the simulation. For example, it is possible to assign probabilities to the selection choices and it is also possible to force a selection to be held for a particular number of time steps. SPANNER allows the creation of a database of sample runs using a set of simulation modules. These modules allow a simulation to be setup, using the constraints mentioned, then simulated, and finally analyzed. The user can analyze the results of the simulation by querying the generated database using an interactive query language, similar to querying the reachability database.

Reachability

Reachability in SPANNER is done in a fairly standard way. Given, a global state of the system (a vector of local states of components), the first step is to generate potential new global states. The is done by checking for all possible transitions that are enabled in each component for that given global state, and then looking at the new local states that result. These new local states are combined to form the next set of global states. This is the *generator* function of reachability. Next, the potential new states are checked to see if they truly are new states. This is done by keeping the reached states in a hash table. This is the *tabulator* function of reachability. Unexplored states are kept on a list of states to be investigated, and they can be explored by either breadth-first or depth-first search.

In this paper, we essentially discuss various ramifications of parallelizing this reachability algorithm. As noted in the introduction, states can be generated in parallel, since the generation of next states from two different initial states can be done independently. Thus, the generator function can be done in parallel. Further, by being careful to handle only parts of the hash table, the tabulator function could also be done in parallel. It should also be noted that the generation of new states from a given initial state can also be made parallel to some extent, since each component process can independently determine the enabled transitions from its initial local state.

3. Implementation

In this section we provide some of the details of our proposed implementation. First, we describe the particular environment in which we intend to construct the parallel version of SPANNER. We then outline the distributed reachability algorithm. The details of the scheduling policy implemented are discussed in section 4.

3.1. Implementation Environment

The system will be implemented on a network of SUN workstations in the Distributed Computing Laboratory of Princeton University. We will use SUN models 2 and 3, which will provide us with some heterogeneity with respect to processor speeds (the model 3 processor is significantly faster than that of the model 2). The machines are connected via a dedicated Ethernet network, which ensures that during our experiments the network is not loaded by extraneous messages.

The machines will be running SUN UNIX version 3.3, which supports a variety of networking protocols [Sun86]. The currently implemented protocols are either *stream* or *datagram* oriented. Stream protocols provide a bidirectional, reliable, sequenced communication channel between processes. The stream protocol implemented by SUN is the TCP protocol defined by DARPA [Po80a]. Datagrams also allow bidirectional communication, but make no delivery guarantees. Datagram packets may be lost, duplicated, etc. SUN's datagram implementation is based on the IP protocol standard [Po80b]. Either type of protocol can support a message rate of somewhat less than 1 megabit per second between two SUN workstations on an otherwise idle Ethernet.

At first glance it would seem that choosing a stream protocol might be the obvious course of action for our work. After all, users of stream protocols do not have to concern themselves with the details of packet formation, dealing with duplicates, ensuring that messages are not lost, etc. However, this functionality comes at the cost of lessened control over the transmission of data. The user view of a communication stream is that of a boundary-less flow of data. That is, users think of streams as if they were inter-processor versions of UNIX pipes [RT78], and are not aware of the details of the underlying communication layers. Typically however, a stream protocol is implemented on top of a datagram protocol (as is the case for our network). System designers who are primarily concerned with efficiency and performance may need to have access to these underlying layers. For example, it may be desirable to control the amount of data in a datagram packet and the time of its transmission.

In practice, communication systems (i.e., the combination of network hardware, protocols and operating system support) have certain packet sizes that they deliver most efficiently; for example, if the networking code in the operating system kernel needs to move the user's data out of the user's address space before shipping it across the network, a packet size equal to the operating system's page size will usually result in the largest throughput. Another fact that must be taken into account in the implementation of a parallel application is that it is usually more efficient to send one large message than many small ones because there is normally an overhead per packet sent.

In light of the above comments, it seems clear that our choice of networking protocol is not an obvious one. Our present approach is to develop the initial code using a stream protocol. Once the debugging stage is complete, we will start using the datagram facility, in order to obtain the maximum performance from our system.

3.2. Software architecture

We have already provided some details of the reachability analysis carried out by SPANNER. The distributed version consists of n generators and m tabulators. Each generator stores the complete description of all component FSMs but keeps no information about the set of reachable states. The "global" hashtable (which would be used by SPANNER in its non-distributed version) is now split into m equal nonoverlapping hashtables to be used by each of the m tabulators. Each tabulator has no information about the FSMs, and will only store the global states which hash in its local hashtable. This implies that each global state can be stored in only one among the m tabulators. One can easily define the function h which maps any global state v to the appropriate tabulator $h(v)$ as follows. Compute the hashvalue of v, and check to which of the local hashtables it corresponds. The index of the corresponding tabulator is the value $h(v)$.

A generator is described in terms of three concurrent processes: a receiver process, which feeds the input queue with unexplored states by unpacking the arriving messages; a next state generator, which given a state produces its successor states; and a sender process, which controls the sending of the resulting states through the network. A tabulator is similarly defined in terms of a receiver, a state tabulator, and a sender process. A more precise description follows:

Generator i, i=1,...,n.
Process Receiver
do until *(end of reachability analysis)* {
    receive message from network;
    break it into states;
    append the states to the generator input queue
    }
Process Next_State_Generator
do until *(end of reachability analysis)* {
    get next state v from the generator input queue;
    compute the set $next(v)$;

```
        for each v' in next(v) do{
                compute j=h(v');
                append <v',v,i> to the generator
                output queue with destination Tabulator j
                }
}
Process Sender
do until (end of reachability analysis) {
        for each output queue with destination Tabulator j, j = 1,...,m, do{
                use the heuristic scheduling policy of section 4
                to pack states into messages sent to Tabulator j

        }
}
```

Tabulator j, j=1,...m.

```
Process Receiver
do until (end of reachability analysis) {
        receive message from network;
        break it into states;
        append these states to the tabulator input queue
}
Process State_Tabulator
variables: U_l is the list of the unacknowledged states
           sent to generator i (in the order sent), i = 1,...n.
do until (end of reachability analysis) {
        get next element <v',v,i> from the tabulator input queue;
        if v is in U_k for some k=1,...n, then
                update U_k=:U_k-(v,v_1,...,v_s),
                where v_1,...,v_s are all states in U_k
                prior to v;
        insert v' in the hash table;
        if v' is a new state, append it to the tabulator output queue
}
Process Sender
do until (end of reachability analysis) {
        Use the heuristic scheduling policy of section 4
        to pack states from the tabulator output queue into messages
        sent to the generators
}
```

What we have not specified yet is how the *end of reachability* condition will be detected by the processes. A simple way to do this is the following. When any processor remains idle for more than time t, it triggers a round where all processors respond about their work status. If all of them have empty input queues, then the above condition is satisfied.

A final point to be made is that at the end of the reachability analysis phase each tabulator will have only a portion of the reachability graph and a final coalescing step will be required in which all the sub-graphs are merged. This coalescing step could also be done in parallel with the tabulation.

4. Scheduling aspects

In this section we examine the scheduling problems which must be addressed by the distributed software described in the previous section. There is a plethora of parameters which are important for the efficient execution of the system. The measure of performance we consider is the total finishing time, i.e., the time at which all the reachable states have been found. Intuitively, this can be minimized if we can keep the work balanced among the various computing resources in the system. Achieving such a balance among the tabulators and the generators constitutes a non-trivial scheduling problem which needs some novel heuristic solution.

We start by examining a simpler system consisting of a single tabulator and n generators. The basic controller of the load of the processors in our system is the tabulator. It is the tabulator's responsibility, once an unexplored (new) state has been found, to ship it to the most appropriate generator among the n available. The following reasons make such a decision very complex. In most LAN's, a message has essentially the same cost (delay) if it contains up to some maximum number of bytes. This implies that sending a message containing one state or some system dependent m_{max} number of states, can be achieved for the same cost. This motivates the batching of a large number of states in the same message. In order to make this possible, states ready to be shipped must be kept in a queue until enough of them accumulate to be batched in a message. The negative side-effect of such a decision is that this can produce idle time for the processors waiting to process these states. On the other hand, if a small number of states per message is sent, this will result in flooding the network with messages and will increase their delay. The reader should be reminded of the size of the problem being on the order of 10^5 to 10^6 states, which makes batching unavoidable. The optimal size of the batch is a crucial parameter to be determined. Note that batching needs to be done from the generator's side as well.

Another consideration is the following. At the beginning of the computation, the number of unexplored states for most problems will grow exponentially fast, and towards the end it will rapidly decrease to zero. If the tabulator ships exceedingly large amounts of work to the generators towards the last phase of the computation, it is likely that during this last phase the workload of the generators will be unbalanced, since there are not enough new unexplored states generated which the tabulator can appropriately distribute to even things out in the generators. A sensible policy should, in the initial phase of the computation, send large amounts of work to the generators to reduce the probability of them staying idle. Towards the end the policy should keep a store of unfinished work in the tabulator's output buffer, from which increasingly smaller batches of work are to be send to the generators in an attempt to keep their outstanding workload balanced, and to reduce the workload uniformly to zero. Such a policy will minimize the total completion time. A key factor in our system that makes such a policy difficult to implement is the random delay with which observations concerning the workload of the generators are made. The tabulator can only estimate the outstanding work in the generator sites from the information in the messages it receives (the newly generated states arrive together with the identity of their generator site). An important characteristic of our system is that the delay of a message through the network is of comparable magnitude to its processing time (time that the destination processor takes to complete the work associated with the states stored in the message). Finding optimal policies with delayed information is in general outside the scope of any realistic analysis; hence, a heuristic solution to the problem is the most for which one can hope.

Following the ideas in the previous discussion, a scheduling policy should define the following decisions for the tabulator: *when* to send a message to a generator, *to which generator* to send it, and *how much* work the message should contain. The available information for such a decision is an estimate of the amount of outstanding work of each generator, its processing rate (estimated), the amount of work stored in the tabulator's queue, and the average delay of messages in the network.

The Queueing Model.

For modeling the system we make the following assumptions. First, the graph constructed by the reachability analysis is characterized by a distribution function f_G of the outdegree of its states, and by some upper bound N_G of its number of states. In our model we assume that each state has d next states, where d is distributed with f_G and is independent for different states. We also assume that each newly generated state at time t has probability $r(t)$ of being already visited. There are many ways to describe $r(t)$ as a function of $x(t)$ and N_G, where $x(t)$ is the number of states found up to time t. Different such functions correspond to different types of graphs. One such choice is $r(t) = x(t)/N_G$, which corresponds to graphs with small diameter.

The model describes n generators and a tabulator connected through a LAN, see Figure 4.1.

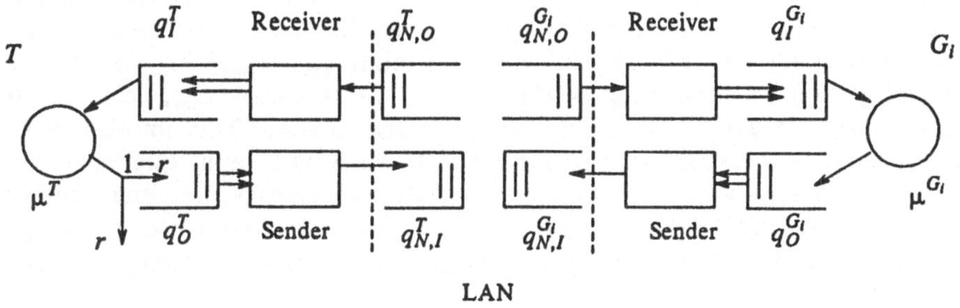

Figure 4.1

There are two types of customers in the system: the "simple" customers (single states) and the "batched" customers (many states batched into one message). Simple customers belong to different classes. Each class describes the origin-destination of the customer (for example, $T-G_i$) and whether or not the customer serves as an acknowledgement (discussed below). The class of a batched customer is the description of the class of each simple customer it contains. Each generator G_i, $i = 1, \cdots, n$, has an input queue $q_I^{G_i}$, an output queue $q_O^{G_i}$, and a server with rate μ^{G_i}. There are two network queues interfacing to the generator, the network output queue $q_{N,O}^{G_i}$, and the network input queue $q_{N,I}^{G_i}$. Customers arriving in the network output queue are of the batched type. Upon arrival, they are automatically "unpacked" (a message containing k states is broken into k single state messages), and the resulting simple customers are placed in the generator input queue. The server serves simple customers from the generator input queue in a FCFS basis, and after each service completion it appends a random number d of simple customers with destination the tabulator, in its output queue. The first among them is tagged as an acknowledgement. As mentioned before, d is distributed with distribution function f_G. A sender processes connects the output queue of the generator to the input queue of the network. Its function consists of making batched customers out of simple customers, and append them at appropriate times in the network queue. Its available information consists of the state of the generator queues, and of some local timer.

The tabulator has an input queue q_I^T, and an output queue q_O^T. These queues interface in a similar fashion to the generators case with a network output queue $q_{N,O}^T$ through an receiver process, and with the network input queue $q_{N,I}^T$ through a sender processes respectively. A batched customer arriving in the network output queue is immediately unpacked, and the resulting simple customers are placed in the tabulator input queue. A server of rate μ^T serves this queue in a FCFS basis. When a customer finishes service at time t, it leaves the network with probability $r(t)$, and with probability $1-r(t)$ it joins the output queue. In this event, the state counter x is incremented by one, to denote that a new state has been found. If a customer finishing service is an acknowledgment and G_l was its origin (class information of the customer), the variable z_l denoting the outstanding work (number of states to be explored) of generator G_l, is decremented by one. The sender process does the packing of simple customers, assigns the destination of the resulting batched customers, and places them at appropriate times in the network input queue. Its available information consists of the values of x, $z_1, \cdots, z_n$, and the state of the tabulator queues.

We are now left with the description of the network. There are $n+1$ input queues and $n+1$ output queues already mentioned before. The model we choose better describes LAN's of the Ethernet type, such as the one in our implementation. Each non-empty input queue is served with rate $\min[\mu_{\max}, \mu_{total}/\ \#of\ non\ \emptyset\ queues]$. To model congestion we choose $\mu_{total}<(n+1)\mu_{\max}$. In this model, $\mu_{\max}$ corresponds to the maximum service rate allocated to any network queue. This implies that the minimum delay of a message being in the front of a network input queue, is on the average $1/\mu_{\max}$. If the number of transmitting stations increases, i.e., the number of non-empty network input queues grows, the service rate allocated to a queue decreases as the total network service rate μ_{total} is being shared equally among the competing queues.

Heuristic Scheduling Policies.

A scheduling policy is defined in terms of the algorithms used by the $n+1$ sender processes of the system. We propose a scheduling policy of the following form:

Generator i: The sender process has a timer of duration τ. While the generator output queue has more than B_G customers, it forms batches of B_G simple customers and delivers them to the network. If there are less than B_G customers, it sets the timer. If the timer times out, and there are still less than B_G customers, it batches them into a message and sends them to the network.

This policy reduces the probability that the tabulator stays idle, while there is work for the tabulator at the generators. It also reduces the number of messages needed.

Tabulator: The sender algorithm uses the following heuristic. The size of the batch is an increasing function of the number of customers in the tabulator output queue q_O^T, starting from zero if the queue is empty, and bounded by some B_T. There is a threshold k^* in the number of customers in q_O^T which affects the operation as follows. If there are more customers than k^*, it continuously makes batches and sends them to the generators, keeping the outstanding work indices $w_l = z_l/\mu^{G_l}$ as close as possible, until an "upper watermark" $W_{\max}$ for each w_l is reached. Then it stops sending, until some w_l drops below, in which case it resumes the sending of work. If the number of customers in q_O^T drops below k^*, it stops sending until for some generator $w_l<W_{min}$, where W_{min} corresponds to some "low watermark". Then it resumes sending to this particular generator, until $w_l \geq W_{min}$. The Readers can convince themselves that in order to achieve an even distribution of customers in the queues of the system, W_{max} and W_{min} have to be increasing functions of the number of customers in q_O^T. Choosing W_{max} to grow appropriately with q_O^T ensures that the tabulator queues do not grow faster than the generator queues. Choosing W_{min} to decrease as q_O^T decreases provides that in the termination phase all queues will decrease uniformly to zero. The appropriate selection of W_{max} and W_{min} remains an open research topic. (A simple queueing theory argument indicates that W_{min} should be proportional to $(q_O^T)^2$.

We can simply define a policy for the general case of m tabulators by having each tabulator use $|U_l|$, as defined in the previous section, in place of z_l.

Some Open Scheduling Problems.

There are two simple versions of the model for which the form of the optimal scheduling policy may be more tractable. Solving these could give a greater insight for how to operate the complete system. These models are derived by reducing to zero the transmission delay of the network in one of the directions from G to T or from T to G. The first model is described in Figure 4.2.

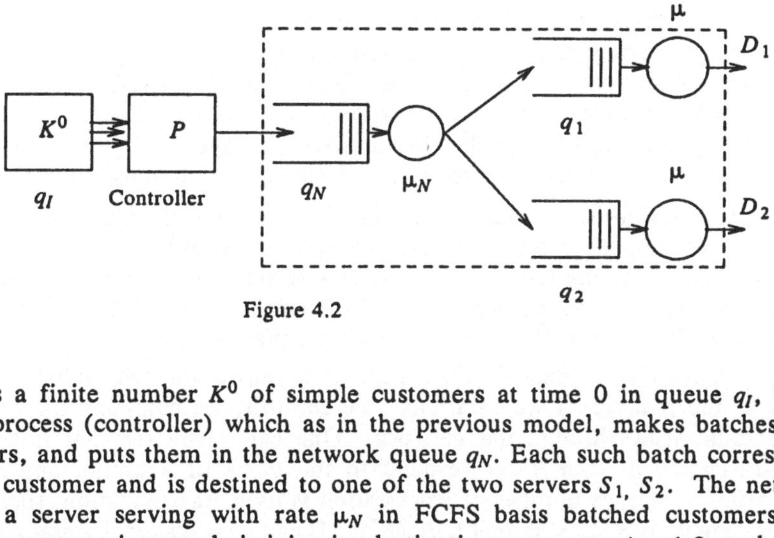

Figure 4.2

There is a finite number K^0 of simple customers at time 0 in queue q_I, There is a sender process (controller) which as in the previous model, makes batches of simple customers, and puts them in the network queue q_N. Each such batch corresponds to a batched customer and is destined to one of the two servers S_1, S_2. The network consists of a server serving with rate μ_N in FCFS basis batched customers. Once a batched customer is served, it joins its destination queue q_i, $i = 1,2$ as the set of its composing simple customers. Each server S_i serves with rate μ from its queue q_i, $i = 1,2$. The information available to the controller is the complete departure process of the system.

The second model is shown in Figure 4.3.
The difference with the previous is that it takes zero delay to append customers in the queues of the two servers, and that the information available to the controller is a delayed picture of the departure process of the system. For both systems we want to minimize the time all customers complete service. One can easily see that a threshold

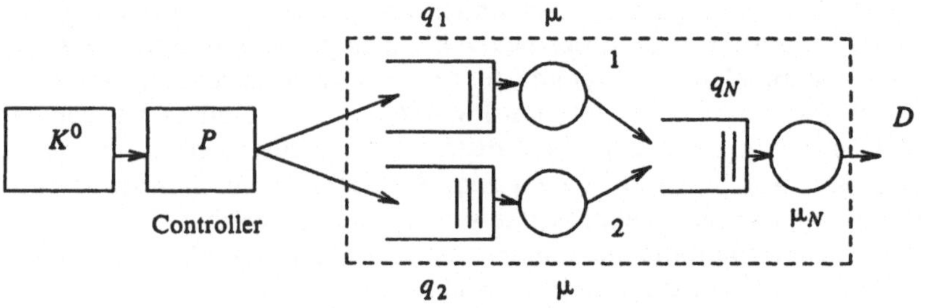

Figure 4.3

type policy of the form described in the previous section should perform well in these models. Although the mathematical analysis of these two models might be prohibitively complex, any progress in this direction can result in a valuable practical contribution.

4.1. Performance Analysis

In this section we provide a simple analytic model in order to predict the approximate performance of our scheme. Let

n = number of generator nodes
m = number of tabulator nodes
N = number of states in the graph to be analyzed
d = average out-degree of vertices in state graph
t_g = average time required to generate a single state (seconds)
t_d = average time required to decide if one state is new (seconds)
B = average number of states batched in each message
$t_m(B)$ = average time required to send one message containing B states (seconds)

Using the above definitions, the reachability analysis performed on a single workstation will take total time $T_1 = Nd(t_g+t_d)$ to complete, since each state is examined d times on average, by both the tabulator and generator software. Assuming that using the current technology, a tool runing on a 1 MIP workstation completes a graph of 10^5 states and $d = 10$ in the order of an hour, we get that t_g+t_g is equal to $5 \cdot 10^{-3}$s.

The first interesting remark is that there is a maximum achievable speed up independent of the number of workstations. To see this, we compute the total time spent in communicating through the network. One can easily see that this time is equal to $N(d+1)t_m(B)/B$. Let B^* correspond to the value of B minimizing $t_m(B)/B$. Then, if an arbitrarily large number of workstation is used, the maximum speed up is aproximately equal to $K_{max} = B^*(t_d+t_g)/t_m(B^*)$. Using $B = 40$ (25 bytes/state, 1024 byte message), and and network bandwith equal to 5 Mbits/s, we get $K_{max} = 100$. (For a large number of two-way conversations the effective bandwidth of an Ethernet is at least 5 Mbits).

We examine now how to choose the m, n. Assuming that we have $m+n < K_{max}$ and we operate in the otimal way (all processors are kept busy until the end), then we must have that $Ndt_d/m = Ndt_g/n = T_1/(n+m)$. From this it follows that the optimal partition is such that $t_d/m = t_g/n$, and the speed up is equal to $m+n$.

5. Conclusions

In this paper we demonstrated that distributed reachability analysis can be easily incorporated into many existing protocol analysis environments and can produce a significant speed up of the analysis. In many research environments there is easy access to LANs with 10-20 workstations, which, according to the results of our performance study, is an ideal environment to implement our method. An important remark which makes our approach even more viable in the future is that fiber optic technology makes communication bandwidth available in a faster rate than the rate of increase of hardware speed. In our method a large number of communicating workstations can utilize this available bandwidth.

We are currently implementing the parallel reachability algorithm in SPANNER. We hope that the results of our implementation will justify the approach presented in this paper.

References

[ABM86] S. Aggarwal, D. Barbara and K. Meth, "Specifying and analyzing protocols with SPANNER", *Proceedings of the IEEE International Conference on Communications*, June 22-25, 1986, Toronto, Canada.

[ABM87] S. Aggarwal, D. Barbara and K. Meth, "SPANNER - A tool for the specification, analysis and evaluation of protocols", to appear in *IEEE Trans. on Soft. Eng.*, 1987.

[ABM88] S. Aggarwal, D. Barbara and K. Meth, "A software environment for the specification and analysis of problems of coordination and concurrency", to appear in the *IEEE Trans. on Software Eng.*, 1988.

[AC85] S. Aggarwal and C. Courcoubetis, "Distributed implementation of a model of communication and computation", *Proc. of the 18th Hawaii Int. Conf. on System Sciences*, January 1985, pp. 206-218.

[ACW86] S. Aggarwal, C. Courcoubetis and P. Wolper, "Adding liveness properties to coupled finite state machines", AT&T Bell Laboratories Technical Memo., 1986.

[AK84] S. Aggarwal and R. Kurshan, "Automated implementation from formal specifications", *Protocol Specification, Testing, and Verification IV*, (Y. Yemini and al. eds.), North Holland, 1984.

[AKS83] S. Aggarwal, R. Kurshan, and K. Sabnani, "A calculus for protocol specification and validation", *Protocol Specification, Testing and Verification III*, H. Rudin and C. West (Eds.), North Holland, 1983.

[An86] J. P. Ansart, et al., "Software tools for Estelle", *Protocol Specification, Testing and Verification VI*, B. Sarikaya and G. Bochman (Eds.), North Holland,1986, pp. 55-62.

[Bo87] G. V. Bochmann, "Usage of protocol development tools: the results of a survey", *Proceedings of the 7th IFIP workshop on Protocol Specification, Testing and Verification*, Zurich, May 5-8, 1987.

[CE82] E. M. Clarke, E. A. Emerson, "Synthesis of synchronization skeletons from branching time temporal logic", *Proc. Logic of Programs Workshop, 1981, Lecture Notes in Comput. Sci.* 131, Springer-Verlag, 1982, 52-71.

[CG86] D. Cohen, B. Gopinath, et al., "*IC**: An environment for specifying complex systems", *Proc. IEEE GLOBECOM Conf.*, Houston, Dec. 1986, pp. 632-637.

[Fl87] A. Fleischmann, "PASS - A Technique for specifying communication protocols", *Proceedings of the 7th IFIP workshop on Prorocol Specification, Testing and Verification*, Zurich, May 5-8, 1987.

[GK82] B. Gopinath and R. Kurshan, "The selection/resolution model for concurrent processes", unpublished.

[GK87] I. Gertner, R. P. Kurshan, "Logical analysis of digital circuits", *Proc. 8th Int'l. Conf. Computer Hardware Description Languages*, 1987, 47-67.

[Po80a] J. Postel, "DOD Standard Transmission Protocol," RFC 761, Information Sciences Institute, January 1980.

[Po80b] J. Postel, "DOD Standard Internet Protocol," RFC 760, Information Sciences Institute, January 1980.

[QS82] J. P. Queille, J. Sifakis, "Specification and verification of concurrent systems in CESAR, *International Symposium in Programming*, LNCS 137, 1982.

[RT78] D. Ritchie and K. Thompson, "UNIX Time-Sharing System," *Bell System Technical Journal*, Vol. 57, Number 6, 1978.

[Su81] C. A. Sunshine, (ed.), *Communication Protocol Modelling*, Artech House, 1981.

[Sun86] "Inter-Process Communication Primer," Sun Microsystems User Documentation, Revision B of February 17, 1986.

[ZWRCB80] P. Zafiropoulo, C. H. West, H. Rudin, D. D. Cowan and D. Brand, "Towards analyzing and synthrsizing protocols", *IEEE Trans. on Comm.*, COM-28, 4 (April 1980), pp. 651-660.

A TOOL FOR THE AUTOMATED VERIFICATION OF ECCS SPECIFICATIONS OF OSI PROTOCOLS

Vincenza Carchiolo and Alberto Faro
Istituto di Informatica e Telecomunicazioni
Facolta' di Ingegneria - Universita' di Catania
Viale Andrea Doria, 6 - 95125 Catania (Italy)
Tel + 39 95 339449 - Telex 970255 UNIVCT I

1. INTRODUCTION

Due to the increasing complexity of the protocols defined by the OSI (Open Systems Interconnections) standards the ability to formally specify communication protocols and services is becoming more and more appreciate. In fact, it is common belief that the use of a formal technique is the only means to obtain a precise and unambiguos specificication of OSI communication protocols and services. Moreover, a formal technique allows one to formally define protocol properties and perform their verification.

The design of OSI systems can be viewed as consisting of two main steps:

i. specifications of services and protocols;

ii. verification (consisting in proving the completeness, correctness and consistency of the specification).

For these reasons the suitability of a formal technique as Formal Description Technique (FDT) for the specification of OSI protocols and services is strongly linked to the existence of automated tools able to aid the design of OSI systems, that is able to support the designer in the effort to provide formal specification and verification.

ECCS language, based on Milner's Calculus of Communicating Systems [Mil 80], provides a framework to formally specify and verify OSI systems as shown in [Car1 86], [Car2 86] and [Car 85]. This paper deals with the ability to provide some tools in order to make ECCS actually useful as FDT.

Two kinds of tools would be provided to implement an automated enviroment for designing OSI systems. The first tool is to make a syntax check and a simulation of the specification [Pap 87] for rapid prototyping; the second one is to perform the verification step.

This paper presents a verification tool for ECCS specifications, written in the logic programming PROLOG [Clo 81]. This tool is based on the simulator for ECCS specifications presented in [Pap 87] and on the bisimulation notion given in [Par 81]. For this reason the tool is named BIsimulation Prover (BIP). The verification algorithm used by this tool is an extension of the Sanderson algorithm [San 82].

The present paper is divided in four main sections: Section 2 contains a description of the used language, Section 3 discusses about the verification concept in the OSI framework, Section 4 gives some information about the algorithm on which the verification tool is based, Section 5 presents the structure of the BIP tool and gives a sort of implementation notes of the BIP core. Finally, some remarks are given on the suitability of PROLOG to implement the proposed verification tool.

2.ECCS: A SPECIFICATION LANGUAGE FOR OSI PROTOCOL

To perform a verification by BIP, the language used to specify a OSI system must be the ECCS language. It is based on Milner's CCS [Mil 80] and it is an attempt to provide an ad-hoc algebraic language for the specification of open systems. Some extensions are introduced in respect to CCS with the purpose to make easy the specification of service and protocol for open systems.

The ECCS (Extended CCS), as said, is the MILNER's Calculus for Communicating Systems, without value passing, and with the addition of the disable operator. Only a few syntatic variants are also introduced with respect to CCS syntax. A brief introduciton of ECCS follows.

ECCS is a calculus for specifying the behaviour of the communicating systems. An ECCS system (called process in the following) can be viewed as a black box interacting with the environment through communication points named 'gates'. The atomic form of communication is the action; the offers of a processes to communicate with the environment are termed 'observable actions'. An observable action, in the case of ECCS without value passing, is defined by:

- the gate where it is offered;
- the direction, input or output from the viewpoint of the process.

An observable action may take one of the following forms:

g?	output at the gate g
g^	input at the gate g

A process may also be capable of internal action denoted by T, for which no agreement with the environment is requested.

An action is the simplest ECCS process expression. An ECCS process expression can be built from others ECCS processes expressions by means of the suitable operators.

Table 1 shows the ECCS syntax, where P and Q are process expressions.

Table. 1. ECCS syntax

Operator	Process expression	Remark
Inaction	nil	
Action	a : P	a is an action
Sum	P + Q	
Composition	P par Q	
Disable	P dis Q	
Relabelling	P {S}	S is a relabelling
Restriction	P \ A	A is a list of gate
Behaviour identifier	p	p is an identifier

In Table 2 the syntax of action and relabelling are given in a BNF form.

Table.2.

```
<action> ::= <gate_name>? | <gate_name>^ | T
<gate_name> ::= <identifier>
<relabelling> ::= <gate_name_list>/<gate_name_list>
<gate_name_list> ::= <identifier> {,<identifier>}
```

The operator precedences are:

composition > disable > sum >
action > (relabbelling = restriction)

The formal language semantics is given in terms of inference rules, i.e. P-a->P' means that P executes the action a and transforms itself into P'. Table 3 presents the formal semantics of ECCS.

In the following an informal interpretation of the ECCS operators is given.

<u>Inaction</u>
nil can execute no action.

<u>Action</u> g? : P   g^ : P   T : P
g? : P can transform itself into P by executing an output action at gate g.
g^ : P can transform itself into P by executing an input action at gate g.
T : P can transform itself into P by executing the internal action T.

Table. 3. ECCS semantics

Operator	Premiss	Conclusions
Action	none	g?:P -g?-> P
		g^:P -g^-> P
		T:P -T-> P
Sum	P -a-> P'	P + Q -a-> P'
	Q -a-> Q'	P + Q -a-> Q'
Composition	P -a-> P'	P par Q -a-> P' par Q
	Q -a-> Q'	P par Q -a-> P par Q'
	P -g?-> P',Q-g^->Q'	P par Q -T-> P' par Q'
Disable	P -a-> P'	P dis Q -a-> P'
	Q -a-> Q'	P dis Q -a-> P dis Q'
Relabelling	P -a-> P'	P{S} -a-> P'{S}
Restriction	a ∉ A P -a->P'	P \ A -a-> P'\A
Behaviour identifier	p := P, P-a->P'	p -a-> P'

Sum P + Q
P + Q may behaves like P or Q.

Composition P par Q
P par Q describes the concurrent behaviour of P and Q with communication through identical gates.

Disable P dis Q
P dis Q may behaves like Q and terminate together with it. Alternatively, at any step, it may start behaving like P

Relabelling P{S}
A relabelling S (expressed by X/Y) is a morphism over gates. If X an Y are gate lists of equal lenght, X/Y denotes that each of gate in X must be mapped with the corresponding gate in Y.

Restriction P\G
The behaviour of P\G is equal to the behaviour of P with the gates in G hidden to the environment.

Behaviour identifier p:=P
Each process expression P can be associated to a behaviour identifier p in the form p:=P

In BIP system to prove the equivalence between two process expressions P and Q, possibly containing some behaviour identifiers, they must be named, i.e. they must be associated with a behaviour identifier and they must be stored in a file.

A relevant concept of a specification language is the notion of equivalence. More than one equivalence relation has been introduced for CCS [Mil 80], [Par 81], [DeN 82]; each of them could be easily introduced in ECCS. In this paper, we shall only introduce the equivalence relation, termed obsevation equivalence, needed for our verification purpose. We shall define this equivalence relation in term of a bisimulation relation like in [Par 81]. To this aim, we define the relation

$$=s=>$$

where s is a string of observable actions (possibly the empty string ε).

If $s=\varepsilon$ then $=s=> \ \equiv (-T->)^*$
If $s=a1\ a2 \ldots aN$ then
$=s=> \ \equiv (-T->)^* -a1-> (-T->)^* -a2-> (-T->)^* -aN-> \ldots (-T->)^*$

P and Q are observation equivalent if there exists a bisimulation relation R such that $\langle P,Q\rangle \in R$. A relation R is a bisimulation if:

$\forall\ s : P=s=>P'$ exists a $Q' : Q=s=>Q' \ \langle P',Q'\rangle \in R$
$\forall\ s : Q=s=>Q'$ exists a $P' : P=s=>P' \ \langle P',Q'\rangle \in R$

3. PROTOCOL VERIFICATION

Protocol verification consists in proving the completeness, correctness and consistency of the protocol specifications. Completeness needs that each protocol entity is able to manage all the inputs coming from the cooperating remote entity and from the adjacent ones. Correctness needs that the protocol satisfies both safety and liveness properties. Consistency needs that the service offered (requested) by the protocol conforms to that expected (given) by the upper (lower) layer.

Incompleteness can produce two undesirable effects:

i) one or both the protocol entities handle as error the arrival of an input which on the contrary is productive for the service to be offered by the protocol;
ii) the protocol does not offer all the service options.

The first situation is not acceptable because it generally leads to deadlock or does not permit the correct execution of the purpose of the protocol, so not satisfying the safety condition. The second situation on the contrary may be acceptable if the user is not interested in all the service options. Completeness analysis is the first step for protocol verification. Generally the completeness is statically proved by checking that there are no unspecified productive receptions for each protocol entity.

Another property to be proved for protocol safety is the absence of overspecification. We can have two main types of overspecifications:

i) the protocol contains some redundant specifications which are never used for offering the service fixed by the designer;

ii) the overspecification constraints determine the impossibility of offering the service fixed by the designer.

We note that the last situation should be surely avoided because it does not conforms to the safety property. On the contrary the former could be acceptable, but it should be avoided because it could lead to unsafe situations. For example the unexecised message receptions specified in the first situation could be execised by erroneous messages, thus leading to bad behaviours. The absence of the first type overspecification can be statically proved by checking that there are no unexecised message receptions for every allowable state pair (protocol global state) of the cooperating entities. The proof of the second type overspecification is part of the proof of the correct execution of the purpose of the protocol (see below).

Other two important properties for protocol safety are: the freeness of state ambiguities and the deadlock freeness. These properties can be proved by respectively checking the absence of entity states shared by two different global stable states and the absence of global states consisting of the pair of the entity states allowing only message receptions.

The above four properties together with the property that the protocol correctly executes its purpose guarantee the protocol safety, i.e. protocol partial correctness. To prove the total correctness of a protocol it is necessary also to prove the liveness properties.

Protocol liveness generally requests the absence of unproductive cycles during the protocol execution (livelocks) due to either an erroneous resource scheduling (starvation condition) or the relative speed of the messages (tempo-blocking condition). In addition liveness requests that the protocol is able to handle faults (i.e. recovery from failure) and to come back to a normal situation after any possible pertubation (i.e. selfsynchronization). Finally liveness requests that the protocol terminates within a finite time (finite termination); however sometimes it may be acceptable an "infinite termination", that is the property that the protocol eventually terminates (fairness). The liveness properties together with the assumed partial correctness allow us to prove that the protocol completes in a finite time the requested service (total correctness).

After proving the protocol completeness and correctness, the protocol verification needs the consistency of the specifications. This can be obtained by proving that the purpose of the protocol (that is the service offered by the protocol) is equivalent to that expected by its users.

Several approaches exist in literature for protocol verification. Generally they use a hybrid approach consisting of both state exploration and assertion proving. Algebraic approach on the contrary allows us to verify protocol by using a single framework. In particular the purpose of the protocol (i.e. the service to be offered to its users) is expressed by an algebraic formula and the verification essentially consists in proving the equivalence between the exepected service and that really offered by the protocol obtained by using algebraic operators from the algebraic specifications of the protocol entities and the intermediate channel. The above verification can be perfomed by the normal equational reasoning for finite state protocol and makes possible to point out the following undesirable situations:

- unspecified message receptions (protocol completeness);
- unexercised message receptions (overspecification);
- state ambiguities;
- states without progress (deadlock).

Therefore algebraic approach allows us to prove partial correctness. Algebraic approach allows us also to prove the liveness properties verifiable by a finite observer. Thus it is not possible to prove that a certain cyclic situation is not a livelock. Livelocks can be treated by extending the algebraic approach by temporal logic constructs. In addition both temporal logic and structural induction should be applied to treat protocol divergence (fairness or infinite state evolution).

4. BISIMULATION PROVING: AN ALGORITHM

As said in the introduction the part of the verification framework that we must take into account in this paper concern with the proof of the consistency of the service offered by the protocol and the one expected by the upper layer.

The proof of other properties could be performed by using ECCS too, but this topic is out of the scope of this paper.

In the spirit of ECCS the proof of consistency consists in proving the observation equivalence between the service expected at the upper layer and the composition of the underlying service with the entities performing the protocol.

Our believe is that a more feasible algoritm for proving the observation equivalence should be one based on the bisimulation definition of Section 2.

This algorithm is based on the definition of observation equivalence in terms of bisimulation relation as introduced in [Par 81]. It aims to show the existence of a bisimulation relation between the two processes. The Sanderson algorithm [San 82] is based on this idea. In this paper we present another algorithm that is an enhancemnt of the Sanderson algorithm.

The key concept of both the algorithms is to build, if any, the set R defined in the following way:

$\langle P,Q\rangle \in R$ iff,

i) $\forall s: P \overset{s}{\Longrightarrow} P'$ exists $Q': Q \overset{s}{\Longrightarrow} Q' \quad \langle P',Q'\rangle \in R$
ii) $\forall s: Q \overset{s}{\Longrightarrow} Q'$ exists $P': P \overset{s}{\Longrightarrow} P' \quad \langle P',Q'\rangle \in R$

where s is a sequence of actions.

This is sufficient to guarantee that a bisimulation

relation between P and Q exists.

If this set does not exist, then the algorithm terminates with a failure. In this case, the behaviours of P and Q are not in a bisimulation relation, that is P and Q are not equivalent under the fixed point definition.

Sanderson algorithm constructs, if any, the set R satisfying the above said conditions (for two behaviours specified by P and Q) under the following restrictive conditions:

1) behaviour P (or Q) must be rigid, i.e. the behaviour of P must be described avoiding the use of internal action T and the non_deterministic choice.
2) the behaviours P and Q must have no derivation that can diverge.

The algorithm used in BIP tool removes the limitation 1) of Sanderson algorithm.

Both methods are based on a recursive algorithm and aiming at building a list containing the behaviour pairs that are in a bisimulation relation.

The condition i) and ii) in the case of Sanderson algorithm can be replaced, as shown in [San 82], by the following simpler conditions:

1) ∀ δ : P -δ-> P′ exists a Q′: Q -δ-> Q′ , <P′,Q′> ε R
or
Q -T-> Q′ , <P,Q′> ε R
2) ∀ δ : Q -δ-> Q′ exists a P′: P -δ-> P′ , <P′,Q′> ε R
3) ∀ T : Q -T->Q′ then <P,Q′> ε R

where δ is a generic visible action and T is the internal action.

In the following we refer to R_list as the list containing the behaviour pairs, and we marking the pairs with a check_mark and unchecked_mark.

Let A and B the behaviour pairs for which the bisimulation relation must be proved, this algorithm can be described as follows:

1. Initially the R_list consists of the only pair <A,B> with the unchecked mark.

2. The first pair of the R_list with a mark unchecked is taken into account. We call this pair <P,Q>; if there is no pair with unchecked mark then the claimed bisimulation is proved.

3. The pair <P,Q> is checked by the condition 1. If there exists the behaviour Q′: Q-δ->Q′ then the pair <P′,Q′> is put in the R_list with the unchecked mark;orif there exists the behaviour Q′: Q-T->Q′ then the pair <P,Q′> is put in the R_list with the unchecked mark; otherwise there is a failure and the bisimulation relation between A and B is disproved.

4. The pair <P,Q> is checked by the condition 2. If there exists the behaviour P′ then the pair <P′,Q′> is put in the

R_list with the unchecked mark; otherwise there is a failure and the bisimulation relation between A and B is disproved.

5. The pair $\langle P,Q\rangle$ is checked by the condition 3 and the pair $\langle P,Q'\rangle$ is put in the R_list with the unchecked mark.

6. Return to step 2

In the case of a no_rigid behaviour pairs the algorithm is symmetric and two main differences in the algorithm can be found.

The first concerns with the fact that condition i) and ii) should be replaced, as shown in the [Car 87], by the following simpler symmetric conditions:

1) $\forall\ \delta$: $P -\delta\to P'$ exists a Q': $Q -\delta\to Q'$, $\langle P',Q'\rangle \in R$
or
$Q -\tau\to Q'$, $\langle P,Q'\rangle \in R$

2) $\forall\ \tau$: $P -\tau\to P'$ exists a Q': $Q -\tau\to Q'$, $\langle P',Q'\rangle \in R$
or
$Q \not\xrightarrow{\tau}$, $\langle P'Q\rangle \in R$

3) $\forall\ \delta$: $Q -\tau\to Q'$ exists a P': $P -\delta\to P'$, $\langle P',Q'\rangle \in R$
or
$P -\tau\to P'$, $\langle P'Q\rangle \in R$

4) $\forall\ \tau$: $Q -\tau\to Q'$ exists a P': $P -\tau\to P'$, $\langle P',Q'\rangle \in R$
or
$P \not\xrightarrow{\tau}$, $\langle P,Q'\rangle \in R$

where δ is a generic visible action and τ is the internal action.

The second one concerns with the need to make a backtraking operation in the construction of the process pair list that are in a bisimulation relation. To manage the backtracing operations two process pair lists are generated, one for the pairs of processes proved equivalent, the second for the pairs of processes assumed temporarily equivalent.

5. THE IMPLEMENTATION CODE

This section gives some information about the implementation code of the BIP system. Our implementation choices and the development of the BIP system are also discussed.

As said in Section 1, BIP system is a PROLOG tool for verifying consistency of distributed systems.

Figure 1 shows the BIP system. It is functionally structured in three modules and uses the information contained in three data base: PDB, DDB and EDB.

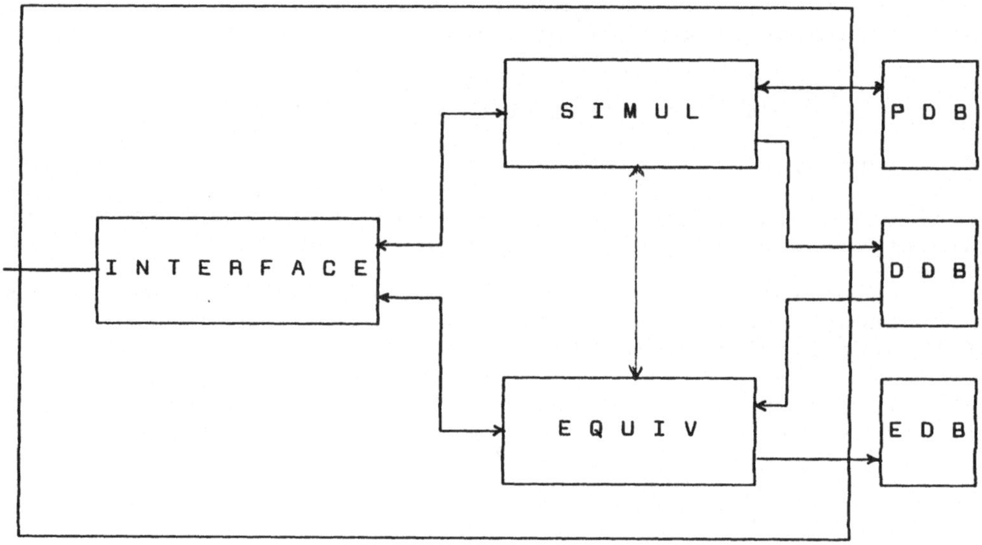

Fig.1 BIP structure

The first data base is a collection of ECCS behaviour; BIP investigates on the consistency of the only processes whose behaviour expressions are associated to behaviour identifiers in the data base. This data base is called Processes Data Base, in the figure it is represented by the box PDB.

The second data base named Derivation Data Base (DDB in the figure) contains the triplet formed by:

- the process P
- the action X
- the process P'

that are in the relation P -X-> P'

The third data base, named Equivalence Data Base (EDB in the figure) contains the behaviour identifier pairs of the processes assigned to be in bisimulation relation.

In addittion, Figure 1 shows three functional module: a user directed module, named INTERFACE, and two operative modules named SIMUL and EQUIV.

INTERFACE module is devoted to manage the interactions between the user and the system and it provides the user interface that making the system interactive. This module contains the predicated devoted to manage the video and to store the processes in the data base PDB.

SIMUL module is devoted to simulate a given process expressed by a ECCS behaviour expression. Let us p the behaviour identifier of a process contained in the data base PDB, the SIMUL module returns, for each possible first action that the process p may perform, the pair X, p'. Where X is the action and p' is the behaviour identifier of the process in which the execution of action X transforms P. Briefly, SIMUL module selects a process of PDB and generates the triplets of DDB to be checked by EQUIV. Moreover, it adds the resulting processes P' to PDB. This module contains the predicates performing the inference rules of table 3. A more deeply description of the structure and functionality of this module can be found in [Pap 87].

EQUIV module is the core of the BIP system. It is the module that actually executes the verification of the consistency between processes. It aims to verify that two processes are observation equivalent, in the sense that there exists a bisimulation relation between the two processes initially inserted in PDB by the BIP user. This module is structured in accord with the bisimulation algorithm presented in the previous section. It contains the predicate bbis(P,Q) devoted to prove the bisimulation relation between two finite state processes having, respectivelly, P and Q, as processes identifiers. The predicate bbis(P,Q) succeeds if P and Q are in a bisimulation relation. The predicate bbis(P,Q) behaves as follows:

i) prove the existence of the bisimulation relation between P and Q.
ii) manage the backtracking operations when the search of a bisimulation relation between P and Q fails.

To attempt at step i), the predicate bbis(P,Q) uses predicate bis(P,Q) in order to check the existence of the bisimulation relation. It succeeds if:

i) the pair P,Q is already in the equivalent behaviour list expressed by the predicate e(P,Q).
ii) the pair P,Q is already in the temporarily equivalent behaviour list expressed by the predicate temp(P,Q).
iii) the conditions 1, 2, 3 and 4 of the previous section, are verified.

Breafly this module starting by the information contained in the data base DDB returns the equivalent behaviour pairs that must be put in the data base EDB.

The advantages of the presented structure of BIP system is the easily up-to-dating of this tool. If we choose another equivalence relation, i.e. the testing equivalence [DeN 82] to adequate the system it is sufficent to modify only the EQUIV module; the other modules can be used without variantions.

6. CONCLUSIONS

A verification tool has been presented aimed at making easy the design of OSI systems. The programming language used, viz. PROLOG, has been shown suitable for a verification tool for two reason:
- the easy mapping of the ECCS semantics and bisimulation definition onto the PROLOG predicates;
- the agreement of the PROLOG with the backtraking operation

needed.

BIP could be easily extended for others FDT having a simiar structure of ECCS like, for example LOTOS [Lot 85], one of the standard FDT defined by ISO.

REFERENCES

[Car 85] V.Carchiolo and G.Pappalardo,"CCS as a Specification and Verification Technique: A Case Study and a Comparison with Temporal Logic", Pacific Computer Communication Symposium, Seoul (Korea), Oct. 1985

[Car1 86] V.Carchiolo et alii,"ECCS and LIPS: two languages for OSI systems specification and verification", Internal Report, Istituto di Informatica e Telecomunicazioni, Catania, Italy, 1986.

[Car2 86] V.Carchiolo et alii,"A LOTOS Specification of the PROWAY Highway Service", IEEE Trans. on Computers, Vol. C-35, No.11, Nov. 1986

[Car 87] V.Carchiolo and A.Faro, "On Bisimulation Relation", Internal Report, University of Catania, 1987

[Clo 81] Clocksin W., Mellish, "Programming in PROLOG, Springer-Verlag, 1981

[DeN 82] R.De Nicola, M.Hennesy, "Testing Equivalences for Processes", Internal Report, University of Edinburgh, CSR-123-82, 1982

[Lot 85] ____, Information Processing Systems - Open Systems Interconnection, LOTOS - A Formal description technique based on temporal ordering of observational behaviour, ISO/TC 97/SC 21/ DP 8807, Jnue 1985

[Mil 80] R.Milner, "A Calculus of Communicating Systems", LNCS 92, Springer Verlag, Berlin 1980

[Pap 87] G.Pappalardo, "Experiences with a verification and simulation tool for behavioural language", proc. of VII IFIP Workshop on Spec., Verif. and Testing, 1987

[Par 81] D. Park, " Concurrency and Automata on Infinite Sequences", in Vol. 104, LNCS, Springer-Verlag, 1981

[San 82] M.T.Sanderson, "Proof Techniques for CCS", Internal Report, University of Edinburgh, CST-19-82, 1982

Supervisory Control of Discrete Event Systems: A Survey and Some New Results

Peter J. Ramadge *

Department of Electrical Engineering
Princeton University, Princeton NJ 08544.

Abstract

We present an overview of the modeling of discrete event systems using formal languages. Some new results on the controllability of sequential behaviors are presented and a standard coordination problem for a product system is shown to be of polynomial complexity.

1 Introduction.

A discrete event system (DES) is a dynamic system that evolves, i.e., changes state, in accordance with the occurrence, at possibly unknown irregular intervals, of discrete events. For example, an event may correspond to the completion of a task or the failure of a machine in a manufacturing system, the arrival or departure of a packet in a communication system, or the occurrence of a disturbance or change of setpoint in a complex control system. Such systems arise in a variety of areas including, for example, computer operating systems, distributed computing, computer networks, data bases, communication networks, manufacturing systems, the start-up and shut-down procedures of industrial plants, and the higher level intelligent control of complex multi-mode processes.

Control problems for DES center on the idea of how to ensure, by control, the orderly and efficient flow of events. Within is overall goal it is possible to recognize a hierarchy of control problems: higher levels dealing with optimization, lower levels with the logical aspects of decision making, information processing, and feedback control, and the lowest levels addressing implementation issues in terms of hardware and real time computer languages.

In this article we are concerned with the second level of the above hierarchy: the logical coordination of DES. We survey the modeling of DES in the framework of [RW1] and report some new results concerning event sequences and the complexity of controller synthesis.

* Research partially supported by the National Science Foundation through grant ECS-8715217 and by an IBM Faculty Development Award.

Numerous models for DES have been proposed in the literature. These models can be roughly grouped as follows: Petri nets [P]; Boolean models [A]; sample path models (traces, runs, languages) [BeN],[CDFV],[H],[MM],[Pa],[RW1]; and models based on temporal [FST], [HO],[MW],[OW],[O],[TW] or modal [Ha],[HZ],[MDH] logic. Although seemingly diverse these models have a common connection through formal languages and sample paths of events. Roughly speaking they are simply different means of specifying and reasoning about the set of system sample paths considered as either a set of finite length strings or as as a set of infinite sequences.

In [RW1],[WR1] Ramadge and Wonham proposed a simple framework for the study of the supervision, i.e., control, of a class of discrete-event systems. This theory uses a simple 'sample path' model for a discrete-event system to study a number of qualitative issues such as the existence, uniqueness, and structure of supervisors for simple control tasks. In addition, algorithms are developed for the synthesis of the desired supervisors. In its use of formal languages the model is similar to the work of [S] and [Sh] on flow expressions and path expressions respectively; the work of [BeN] on using automata models to study process synchronization; and there are certain points of similarity with the linguistic approach of Hoare to the specification of concurrent processes [H]. The framework has proved useful in the theoretic analysis of a number of basic supervisory control problems [RW1],[WR1]; has motivated investigations using related models in database systems [LaW], and manufacturing systems [MT]; and more recently has been extended to cover modular [RW2],[WR2] and distributed [CDFV],[LW1],[LW2] control.

The remainder of the paper is organized as follows. In Section 2 we describe the the modeling of DES in terms of languages, Section 3 introduces the concept of a controllable language and discusses a basic control problem, and in Section 5 we consider the issue of the complexity of supervisor synthesis in the context of a standard coordination problem. Space limitations preclude the inclusion of proofs - for these the interested reader is referred to the appropriate literature.

2 Discrete Event Systems

Intuitively a DES consists of a set of elementary events together with a specification of the possible orders in which these events can occur. To formalize this notion let Σ denote a finite set of events, and Σ^* denote the set of all finite strings of elements of the set Σ, including the empty string 1. We say that $u \in \Sigma^*$ is a *prefix* of $v \in \Sigma^*$, denoted $u \preceq v$, if for some string $w \in \Sigma^*$, $v = wu$, and a *proper prefix*, denoted $u \prec v$, if $w \neq 1$. The *prefix closure* of $L \subseteq \Sigma^*$ is the subset $\bar{L} \subset \Sigma^*$ defined by

$$\bar{L} = \{u : uv \in L \text{ for some } v \in \Sigma^*\}$$

and L is *prefix closed* if $\bar{L} = L$.

The behavior of a DES can then be modeled as a prefix closed language $L \subseteq \Sigma^*$. Here L represents the possible (finite) strings of events that the DES can generate.

A natural extension of the above model is to consider infinite sequences of events in addition to finite strings. For this let $N = \{1, 2, 3, \ldots\}$ denote the set of positive integers, and Σ^ω denote the set of all sequences of elements of Σ, i.e.,

$$\Sigma^\omega = \{e \colon \; e \colon N \to \Sigma\}$$

For $e \in \Sigma^\omega$ and $j \in N$, let $e(j)$ denote the jth element of e, and $e^j = e(1)e(2)\ldots e(j)$ denote the string consisting its first j elements. A subset $B \subseteq \Sigma^\omega$ is termed an *ω-language*, and the *prefix* of B is the subset $pr(B)$ of Σ^* defined by

$$pr(B) = \cup_{j \geq 1} \cup_{e \in B} e^j$$

For an increasing sequence $u_1 \prec u_2 \prec u_3 \prec \ldots$ of elements of Σ^*, there is a unique element $e \in \Sigma^\omega$ such that $e^j = u_k$ for $j = |u_k|$, $k \in N$. We call e the *limit* of $\{u_k\}$. The *adherence* [(1)] or *limit* of $L \subseteq \Sigma^*$ is the ω-language

$$L^\infty = \{e \colon e \in \Sigma^\omega \;\&\; e^j \in L \text{ for infinitely many } j \in N\}$$

Thus $e \in L^\infty$ if and only if there exists an increasing sequence $u_1 \prec u_2 \prec \ldots$ of elements of L such that $e = \lim u_k$. Note that if L is prefix closed, then $e \in L^\infty$ if and only if $e^j \in L$ for all $j \in N$.

We incorporate sequences into the DES model by modeling a DES as a pair $A = (L, S)$ where L is a prefix closed subset of Σ^* and S is a subset of L^∞. In general it need not be the case that $pr(S) = L$. Equality implies that every string in L is a prefix of some sequence in S. Roughly this can be interpreted to mean that the system is never 'blocked' or 'deadlocked' and thus unable to produce a string in S. Hence when $pr(S) = L$ we say that A is *nonblocking*.

The language based model defined above is representation independent. The languages L and S could be specified, for example, by finite automata, Petri nets, fixed point equations, Turing machines, etc.. We make no assumption or restriction at this point about specific representations. Of course at some point it may be interesting (or necessary) to specialize to particular classes of languages for which a deeper analysis is possible.

2.1 Control

We assume that Σ is partitioned into *uncontrollable* and *controllable* events: $\Sigma = \Sigma_u \cup \Sigma_c$. An admissible input for A consists of a subset $\gamma \subseteq \Sigma$ satisfying $\Sigma_u \subseteq \gamma$. Let $\Gamma \subseteq 2^\Sigma$ denote the set of admissible inputs; note that Γ is closed under set union and set intersection. If $\gamma \in \Gamma$ and $\sigma \in \gamma$, then we say σ is *enabled* by γ, otherwise we say σ is *disabled* by γ. Disabled events are prevented from occurring while enabled events can occur when permitted by the prescribed dynamics; thus γ represents the allowed 'next' events. The condition $\Sigma_u \subseteq \gamma$ means that the uncontrollable events are always enabled.

A *supervisor* for the controlled DES (CDES) $A = (L, S)$ is a map

$$f \colon L \to \Gamma$$

---

(1) After [BoN]. See also [Mc],[E].

specifying for each possible (finite) string of generated events the next input to be applied. The closed loop DES consisting of f supervising A is denoted by (A, f), and the closed loop behaviors, denoted L_f and S_f, are defined as follows:

(i) $1 \in L_f$; and

(ii) $w\sigma \in L_f$ if and only if $w \in L_f$ & $\sigma \in f(w)$.

(iii) $S_f = L_f^{\infty} \cap S$

Note that it is sufficient for f to be specified on a subset K of L containing L_f.

From the above definition it follows that

$$pr(S_f) \subseteq L_f$$

In general there need not be equality in this expression. Equality implies that the system (A, f) is nonblocking in which case we say that f is *nonblocking* for A.

3 Controllable Languages

The basic control problem in the above framework is the following: given a language $K \subseteq L$ (resp. an ω-language $B \subseteq S$) does there exist a nonblocking supervisor f such that $L_f = K$ (resp. $S_f = B$). The answer in the case of string languages was given in [RW1] in terms of the concept of a controllable language. This is a language $K \subseteq L$ satisfying the following invariance property:

$$\bar{K}\Sigma_u \cap L \subseteq \bar{K}$$

It can be shown that for nonempty $K \subseteq L$ there exists a supervisor f such that $L_f = K$ if and only if K is both prefix closed and controllable [RW1, Prop. 5.1]. We show below that a similar result holds in the case of sequential behaviors.

A metric ρ can be defined on Σ^ω by

$$\rho(e_1, e_2) = \begin{cases} 1/n, & \text{if } e_1^{n-1} = e_2^{n-1} \text{ and } e_1(n) \neq e_2(n); \\ 0, & \text{if } e_1 = e_2. \end{cases}$$

The topolgical closure of a set $B \subset \Sigma^\omega$ with respect to the above metric is denoted $\bar{B}$, and $B \subseteq S$ is said to be *closed relative to S* if $\bar{B} \cap S = B$.

Proposition 3.1.

If $B \subseteq S$ is nonempty, then there exists a nonblocking supervisor f such that $S_f = B$ if and only if

(1) $pr(B)$ is controllable, i.e., $pr(B)\Sigma_u \cap L \subseteq pr(B)$; and

(2) B is closed relative to S, i.e., $\bar{B} \cap S = B$.

A subset $B \subseteq S$ satisfying the two conditions of the above proposition is said to be a *controllable sequential behavior.*

It was shown in [RW1] that the family of prefix closed and controllable sublanguages of L is closed under set union and set intersection, and hence forms a lattice under subset inclusion. Since the empty set and L are controllable it follows that and for any closed $K \subseteq L$ there exists a unique largest closed and controllable language $K^{\uparrow}$ and a unique smallest closed and controllable language $K^{\downarrow}$ such that $K^{\uparrow} \subseteq K \subseteq K^{\downarrow}$. These can be thought of as the best controllable approximations to the language K.

The set of controllable ω-languages is closed under arbitrary intersections and under finite unions, but not in general under countable unions. Nevertheless in certain situations it is still the case that there exists a unique maximal controllable ω-language contained in a prescibed ω-language B.

Proposition 3.2.

If $B \subseteq S$ is closed relative to S, then there exists a unique maximal controllable ω-language $B^{\uparrow}$ contained in B.

4 Finite Representations and Computation

For purposes of computation it is necessary to select finite representations for the languages L and S. One way in which this can be done is as follows. A *generator* G is a dynamic system consisting of a state set Q, an initial state q_0, and transition function $\delta : \Sigma \times Q \to Q$ (in general a partial function). Without loss of generality we assume that every state of G is reachable from the initial state, i.e., that G is accessible. δ is extended to a (partial) function on $\Sigma^* \times Q$ in the standard fashion [HU, p.17], and we write $\delta(w,q)!$ as an abbreviation for the phrase '$\delta(w,q)$ is defined'. Then the language *generated* by G is defined to be the subset

$$L(G) = \{w : w \in \Sigma^* \ \&\ \delta(w, q_0)!\}$$

Every closed language $L \subseteq \Sigma^*$ has such a representation. However, as is well known, L has a finite state representation if and only if it is a closed regular language.

The limit behavior S can be specified as follows. Adjoin to $G = (\Sigma, Q, \delta, q_0)$ one or more subsets of states $Q_m \subseteq Q$ (2) . To each sequence of events $e \in L(G)^{\infty}$ there corresponds a unique state trajectory $s_e: N \to Q$ satisfying

$$s_e(j) = \delta(e^j, q_0)$$

The sequence e and trajectory s_e are said to be *admissible* if s_e visits the set Q_m infinitely often. The set of event sequences generated by $G = (\Sigma, Q, \delta, q_0, Q_m)$ is then defined to be (3)

$$S(G) = \{e : e \in L(G)^{\infty}, \text{ and } s_e \text{ is admissible } \}$$

It is well known that an ω-language S can be represented in this fashion if and only if S is the adherence of a regular language. This comprises a proper subset of the regular ω-languages [E].

(2) For simplicity we restrict attention at this point to one subset.

(3) This is a deterministic Büchi automaton [B].

It is clear that in the above representation

$$S(G) \subseteq L(G)^{\infty} \tag{G1}$$

with equality if $Q_m = Q$; and that

$$pr(S(G)) \subseteq L(G) \tag{G2}$$

Similarly for computation and implementation purposes one must select a finite representation for a supervisor. One possibilty is to realize a supervisor in terms of a state machine together with an output map [RW1],[WR1]. For this let $S = (\Sigma, X, \xi, x_0)$ be an automaton and $\phi : X \to \Gamma$. We say that the pair (S, ϕ) *realizes* the supervisor f if for each $w \in L(G, f)$

$$\phi(\xi(w, x_0)) = f(w)$$

We interpret S as a standard automaton whose state transitions are driven by the events in Σ. In turn the state feedback map ϕ determines the input for G as a function of the state of S. We say that f is a *finite state supervisor* if it has a finite state realization.

Let G_1 be a finite state generator for the DES $A = (L, S)$, and let $K \subseteq L$ be a regular language represented by a finite state automaton G_2. An algorithm for computing the supremal controllable sublanguage of K, based on a lattice fixpoint characterization of $K^{\uparrow}$, is given in [WR1]. This algorithm requires a time bounded by a polynomial in the number of states of G_1 and G_2, and produces a finite state generator for $K^{\uparrow}$. A finite state supervisor realization that implements $K^{\uparrow}$ can then be synthesized directly from the generator for $K^{\uparrow}$. This algorithm also provides a polynomial time decision procedure for testing the controllability of a given language $K \subset L$.

5 Product Systems

Our main interest is in a class of structured DES which we call product systems. These are DES composed of a finite set of asynchronous interacting components. Such systems arise naturally when modeling the concurrent operation of several asynchronous, or partially synchronous discrete dynamical systems. One of the principal difficulties in dealing with product systems is that the number of states increases exponentially with the number of components. Thus synthesis methods based on searching over the product state space are not computationally feasible. For example, the general supervisor synthesis problems posed and solved in [RW1], [WR1], although known to be of polynomial complexity when the size of a problem instance is measured in terms of the number of system states [RW3], cannot be regarded as computationally tractable for product systems. We regard a decision or synthesis problem for a product system as computationally feasible if it can be solved in a time bounded by a polynomial in the size of the component subsystems n and the number of components p.

Let $A_i = (L_i, S_i)$ be be p finite state DES over disjoint alphabets $\Sigma_1, \ldots, \Sigma_p$, with control partitions $\Sigma_i = \Sigma_{ci} \cup \Sigma_{ui}$. For each A_i assume

$$S_i \neq \emptyset$$

$$pr(S_i) = L_i$$

Let $\Sigma = \cup_{i=1}^{p} \Sigma_i$, and define the projection $p_i : \Sigma^* \to \Sigma_i^*$ of Σ^* onto Σ_i^* by

$$p_i(\sigma) = \begin{cases} \sigma, & \text{if } \sigma \in \Sigma_i; \\ 1, & \text{if } \sigma \in \Sigma_j \text{ with } i \neq j. \end{cases}$$
$$p_i(w\sigma) = p_i(w)p_i(\sigma) \quad w \in \Sigma^*,\ \sigma \in \Sigma$$

A sequence $e \in \Sigma^\omega$ is Σ_i-*recurrent* if $e(j) \in \Sigma_i$ infinitely often. In this case let e_i denote the unique subsequence of e consisting of the elements of Σ_i, and extend the projection p_i to a partial function $p_i : \Sigma^\omega \to \Sigma_i^\omega$ by defining

$$p_i(e) = \begin{cases} e_i, & \text{if } e \text{ is } \Sigma_i\text{-recurrent;} \\ \text{undefined}, & \text{otherwise.} \end{cases}$$

The product system $A = ||_{i=1}^{p} A_i$ is defined to be the DES (L, S) with [4]

$$L = \{w : w \in \Sigma^* \ \&\ p_i(w) \in L_i, i = 1, \ldots, p\}$$

and

$$S = \{e : e \in \Sigma^\omega \ \&\ p_i(e) \in S_i,\ i = 1, \ldots, p\}$$

Assume that for $i = 1, \ldots, p$ the component DES A_i has a finite state realization

$$G_i = (\Sigma_i, Q_i, \delta_i, q_{0i}, Q_{mi})$$

Let $|Q_i|$ denote the cardinality of Q_i and set $n = max\{|Q_i| : 1 \le i \le p\}$.

The product generator $G = ||_{i=1}^{p} G_i$ is defined according to

$$G = (\Sigma, Q, \delta, q_0)$$

with

$$\Sigma = \cup_{i=1}^{p} \Sigma_i \quad (\Sigma_c = \cup_{i=1}^{p} \Sigma_{ci})$$
$$Q = \Pi_{i=1}^{p} Q_i$$
$$q_0 = (q_{01}, \ldots, q_{0p})$$

and for $\sigma \in \Sigma_i$

$$\delta(\sigma, (q_1, \ldots, q_i, \ldots, q_p)) = (q_1, \ldots, \delta_i(\sigma, q_i), \ldots, q_p)$$

provided $\delta_i(\sigma, q_i)!$.

For $i = 1, \ldots, p$ let

$$Y_{mi} = \{q : q \in Q, q_i \in Q_{mi}\}$$

These sets will pay the role of the set Q_{mi} for the generator G_i, except for G there are p such recurrent sets - one for each of the component generators.

(4) This assumes a fair shuffling of the component behaviors. If this is not acceptable then the sequential behavior can be enlarged to include unfair shufflings.

The language generated by G is defined in the usual fashion, i.e.,

$$L(G) = \{w : w \in \Sigma^*, \delta(w, q)!\}$$

To each event sequence $e \in L(G)^\infty$ there corresponds a unique state trajectory s_e. The sequence e and trajectory s_e are *admissible* if s_e visits each of the sets Y_{mi}, $i = 1, \ldots, p$, infinitely often. The sequential behavior of G is then defined to be the set

$$S(G) = \{e : e \in L(G)^\infty, \text{ and } s_e \text{ is admissible }\}$$

It is readily verified that for each i, $1 \leq i \leq p$,

$$L(G) = \{w : w \in \Sigma^* \ \& \ p_i(w) \in L(G_i), i = 1, \ldots, p\}$$

and that

$$S(G) = \{e : \ e \in \Sigma^\omega \ \& \ p_i(e) \in S(G_i), \ i = 1, ..., p\}$$

Thus G is a representation of the product system A.

Note that if each G_i has n states, then G has n^p states. If p is bounded, then the size of G is bounded by a polynomial in n. It follows from our previous remarks that control problems for G formulated in the framework of [RW1] are decidable in a time bounded by a polynomial in n. Here, however, we are interested in the case when both p and n are variable and both are to be taken as a measure of problem size.

6 A Coordination Problem

In what follows $A = \|_{i=1}^p A_i$ will be a product system with components $A_i = (L_i, S_i)$, $i = 1, \ldots, p$. In order to discuss the complexity of decision and synthesis problems for A we need to have a finite representation for the product system. For this we assume that finite state realizations $G_i = (\Sigma, Q_i, \delta_i, q_{0i}, Q_{mi})$, $i = 1, \ldots, p$, are provided for each of the components A_i, and let G denote the corresponding product generator.

A supervisor f for the product system is a *coordinator* if for every set of p event sequences $e_1, e_2, \ldots, e_p$ with $e_i \in S_i$, $i = 1, ..., p$, there exists a sequence e in the closed loop behavior S_f such that for $i = 1, ..., p$

$$p_i(e) = e_i$$

i.e., the supervisor does not modify the open loop behaviors of the individual DES; it only constrains how they interact by controlling the relative order of events.

A subset $\bar{Q}$ of the state set of the generator G is said to be *nontransient* if there exists an admissible state trajectory for G that visits $\bar{Q}$ infinitely often.

We now analyze the following standard problem:

MUTUAL EXCLUSION (MEX): *Let $\bar{Q}_i \subseteq Q_i$ be p given nontransient subsets and k be a fixed integer with $1 \leq k < p$. Synthesize (if possible) a nonblocking supervisor f for A satisfying the following two conditions:*

(1) f is a coordinator; and

(2) For each $e \in S_f$, and each $j \geq 1$, after e^j at most k of the generators G_i satisfy $q_i \in \bar{Q}_i$.

The problem requires the A_i to be coordinated so that at most k of the generators G_i are in the designated subsets of states at any one time. For $k = 1$ this is the traditional mutual exclusion problem. When $k = p - 1$ the problem is equivalent to ensuring that the state of G never enters the subset $\bar{Q} = \Pi_{i=1}^{p} \bar{Q}_i$, or equivalently that $Q - \bar{Q}$ is an invariant set.

Let B denote the subset of S consisting of all sequences in the open loop behavior that satisfy the mutual exclusion constraint, i.e., that satisfy item (2) above. If $\{e_n\}$ is a sequence in B that converges in the ρ-topology to a sequence e, then it is clear that e also satisfies the mutual exclusion constraint. So if $e \in S$, then $e \in B$. Thus B is closed relative to S. It follows from Prop. 3.2 that there exists a unique largest controllable sequential behavior contained in B. If MEX is solvable, then a supervisor f that solves MEX and that implements $B^{\dagger}$ is said to be a minimally restrictive solution.

Our main result on MEX is

Theorem 6.1.

MEX is polynomially decidable and polynomially solvable. Furthermore, when MEX is solvable it is possible to synthesize a minimally restrictive solution in polynomial time.

That the problem is polynomial is due to the fact that it can be decoupled and analyzed in terms of the component DES. To show this it will be helpful to introduce the following notation. Let $\Sigma_{ui} = \Sigma_i - \Sigma_{ci}$ be the set of uncontrolled events of G_i, and D_i denote the set of states of G_i from which is possible to reach $\bar{Q}_i$ via uncontrollable events:

$$D_i = \{q_i : q_i \in Q_i \ \& \ \delta_i(w, q_i) \in \bar{Q}_i, \text{ for some } w \in \Sigma_{ui}^*\}$$

Necessary and sufficient conditions for the solvability of MEX are readily determined in terms of the sets D_i:

Proposition 6.1.

MEX is solvable if and only if the following conditions are satisfied:

(1) For at most k of the G_i, $q_{0i} \in D_i$; and

(2) There exist $p - k + 1$ generators with the property that every admissible state trajectory of G_i enters $Q_i - D_i$ infinitely often.

The second condition of the previous proposition can be further resolved as follows:

Proposition 6.2.

Every admissible state trajectory of the generator G_i enters the set $Q_i - D_i$ infinitely often if and only if G_i has no cycles in D_i that intersect Q_{mi}.

Using these results it is straightforward to prove Theorem 6.1.

7 Conclusion

The modeling of DES in terms of formal languages provides a setting for the study of the logical coordination of the components of a DES, and, as we have shown in the context of a simple example, can lead to computationally feasible synthesis methods for certain classes of systems. The model has some limitations particularly in terms of its modeling scope. Extensions to include quantitative aspects of system behavior is a subject of current research.

8 References

[A] Aveyard, R., *A boolean model for a class of discrete event systems. IEEE Trans. Sys. Man. and Cyb.*, SMC-4, 249-258, 1974.

[BeN] Beauquier, J., and M. Nivat, *Application of formal language theory to problems of security and synchronization.* In R.V. Book (Ed.), *Formal Language Theory - Perspective and Open Problems,* Academic Press, New York; pp. 407-454, 1980.

[BoN] Boasson, L., and M. Nivat, *Adherences of languages, Journal of Computer and System Sciences,* **20**, 285–309, 1980.

[B] Buchi, J.R., *On a decision method in restricted second order arithmetic, International Congress Logic Methodology and Philosophy of Science,* Stanford, Calif., 1960.

[CDFV] Cieslak, R., C. Desclaux, A. Fawaz, and P. Varaiya, *Supervisory control of discrete event processes with partial observations,* Memo no. UCB/ERL M86/63, Electronics Research Lab., College of Eng., Univ. of Calf., Berkeley, 1986.

[E] Eilenberg, S., *Automata, Languages, and Machines Volume A,* Academic Press, New York, NY, 1974.

[FST] Fusaoko, A., H. Seki, and K. Takahashi, *A description and reasoning of plant controllers in temporal logic, Proc. 8th International Joint Conference on Artificial Intelligence,* 405–408, August 1983.

[Ha] Halpern, J.Y., *Using reasoning about knowledge to analyze distributed systems,* To appear: *Comp. Science Annual Review,* 1987.

[HZ] Halpern, J.Y., and L.D. Zuck, *A little knowledge goes a long way: simple knowledge-based derivations and proofs for a family of protocols,* Extended abstract, IBM Almaden Research Center, Dept. K53/801, 650 Harry Rd., San Jose, CA 95120, February 1987.

[HO] Hailpern, B.T., and S.S. Owicki, *Modular verification of computer communication protocols, IEEE Trans. Commun.,* COM-31, 56-68, 1983.

[H] Hoare, C.A.R., *Communicating Sequential Processes,* Prentice-Hall, Englewood Cliffs, New Jersey, 1985.

[HU] Hopcroft, J.E., and Ullman, J.D., *Introduction to Automata Theory, Languages and Computation,* Addison-Wesley Pub. Co., Reading, MA., 1979.

[LaW] Lafortune, S. and E. Wong, *A state model for the concurrency control problem in data base management systems,* Memorandum no. UCB/ERL M85/27 Electronic Systems Laboratory, College of Engineering, University of California Berkeley, CA 94720., April 1985.

[LW1] Lin, F., and W.M. Wonham, *Decentralized supervisory control of discrete event systems, Tenth World Congress, International Federation of Automatic Control (IFAC),* Munich,

W. Germany, July 1987; also to appear: *Information Sciences,* 1987. See also: Systems Control Group Report No. 8612, Department of Electrical Engineering, University of Toronto, July 1986.

[LW2] Lin, F., and W.M. Wonham, *On observability of discrete-event systems,* Systems Control Group Report #8701, Department of Elect. Eng., University of Toronto, 1987.

[MT] Maimon, O., G. Tadmor, "Efficient low level control of FMS," LIDS-P-1571, Laboratory for Information and Decisions Systems, MIT, Cambridge, MA, 02139, June 1986.

[MW] Manna, Z., A. Wolper, *Synthesis of Communicating processes from temporal logic specifications, ACM Trans. on Programming Languages and Systems,* **6**, 68–93, 1984.

[MM] Milne, G., and R. Milner, *Concurrent processes and their syntax, J. Assoc. Comp. Mach.,* **26**, 302-321, 1979.

[MDH] Moses, Y., D. Dolev, and J. Halpern, *Cheating husbands and other stories: a case study of knowledge, action, and communication, Distributed Computing,* **1**, 167–176, 1986.

[Mc] McNaughton, R., *Testing and generating infinite sequences by finite automata, Inform. Contr.,* **9**, 521–530, 1966.

[OW] Ostroff, J.S., and W.M. Wonham, *A temporal logic approach to real time control, Proc. 24th IEEE Conference on Decision and Control,* Florida, December 1985.

[O] Ostroff, J.S., *Real Time Computer Control of Discrete Event Systems Modelled by Extended State Machines: A Temporal Logic Approach,* Report no. 8618 Department of Electrical Engineering, University of Toronto, September 1986.

[Pa] Park, D., *Concurrency and automata on infinite sequences, Theoretical Computer Science,* 5th GI-Conference, Karlsruhe, March 1981, *Lecture Notes in Computer Science,* **104**, 167–183, 1981.

[P] Peterson, J.L., *Petri Net Theory and the Modeling of Systems,* Prentice-Hall, Inc., Englewood Cliffs, NJ., 1981.

[RW1] Ramadge, P.J., and W.M. Wonham, *Supervisory control of a class of discrete-event processes, SIAM J. on Contr. and Optimization,* **25** (1), 206–230, January 1987.

[RW2] Ramadge, P.J., and W.M. Wonham, *Modular feedback logic for discrete event systems,* To appear: *SIAM J. on Contr. and Optimization,* 1987; see also: *Proc. 4th IFAC IFORS Symposium Large Scale Systems: Theory and Applications,* Zurich, Switzerland, August 1986.

[RW3] Ramadge, P.J., and W.M. Wonham, *Modular supervisory control of discrete event systems, Proc. of the Seventh International Conference on Analysis and Optimization of Systems,* Antibes, June, 1986.

[S] Shaw, A.C., *Software descriptions with flow expressions, IEEE Trans. on Software Engineering,* SE-4 (3), 242-254, 1978.

[Sh] Shields, M.W., "COSY Train Journeys," Rpt. ASM/67, Computing Laboratory, Univ. of Newcastle-upon-Tyne, 1979.

[TW] Thistle, J.G., and W.M. Wonham, *Control problems in a temporal logic framework,* Systems Control Group Report No. 8510, Department of Electrical Engineering, University of Toronto, Toronto, Canada, M5S1A4, August 1985.

[WR1] Wonham, W.M., and P.J. Ramadge, *On the supremal controllable sublanguage of a given language, SIAM J. on Contr. and Optimization,* **25** (3), 637–659, May 1987.

[WR2] Wonham, W.M., and P.J. Ramadge, *Modular supervisor control of discrete event systems,* to appear: *Mathematics of Control, Signals and Systems,* 1987; see also Information Sciences and Systems Report No. 49, Department of Electrical Engineering, Princeton University, June 1986, revised February 1987.

Using trace theory to model discrete events

Rein Smedinga
Department of computing science
University of Groningen
p.o.box 800
Groningen, the Netherlands

August 1987

Abstract

In this paper discrete processes are defined by means of trace structures. Every symbol in a trace denotes (the occurrence of) some discrete event. The trace alphabet is split into two disjoint sets, one denoting the communication events, the other denoting the exogenous events. Control of a discrete process means constructing a second discrete process having as alphabet the communication events only, so that the connection of the two discrete processes results in a desired exogenous trace set. Connection of discrete processes means blending of the corresponding trace structures.

An algorithm is derived to construct a controller, given a process to be controlled and specifications of the desired exogenous behavior.

Two examples of the use of this algorithm are presented.

1 Introduction

A number of possibilities exists to model discrete events. Most of them however lack the possibility to model plant and controller separately (for example Petri Nets). The controller has to be known beforehand and plant and controller are handled as one system. There are no ways to find a controller in a systematic way. Intuition and clever thinking are the only possibilities in finding a controller.

Other theories exists in which plant and controller are handled separately (for example the supervisory control theory of Wonham (see [RaWo])). Given a model of the plant it is possible to compute a controller (a supervisor) and the behavior of plant, controller and the closed loop system can be studied. However plant and controller need different interpretations. The plant is given in the form of a generator of events, while the controller (the supervisor) acts as an observer and generates enable/disable-strings. This different interpretation makes it hard to connect more than two processes or, for example, to supervise a supervisor.

In this paper trace theory is used to model discrete processes. It turns out that plant and controller can be described in exactly the same way. Furthermore a nice algorithm is developed to construct a controller.

Throughout this paper we use the following notation:

$(\forall x : B(x) : C(x))$ is true if $C(x)$ holds for every x that satisfies $B(x)$, e.g. $(\forall x : x \in \mathbb{N} : x \geq 0)$

$(\exists x : B(x) : C(x))$ is true if there exists an x satisfying $B(x)$ for which $C(x)$ holds, e.g. $(\exists x : x \in \mathbb{N} : x = 10)$

$\{x : B(x) : y(x)\}$ is the set constructor and denotes the set of all elements $y(x)$ constructed using elements x satisfying $B(x)$, e.g. $\{n : n \in \mathbb{N} : a^n b^n\} = \{\epsilon, ab, aabb, aaabbb, \ldots\}$

2 Discrete processes

Describing a discrete process means:

- defining all possible events
- defining the behavior of the process

All possible events are collected in an event set (which we call the *alphabet*). Events are denoted by small letters near the beginning of the Latin alphabet, like a, b, c, etc.

The behavior of the process is given as a collection of sequences of events (which is called the *trace set*). A sequence of events is denoted by a string of letters, like abc, meaning, that the events a, b, and c may appear in that order (first a, then b and at last c). Such a string is called a *trace*. Sometimes we use small letters near the end of the Latin alphabet for traces, like x and y. With ϵ we denote the empty string (a sequence of no events).

The set of all possible traces, together with the set of all events that a given system can produce, is called a *trace structure* (TS for short) and is denoted by

$$P = < \mathbf{t}P, \mathbf{a}P >$$

where $\mathbf{a}P$ stands for the alphabet and $\mathbf{t}P$ for the trace set.

3 Trace theory

In this paper we discuss the notion of *control* of discrete processes. Therefore we introduce *connections* between discrete processes. It turns out that connections can be defined using existing operators from trace theory. Before we can define the notions *discrete process* and *connection of discrete processes* we first introduce these operators and restrict our attention to trace structures.

A connection of trace structures is in fact a shuffling, where identical events have to occur simultaneously, i.e. occur in both processes at the same time. This means that a common event can only occur if it occurs in both processes simultaneously. This kind of operation is called *weaving* and is defined as follows:

Definition 1 *The weaving* w *of two TS's P and R is defined to be*

$$\begin{array}{ll} & P \,\mathrm{w}\, R \\ = & \\ & < \{x : x \in (\mathbf{a}P \cup \mathbf{a}R)^* \;\wedge\; x\lceil\mathbf{a}P \in \mathbf{t}P \;\wedge\; x\lceil\mathbf{a}R \in \mathbf{t}R : x\} \\ & ,\ \mathbf{a}P \cup \mathbf{a}R \\ & > \end{array}$$

The symbol $\lceil$ stands for the *restriction* of a trace to some alphabet, meaning that all events in the trace, that do not belong to the alphabet are deleted.

Sometimes we are only interested in those events, that are not common, i.e. belong to only one of the TS's. Then we use *blending*:

Definition 2 *The blending* b *of two TS's P and R is defined to be*[1]

$$\begin{array}{ll} & P \,\mathrm{b}\, R \\ = & \\ & < \{x : x \in (\mathbf{a}P \cup \mathbf{a}R)^* \;\wedge\; x\lceil\mathbf{a}R \in \mathbf{t}P \;\wedge\; x\lceil\mathbf{a}R \in \mathbf{t}R : x\lceil(\mathbf{a}P \div \mathbf{a}R)\} \\ & ,\ \mathbf{a}P \div \mathbf{a}R \\ & > \end{array}$$

[1]The operator $\div$ stands for symmetric set difference.

In the sequel we use the following partial ordering:

Definition 3 *For two trace structures P and R the ordering $P \sqsubseteq R$ is defined as:*

$$\mathbf{a}P = \mathbf{a}R \ \wedge\ \mathbf{t}P \subseteq \mathbf{t}R$$

Property 4 *For TS's P and R, with $\mathbf{a}R \subseteq \mathbf{a}P$ the following properties hold:*

$$\begin{array}{ll}
(1) & P \mathbin{\mathrm{w}} R = <\{x : x \in \mathbf{t}P \ \wedge\ x\lceil \mathbf{a}R \in \mathbf{t}R : x\}, \mathbf{a}R> \\
(2) & P \mathbin{\mathrm{b}} R = <\{x : x \in \mathbf{t}P \ \wedge\ x\lceil \mathbf{a}R \in \mathbf{t}R : x\lceil(\mathbf{a}P \setminus \mathbf{a}R)\}, \mathbf{a}P \setminus \mathbf{a}R> \\
(3) & P \mathbin{\mathrm{b}} R = <\{z : z \in \mathbf{t}P\lceil(\mathbf{a}P \setminus \mathbf{a}R) \ \wedge \\
 & \qquad (\exists x : x \in \mathbf{t}P \ \wedge\ x\lceil(\mathbf{a}P \setminus \mathbf{a}R) = z : x\lceil \mathbf{a}R \in \mathbf{t}R) : z\} \\
 & \quad , \mathbf{a}P \setminus \mathbf{a}R> \\
(4) & (P \mathbin{\mathrm{w}} R)\lceil \mathbf{a}R = (P\lceil \mathbf{a}R) \cap R
\end{array}$$

and for TS's P, R_1, and R_2 we have:

$$(5) \quad R_1 \sqsubseteq R_2 \Rightarrow (P \mathbin{\mathrm{b}} R_1) \sqsubseteq (P \mathbin{\mathrm{b}} R_2)$$

4 Regular trace structures

If the trace set of a TS is regular[2], it is possible to describe such a set by means of a finite state machine and also by means of so called regular expressions.

A regular expression (RE) is defined (recursively) as follows:

Definition 5 *The empty string (ϵ) and every single symbol is a RE and if x and y are RE's, then also:*

$x;y$	*concatenation*	*first x, then y*
$x\|y$	*union*	*x or y*
x^*	*repetition*	*zero or more concatenations of x*
(x)		*to change priority in evaluation*
x,y	*weaving*	

The corresponding trace structures are:

$$\begin{array}{lcl}
TR(\epsilon) & = & <\{\epsilon\}, \emptyset> \\
TR(a) & = & <\{a\}, \{a\}> \\
TR(x;y) & = & <\{t, u : t \in \mathbf{t}TR(x) \ \wedge\ u \in \mathbf{t}TR(y) : tu\} \\
 & & , \mathbf{a}TR(x) \cup \mathbf{a}TR(y) \\
 & & > \\
TR(x|y) & = & TR(x) \cup TR(y) \\
TR(x^*) & = & <\{t : t \in \mathbf{t}TR(x) : t^*\}, \mathbf{a}TR(x)> \\
TR(x,y) & = & TR(x) \mathbin{\mathrm{w}} TR(y)
\end{array}$$

A more detailed introduction to this notation and terminology can be found in [JvdS].

A finite state machine (FSM for short) is defined as:

[2]A trace set is called regular if the number of equivalence classes is finite, where the equivalence relation on (prefixes of) traces of a TS T is defined as:

$$x \equiv y = (\forall z : z \in (\mathbf{a}T)^* : xz \in \mathbf{t}T = yz \in \mathbf{t}T)$$

Figure 1: Representation of a FSM

Definition 6 *A Finite state machine is a tuple $M = (A, Q, \delta, q, F)$ with:*

A	*a finite set of events, called the alphabet*
Q	*a finite set of states*
$q \in Q$	*initial (or start) state*
$F \subseteq Q$	*set of final (or marker) states*
$\delta : Q \times A \to Q$	*the transition function*

From the state transition function δ we derive a path x from $p_1 \in Q$ to $p_2 \in Q$ if $p_2 = \delta^*(p_1, x)$, with δ^* the closure of δ, defined as:

$$\begin{array}{ll} \delta^*(p, \epsilon) = p & \text{if } x = a \\ \delta^*(p, x) = \delta^*(\delta(p, a), y) & \text{if } x = ay \end{array}$$

The corresponding trace structure is:

$$TR(M) = < \{x : \delta^*(q, x) \in F : x\}, A >$$

In figure 1 we have given the representation of a FSM in a drawing.

In the sequel we consider FSM's that are *minimal* (i.e. contain a minimal number of states in order to represent a certain behavior), *deterministic* (i.e. δ is a function), and *complete* (i.e. δ is defined for all pairs (p, x)). If a FSM M does not represent A^* but only a subset of A^* then M contains an *error state*[3], denoted by $[\emptyset]$, with the property:

$$(\forall a : a \in A : \delta([\emptyset], a) = [\emptyset]) \wedge [\emptyset] \notin F$$

Once we have reached an error state using path x we are unable to reach a final state. This means, that x is a trace that is no part of the corresponding trace set. So an error state is the endpoint of all traces that do not belong to the corresponding trace structure.

A more detailed introduction to FSM's can be found in [HoUl] and many other books.

5 Discrete processes

We now return to the notion of control of discrete processes. First we split the alphabet while introducing two kinds of events:

- exogenous events
- communication events

The exogenous events are used to model *own* actions of the process. This means that exogenous events do not appear in other processes[4].

[3]In drawings of FSM's we omit this error state and all transitions going to it.

[4]One can argue about the name *exogenous*, endogenous events are perhaps more convenient. However it turns out, that exogenous events are retained in a connection while communication events disappear. So communication events are internal and exogenous events are external.

The communication events are used to model actions of a process, that may be common to other processes. This kind of events is of interest in communication with other processes. Furthermore we like to use communication events to control the exogenous events.

Such a TS, in which the alphabet is split up, is called a *discrete process* (DP for short) and denoted as:

$$P = < \mathbf{t}P, \mathbf{e}P, \mathbf{c}P >$$

where $\mathbf{t}P$ is the trace set of the process (the set of all possible sequences of occurring events), $\mathbf{e}P$ the set of exogenous events and $\mathbf{c}P$ the set of communication events.

Notice, that such a DP P is only well defined if

$$\mathbf{e}P \cap \mathbf{c}P = \emptyset$$

We are not concerned with how this communication is actually performed, i.e. which process *generates* the event and which process *receives* it. In other words, we do not make any distinction between input and output events here.

In the sequel we sometimes look at a DP as being a TS, i.e. use P as if it was defined as a TP $P = < \mathbf{t}P, \mathbf{e}P \cup \mathbf{c}P >$. Then we use $\mathbf{a}P$ as abbreviation for $\mathbf{e}P \cup \mathbf{c}P$.

If we restrict our attention to the exogenous events we have, what is called the exogenous behavior or external behavior $\mathbf{t}P \lceil \mathbf{e}P$. If we restrict our attention to the communications we have the communication behavior or internal behavior $\mathbf{t}P \lceil \mathbf{c}P$.

Control of discrete processes can be described as using the communication events to establish some predefined behavior of the exogenous events. Therefore we need:

- the uncontrolled exogenous behavior: $\mathbf{t}P \lceil \mathbf{e}P$.
- one or more other DP's communicating with P by means of the events from $\mathbf{c}P$.
- the resulting controlled exogenous behavior: $\mathbf{t}(P \otimes R) \lceil \mathbf{e}P$ (where $\otimes$ for this moment denotes the connection of two DP's and R is the controller).

In order to be able to discuss this topic we have to define what is meant by connection of discrete processes.

6 Connections

In this section we define the notion *connection* of discrete processes.

Definition 7 *Given two DP's $P = < \mathbf{t}P, \mathbf{e}P, \mathbf{c}P >$ and $R = < \mathbf{t}R, \mathbf{e}R, \mathbf{c}R >$ with $\mathbf{e}P \cap \mathbf{a}R = \emptyset$ and $\mathbf{e}R \cap \mathbf{a}P = \emptyset$ then the connection of P and R is defined as*

$$< \mathbf{t}(P \ \mathrm{b} \ R), \mathbf{e}P \cup \mathbf{e}R, \mathbf{c}P \div \mathbf{c}R >$$

and denoted[5] *by $P \ \mathrm{b} \ R$.*

Note that all exogenous events of P and R are exogenous events of the connection. From the communication events only those that do not belong to both processes are retained. This guarantees, that

$$\mathbf{a}(P \ \mathrm{b} \ R) = \mathbf{a}P \div \mathbf{a}R$$

so that the blend is well defined.
In the sequel we write $P \ \mathrm{b} \ R$ only if $\mathbf{e}P \cap \mathbf{a}R = \emptyset$ and $\mathbf{e}R \cap \mathbf{a}P = \emptyset$.

It is not difficult to prove the following properties:

[5] So we use the same notation b for blending of TS's as well as for connection of DP's.

Property 8 *For the connection* b *the following properties hold:*
(1) $P \,\mathrm{b}\, R = R \,\mathrm{b}\, P$
(2) $P \,\mathrm{b} < \{\epsilon\}, \emptyset, \emptyset > = P$
(3) $\mathbf{c}P \cap \mathbf{c}R \cap \mathbf{c}S = \emptyset \Rightarrow (P \,\mathrm{b}\, R) \,\mathrm{b}\, S = P \,\mathrm{b}\, (R \,\mathrm{b}\, S)$

For part (3) we have to remark that $(P \,\mathrm{b}\, R) \,\mathrm{b}\, S = P \,\mathrm{b}\, (R \,\mathrm{b}\, S)$ only holds, if $\mathbf{a}P \cap \mathbf{a}R \cap \mathbf{a}S = \emptyset$, which is established through the condition $\mathbf{c}P \cap \mathbf{c}R \cap \mathbf{c}S = \emptyset$. This last property allows to write multi-connections without parentheses.

The DP $< \{\epsilon\}, \emptyset, \emptyset >$ is the *unit element* of the operator b . Notice, that it is necessary to have at least the empty trace ϵ in the trace set, because for every DP P we have that

$$P \,\mathrm{b} < \emptyset, \emptyset, \emptyset > = < \emptyset, \emptyset, \emptyset >$$

(thus $< \emptyset, \emptyset, \emptyset >$ is not a unit element).

Sometimes we are interested in the total behavior of the connected system. Therefore we introduce the *overall* or *total connection* as well.

Definition 9 *Given two DP's* $P = < \mathbf{t}P, \mathbf{e}P, \mathbf{c}P >$ *and* $R = < \mathbf{t}R, \mathbf{e}R, \mathbf{c}R >$ *with* $\mathbf{e}P \cap \mathbf{a}R = \emptyset$ *and* $\mathbf{e}R \cap \mathbf{a}P = \emptyset$ *then the total connection of* P *and* R *is defined as:*

$$< \mathbf{t}(P \,\mathrm{w}\, R), \mathbf{e}P \cup \mathbf{e}R \cup (\mathbf{c}P \cap \mathbf{c}R), \mathbf{c}P \div \mathbf{c}R >$$

and denoted[6] *by* $P \,\mathrm{w}\, R$.

Notice, that we have put all common communication events of P and R in the exogenous event set of the total connection. This guarantees that $\mathbf{a}(P \,\mathrm{w}\, R) = \mathbf{a}P \cup \mathbf{a}R$, that communication events can be observed outside the connection and that communication events can not be used in other connections for communication purposes. A communication event can thus serve as communication between exactly two DP's.

We have the following properties for the total connection:

Property 10 *For the total connection* w *the following properties hold:*
(1) $P \,\mathrm{w}\, R = R \,\mathrm{w}\, P$
(2) $P \,\mathrm{w} < \{\epsilon\}, \emptyset, \emptyset > = P$
(3) $(P \,\mathrm{w}\, R) \,\mathrm{w}\, S = P \,\mathrm{w}\, (R \,\mathrm{w}\, S)$

7 Introduction to control

Before we give an exact description of our control problem we give a illustrative example first.

Suppose a shop sells two kinds of articles and in order to get an article one has to pay for it. Paying for an article is supposed to be a communicating action. So we have:

a_1 article 1 is sold
a_2 article 2 is sold
p_1 pay for article 1
p_2 pay for article 2

The complete process becomes:

$$P = < ((p_1; a_1)|(p_2; a_2))^*, \{a_1, a_2\}, \{p_1, p_2\} >$$

A customer can now be described as "letting event p_i occur for every article (number i) wanted," for example:

[6] Again the same notation for weaving of TS's and for the total connection of DP's.

$R =< ((p_1; p_1), p_2), \emptyset, \{p_1, p_2\} >$

Connecting these two processes results in:

$$P \,\mathrm{b}\, R$$
$$=$$
$$< ((a_1; a_1), a_2), \{a_1, a_2\}, \emptyset >$$

Notice that the uncontrolled behavior of the shop P equals

$$\mathbf{t}P\lceil \mathbf{e}P = (a_1|a_2)^*$$

while the controlled behavior is

$$\mathbf{t}(P \,\mathrm{b}\, R) = (a_1; a_1), a_2$$

Notice, that the customer has in fact controlled the exogenous behavior of the shop.

8 A control problem

Using the previous definitions we are now able to state our control problem in a formal way:

Given are a DP $P =< \mathbf{t}P, \mathbf{e}P, \mathbf{c}P >$ and two trace structures L_{min} and L_{max} with

$$L_{min} =< \mathbf{t}L_{min}, \mathbf{e}P, \emptyset >$$
$$L_{max} =< \mathbf{t}L_{max}, \mathbf{e}P, \emptyset >$$
$$L_{min} \subseteq L_{max}$$

L_{min} and L_{max} specify the range of resulting exogenous traces, that are acceptable.
The problem is to find, if possible, a DP R with:

$$R =< \mathbf{t}R, \emptyset, \mathbf{c}R > \quad \text{with: } \mathbf{c}R \subseteq \mathbf{c}P$$

such that

$$L_{min} \subseteq (P \,\mathrm{b}\, R)\lceil \mathbf{e}P \subseteq L_{max}$$

This last condition is called the *minmax condition.*

In words: construct, given a DP P, a second DP R, that controls the exogenous events of P as specified by L_{min} and L_{max}, and whose possible events are at most all communication events of P.

The restriction of $\mathbf{t}L_{max}$ to be a subset of $(\mathbf{e}P)^*$ is needed because we can only give restrictions on the existing exogenous events. Without loss of generality we assume:

$$L_{min} \subseteq L_{max} \subseteq P\lceil \mathbf{e}P$$

(so we can give restrictions to existing exogenous traces only).

This problem is called *control of discrete events* (CODE for short).

In our previous example we had:

$$\mathbf{t}L_{min} = \mathbf{t}L_{max} = (a_1; a_1), a_2$$

In the following section we give a general solution for CODE.

9 Solution for CODE

In the sequel we deal (without loss of generality) only with the situation that $R =< \mathrm{t}R, \emptyset, \mathrm{c}P >$, i.e. $\mathrm{c}R = \mathrm{c}P$. So we use *all* communication events to control the process.

Furthermore we use DP's P, R, and S and assume[7]:

$$P =< \mathrm{t}P, \mathrm{e}P, \mathrm{c}P >$$
$$R =< \mathrm{t}R, \emptyset, \mathrm{c}P >$$
$$L =< \mathrm{t}L, \mathrm{e}P, \emptyset >$$

In this section we give an algorithm to construct a solution for CODE that also can be used to investigate if CODE has a solution at all. For the algorithm we need two functions:

Definition 11 *With the CODE problem we associate the functions:*

$$F(L) = (P \mathrm{\,b\,} L) \setminus (P \mathrm{\,b\,} (P \lceil \mathrm{e}P \setminus L))$$

called the friend *of L, and*

$$G(L) = P \mathrm{\,b\,} F(L)$$

called the guardian *of L.*

In most cases we only need the trace set of F or G, so we introduce:

$$f(L) = \mathrm{t}F(L)$$
$$g(L) = \mathrm{t}G(L)$$

Notice that:

$$\mathrm{e}F(L) = \emptyset \quad \mathrm{c}F(L) = \mathrm{c}P$$
$$\mathrm{e}G(L) = \mathrm{e}P \quad \mathrm{c}G(L) = \emptyset$$

The algorithm (called the *deCODEr*) is described as follows:

forall L **such that** $L_{min} \subseteq L \subseteq L_{max}$:
  if $L_{min} \subseteq G(L) \subseteq L_{max}$
  then $F(L)$ is a solution

Of course, it is not immediately clear that this algorithm works. In the next section we try to make it convincing. In the section thereafter we give a proof of the algorithm and, more important, give a necessary and sufficient condition under which the problem is solvable.

10 Outline of the algorithm

First let us try

$$F'(L) = P \mathrm{\,b\,} L$$

So, if $G'(L) = P \mathrm{\,b\,} F'(L) = P \mathrm{\,b\,} (P \mathrm{\,b\,} L)$ satisfies the minmax condition then $R = F'(L)$ should be a solution.

However, starting with an L such that $L_{min} \subseteq L \subseteq L_{max}$ does not guarantee, that $L_{min} \subseteq G'(L) \subseteq L_{max}$. For example:

$$P =< \{ac, ad, bc\}, \{c, d\}, \{a, b\} >$$
$$L_{min} = L_{max} =< \{c\}, \{c, d\}, \emptyset >$$

[7] In the sequel checking the alphabets in proofs is omitted, if this is clear from this context.

results (with $L = L_{max}$) in:

$$f'(L) = \mathrm{t}(P \;\mathbf{b}\; R) = \{a, b\}$$
$$g'(L) = \mathrm{t}(P \;\mathbf{b}\; F'(L)) = \{c, d\} \neq \mathrm{t}L$$

In $f'(L)$ the trace a does not lead to the desired solution because it also allows exogenous event d to occur and d does not satisfy the minmax condition.
We conclude the following:

- First compute $P \;\mathbf{b}\; L$ to get all possible control traces.
- Next compute $P \;\mathbf{b}\; \neg L$ to get all control traces, that give undesired results[8].
- At last take $(P \;\mathbf{b}\; L) \setminus (P \;\mathbf{b}\; \neg L)$ to find exactly the right control traces.

This is precisely, what $F(L)$ does:

$$F(L) = \underbrace{\underbrace{(P \;\mathbf{b}\; L)}_{\text{possible controls}} \setminus \underbrace{(P \;\mathbf{b}\; \underbrace{(P\lceil \mathrm{e}P \setminus L)}_{\neg L})}_{\text{undesired controls}}}_{\text{desired controls}}$$

11 Proof of the algorithm

First a number of properties of the friend and the guardian (and the operators used) are listed.

Lemma 12 *The friend and the guardian of L satisfy:*

$$\begin{array}{lcl} f(L) & = & \{z : z \in \mathrm{t}P\lceil \mathrm{c}P \wedge (\forall x : x \in \mathrm{t}P \wedge x\lceil \mathrm{c}P = z : x\lceil \mathrm{e}P \in \mathrm{t}L) : z\} \\ g(L) & = & \{x : x \in \mathrm{t}P \wedge (\forall y : y\lceil \mathrm{c}P = x\lceil \mathrm{c}P : y\lceil \mathrm{e}P \in \mathrm{t}L) : x\lceil \mathrm{e}P\} \end{array}$$

proof: We have:

$$\begin{array}{ll} & \mathrm{t}(P \;\mathbf{b}\; (P\lceil \mathrm{e}P \setminus L)) \\ = & \quad [\text{ property 4 (3) }] \\ & \{z : z \in \mathrm{t}P\lceil \mathrm{c}P \wedge (\exists x : x \in \mathrm{t}P \wedge x\lceil \mathrm{c}P = z : x\lceil \mathrm{e}P \in \mathrm{t}(P\lceil \mathrm{e}P \setminus L)) : z\} \\ = & \quad [\; x\lceil \mathrm{e}P \in \mathrm{t}P\lceil \mathrm{e}P \;] \\ & \{z : z \in \mathrm{t}P\lceil \mathrm{c}P \wedge (\exists x : x \in \mathrm{t}P \wedge x\lceil \mathrm{c}P = z : x\lceil \mathrm{e}P \notin \mathrm{t}L) : z\} \end{array}$$

Hence:

$$\begin{array}{ll} & f(L) \\ = & \\ & \mathrm{t}((P \;\mathbf{b}\; S) \setminus (P \;\mathbf{b}\; (P\lceil \mathrm{e}P \setminus L))) \\ = & \quad [\text{ definition of } \mathbf{b} \text{ and previous equation }] \\ & \{z : z \in \mathrm{t}P\lceil \mathrm{c}P \wedge (\exists x : x \in \mathrm{t}P \wedge x\lceil \mathrm{c}P = z : x\lceil \mathrm{e} \in \mathrm{t}L) : z\} \\ & \setminus \{z : z \in \mathrm{t}P\lceil \mathrm{c}P \wedge (\exists x : x \in \mathrm{t}P \wedge x\lceil \mathrm{c}P = z : x\lceil \mathrm{e}P \notin \mathrm{t}L) : z\} \\ = & \quad [\; x\lceil \mathrm{e}P \in \mathrm{t}P\lceil \mathrm{e}P \;] \\ & \{z : z \in \mathrm{t}P\lceil \mathrm{c}P \wedge (\forall x : x \in \mathrm{t}P \wedge x\lceil \mathrm{c}P = z : x\lceil \mathrm{e}P \in \mathrm{t}L) : z\} \end{array}$$

$g(L)$ now easily follows from $f(L)$.
(end of proof)

[8]The notation $\neg L$ is an abbreviation of $P\lceil \mathrm{e}P \setminus L$. It stands for the trace structure containing all traces (over the same alphabet) that do not belong to L.

The expression of $f(L)$ in this lemma is useful in the proofs of the next three lemmas. These lemmas together make up the proof of the correctness of the algorithm and result in theorem 16.

Lemma 13 $G(L) \subseteq L$

proof:

$$\begin{array}{ll} & z \in g(L) \\ = & \quad [\text{ definition of the guardian }] \\ & z \in \mathbf{t}(P \,\mathrm{b}\, F(L)) \\ = & \quad [\text{ definition of } \mathrm{b} \;] \\ & (\exists x : x \in \mathbf{t}P \wedge x\lceil \mathbf{c}P \in f(L) : x\lceil \mathbf{e}P = z) \\ \Rightarrow & \quad [\text{ lemma 12: } x \in \mathbf{t}P \wedge x\lceil \mathbf{c}P \in f(L) \Rightarrow x\lceil \mathbf{e}P \in \mathbf{t}P \;] \\ & z \in \mathbf{t}L \end{array}$$

(end of proof)

Lemma 13 implies that by choosing $L = L_{max}$ (the largest possible choice) the solution found by the deCODEr still satisfies the right part of the minmax condition of CODE. Notice that $G(L) = L$ does not hold in general.

Lemma 14 $R \subseteq P\lceil \mathbf{c}P \wedge P \,\mathrm{b}\, R \subseteq L \Rightarrow R \subseteq F(L)$

proof:

$$\begin{array}{ll} & z \in \mathbf{t}R \\ \Rightarrow & \quad [\; R \subseteq P\lceil \mathbf{c}P \wedge \mathbf{t}(P \,\mathrm{b}\, R) \subseteq L \;] \\ & z \in \mathbf{t}P\lceil \mathbf{c}P \wedge (\forall x : x \in \mathbf{t}P \wedge x\lceil \mathbf{c}P = z : x\lceil \mathbf{e}P \in \mathbf{t}L) \\ \Rightarrow & \quad [\text{ lemma 12 }] \\ & z \in f(L) \end{array}$$

(end of proof)

Lemma 14 is a very important lemma. It implies, that (take $L = L_{max}$) every solution of CODE, that is contained in $P\lceil \mathbf{c}P$, is contained in $R_{max} = F(L_{max})$. Using lemma 13 we see that R_{max} therefore is the greatest possible solution of CODE and can be constructed using the algorithm[9].

Notice, that we cannot prove:

$$R \subseteq P\lceil \mathbf{c}P \wedge L \subseteq P \,\mathrm{b}\, R \Rightarrow F(L) \subseteq R$$

and therefore, we cannot find in general a smallest possible solution. In other words, as we shall see, if a solution exists, it is a most *liberal* solution. There may not be a most *conservative* one.

Lemma 15 $L_1 \subseteq L_2 \Rightarrow G(L_1) \subseteq G(L_2)$

proof:

$$\begin{array}{ll} & L_1 \subseteq L_2 \\ \Rightarrow & \quad [\text{ lemma 12 }] \\ & F(L_1) \subseteq F(L_2) \\ \Rightarrow & \quad [\text{ property 4 (5) }] \\ & P \,\mathrm{b}\, F(L_1) \subseteq P \,\mathrm{b}\, F(L_2) \\ = & \\ & G(L_1) \subseteq G(L_2) \end{array}$$

(end of proof)

[9] Notice, that it is always possible to add to R_{max} traces that have no influence when they are blended with P. So it is only possible to find a greatest solution that is contained in $P\lceil \mathbf{c}P$.

This lemma states, that an increasing set of choices of L leads to an increasing set of resulting exogenous traces P b $F(L)$ of the CODE problem (monotonicity).

We are able now to formulate the following important theorem:

Theorem 16 *CODE has a solution if and only if* $L_{min} \subseteq G(L_{max})$

proof: Suppose CODE has a solution, say R, then write

$$R = R_{int} \cup R_{ext}$$

with:

$$R_{int} = P\lceil cP \cap R$$
$$R_{ext} = R \setminus R_{int}$$

(which gives P b $R = P$ b R_{int}). Then we have:

$$\begin{array}{ll} & L_{min} \\ \subseteq & \quad [\ R \text{ is a solution: } L_{min} \subseteq \mathrm{t}(P \text{ b } R_{int}) = \mathrm{t}(P \text{ b } R)\ ] \\ & P \text{ b } R_{int} \\ \subseteq & \quad [\ R_{int} \subseteq P\lceil cP \ \wedge\ \mathrm{t}(P \text{ b } R_{int}) \subseteq L_{max} \Rightarrow R_{int} \subseteq F(L_{max})\ ] \\ & P \text{ b } F(L_{max}) \\ = & \\ & G(L_{max}) \end{array}$$

Next, suppose CODE has no solution:

$$\begin{array}{ll} & (\forall R : P \text{ b } R \subseteq L_{max} : L_{min} \not\subseteq P \text{ b } R) \\ \Rightarrow & \quad [\text{ choose } R = F(L_{max}) \text{ (such an } R \text{ exists, see lemma 13) }] \\ & L_{min} \not\subseteq P \text{ b } F(L_{max}) \\ = & \\ & L_{min} \not\subseteq G(L_{max}) \end{array}$$

(end of proof)

This theorem implies, that if the deCODEr does not abort, the constructed R is a solution of CODE. If the deCODEr aborts with $L = L_{max}$ we may conclude that no solution exists.

12 Some properties of the deCODEr

The next lemma gives a property of solutions of CODE, as constructed using the friend and the guardian.

Lemma 17

$$\begin{array}{ll} & L_{min} \subseteq G(L_1) \subseteq L_{max} \ \wedge\ L_{min} \subseteq G(L_2) \subseteq L_{max} \\ \Rightarrow & \\ & L_{min} \subseteq G(L_1 \cup L_2) \subseteq L_{max} \end{array}$$

proof: trivial, using:

$$\begin{array}{ll} & G(L_1 \cup L_2) \\ = & \\ & P \text{ b } F(L_1 \cup L_2)) \\ = & \quad [\ F(L_1 \cup L_2) = F(L_1) \cup F(L_2), \text{ see lemma 12 }] \\ & P \text{ b } (F(L_1) \cup F(L_2)) \\ = & \quad [\text{ property 1.34 in [JvdS] }] \\ & (P \text{ b } F(L_1)) \cup (P \text{ b } F(L_2)) \\ = & \\ & G(L_1) \cup G(L_2) \end{array}$$

(end of proof)

This lemma states, that if R_1 and R_2 are both solutions of CODE (and of the form $F(L)$ for some L), then also $R_1 \cup R_2$ is a solution. This lemma implies that a maximal solution (contained in $P\lceil cP$) exists.

In general, however, $R_1 \cap R_2$ and R_1 w R_2 need not be solutions, which prevents the existence of a minimal solution, as is shown in the following example: Let

$$P =< \{abc, bad\}, \{a\}, \{b,c,d\} >$$
$$R_1 =< \{bc\}, \emptyset, \{b,c,d\} >$$
$$R_2 =< \{bd\}, \emptyset, \{b,c,d\}$$

then:

$$\mathrm{t}(P \text{ b } R_1) = \{a\} \text{ and } \mathrm{t}(P \text{ b } R_2) = \{a\}$$

(so both R_1 and R_2 are solutions of CODE), but:

$$\mathrm{t}(P \text{ b } (R_1 \cap R_2)) = \emptyset \not\supseteq \mathrm{t}L_{min}$$

(hence $R_1 \cap R_2$ is no solution of CODE).
This is due to the fact, that (according to property 1.34 in [JvdS]):

$$P \text{ b } (R_1 \cap R_2) \subseteq (P \text{ b } R_1) \cap (P \text{ b } R_2)$$

Next we like to investigate if every solution of CODE can be written in terms of a friend of some L satisfying the minmax condition. Because every solution of CODE can be extented with traces that do not have any influence on the result (take $R_{new} = R \cup R_{ext}$ for an R_{ext} with $\mathrm{t}(P \text{ b } R_{ext}) = \emptyset$, then R_{new} is also a solution), we can only hope that every solution R with $R \subseteq P\lceil cP$ can be written in terms of a certain friend. Suppose R is a solution of CODE with $R \subseteq P\lceil cP$, then P b R satisfies the minmax condition. From lemma 14 we have:

$$R \subseteq F(P \text{ b } R)$$

In general, we have no equality here.

It is easily seen, that if all exogenous traces can be found by applying a unique communication trace only, all solutions of CODE can be found by applying the deCODEr (i.e. are of the form $F(L)$):

Lemma 18

$$\begin{array}{ll} & (\forall x, y : x \in \mathrm{t}P \ \wedge \ y \in \mathrm{t}P : x\lceil cP \neq y\lceil cP \Rightarrow x\lceil eP \neq y\lceil eP) \\ \Rightarrow & \\ & (\forall R :\ R \textit{ is a solution of CODE } \ \wedge \ R \subseteq P\lceil cP : F(P \text{ b } R) = R) \end{array}$$

proof: We only have to prove that $R \supseteq F(P \,\mathbf{b}\, R)$:

$$
\begin{array}{ll}
 & z \in f(P \,\mathbf{b}\, R) \\
\Rightarrow & \\
 & z \in \mathbf{t}P\lceil \mathbf{c}P \ \wedge\ (\forall x : x \in \mathbf{t}P \ \wedge\ x\lceil \mathbf{c}P = z : x\lceil \mathbf{e}P \in \mathbf{t}(P \,\mathbf{b}\, R)) \\
\Rightarrow & \quad [\ \mathbf{t}(P \,\mathbf{b}\, R) = \{y : y \in P \ \wedge\ y\lceil \mathbf{c}P \in \mathbf{t}R : y\lceil \mathbf{e}P\}\ ] \\
 & z \in \mathbf{t}P\lceil \mathbf{c}P \ \wedge\ (\forall x : x \in \mathbf{t}P \ \wedge\ x\lceil \mathbf{c}P \in z : \\
 & \qquad\qquad (\exists y : y \in \mathbf{t}P \ \wedge\ y\lceil \mathbf{c}P \in \mathbf{t}R : x\lceil \mathbf{e}P = y\lceil \mathbf{e}P)) \\
\Rightarrow & \quad [\ \text{assumption implies } y = x\ ] \\
 & z \in \mathbf{t}R
\end{array}
$$

(end of proof)

If P has the property that

$$(\forall x, y : x \in \mathbf{t}P \ \wedge\ y \in \mathbf{t}P : x\lceil \mathbf{c}P \neq y\lceil \mathbf{c}P \Rightarrow x\lceil \mathbf{e}P \neq y\lceil \mathbf{e}P)$$

we call P *observable.*

From lemma 18 it is clear, that every solution of CODE for an observable DP P has the form $F(L)$. In that case, the deCODEr gives exactly all possible solutions.

If P is observable we also have, that

$$G(P \,\mathbf{b}\, R) = P \,\mathbf{b}\, R$$

This property however holds for every P and every solution R:

Lemma 19 $G(P \,\mathbf{b}\, R) = P \,\mathbf{b}\, R$

proof: From lemma 14 we have (take $L = P \,\mathbf{b}\, R$) that $G(P \,\mathbf{b}\, R) \subseteq P \,\mathbf{b}\, R$. So it remains to prove $G(P \,\mathbf{b}\, R) \supseteq P \,\mathbf{b}\, R$. We have:

$$
\begin{array}{ll}
 & x\lceil \mathbf{c}P \in \mathbf{t}R \\
\Rightarrow & \quad [\ \text{take } u = y\ ] \\
 & (\forall y : y \in \mathbf{t}P \ \wedge\ y\lceil \mathbf{c}P = x\lceil \mathbf{c}P : \\
 & \qquad (\exists u : u \in \mathbf{t}P \ \wedge\ u\lceil \mathbf{c}P \in \mathbf{t}R : y\lceil \mathbf{e}P = u\lceil \mathbf{e}P)) \\
= & \\
 & (\forall y : y \in \mathbf{t}P \ \wedge\ y\lceil \mathbf{c}P = x\lceil \mathbf{c}P : y\lceil \mathbf{e}P \in \mathbf{t}(P \,\mathbf{b}\, R))
\end{array}
$$

Hence:

$$
\begin{array}{ll}
 & z \in \mathbf{t}(P \,\mathbf{b}\, R) \\
= & \\
 & (\exists x : x \in \mathbf{t}P \ \wedge\ x\lceil \mathbf{c}P \in \mathbf{t}R : z = x\lceil \mathbf{e}P) \\
= & \quad [\ \text{above implication}\ ] \\
 & (\exists x : x \in \mathbf{t}P \ \wedge\ (\forall y : y \in \mathbf{t}P \ \wedge\ y\lceil \mathbf{c}P = x\lceil \mathbf{c}P : y\lceil \mathbf{e}P \in \mathbf{t}(P \,\mathbf{b}\, R)) \\
 & \qquad : z = x\lceil \mathbf{e}P) \\
\Rightarrow & \quad [\ \text{see lemma 12}\ ] \\
 & z \in g(P \,\mathbf{b}\, R)
\end{array}
$$

(end of proof)

We end this section with a summary of the founded results:

Theorem 20 *Associated with CODE the following conclusions hold:*

- *If CODE has a solution, we can find one by constructing* $R = F(L)$, *with* L *satisfying:* $L_{min} \subseteq L \subseteq L_{max}$

- *If CODE has a solution and P is observable, all solutions are of the form $R = F(L)$, with L satisfying: $L_{min} \subseteq L \subseteq L_{max}$*
- *If no solutions of the form $R = F(L)$ exists (i.e. if $F(L_{max})$ does not lead to a solution), no solution of CODE exists.*

13 An example: a ship lock

As an example of the use of the deCODEr we look at the following situation. Consider a ship lock with two doors in which ships can pass from west to east (see figure 2). The lock is given as:

$$P =< \mathbf{t}P, \mathbf{e}P, \mathbf{c}P >$$

with

$$\mathbf{e}P = \{p_1, p_2\}$$
$$\mathbf{c}P = \{o_1, o_2, c_1, c_2\}$$

The behavior is given in figure 3.

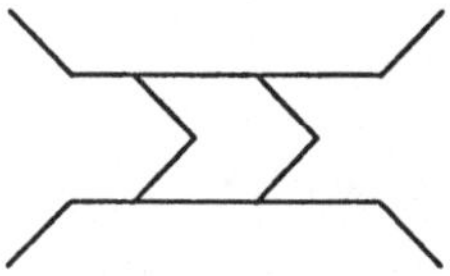

Figure 2: A ship lock

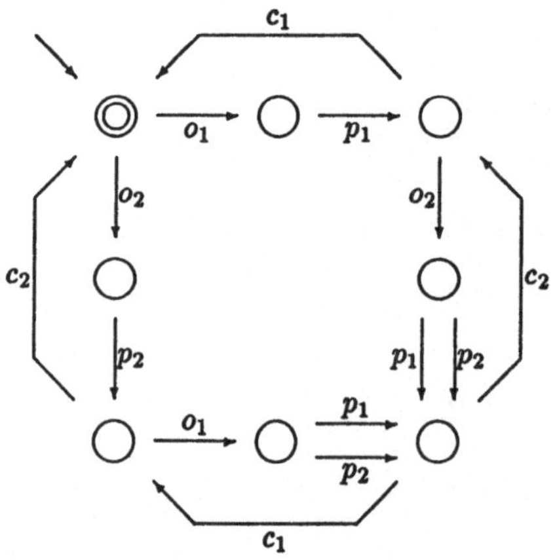

Figure 3: Behavior of the lock

The meaning of the events is given in table 1. The lock can contain one ship at the time. Our desired behavior therefore is:

$$L =< (p_1; p_2)^*, \mathbf{e}P, \emptyset >$$

Using the deCODEr[10] we find the controller as in figure 4. This controller does precisely

[10]We use a computer program here, so we do not give any calculations.

event	meaning
p_1	a ship passes through door 1
p_2	a ship passes through door 2
o_1	open door 1
o_2	open door 2
c_1	close door 1
c_2	close door 2

Table 1: Meaning of the events of the lock

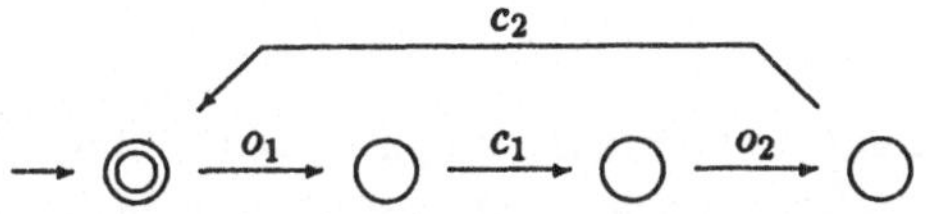

Figure 4: Controller for the lock

what we expected he should do: first let a ship in by opening and closing door 1, next let the ship go out by opening and closing door 2.

In figure 5 we have given P b L. Just computing P b L in general does not give the right controller: in P b L for example the behavior

$o_1; o_2; c_1; c_2$

is possible. This may lead to

$p_1; p_2$

but also

$p_1; p_1$

is possible and this last exogenous behavior is certainly not desired.

It can easily be verified, that the exogenous behavior of the connection equals:

$\mathsf{t}(P \text{ b } R) = (p_1; p_2)^*$

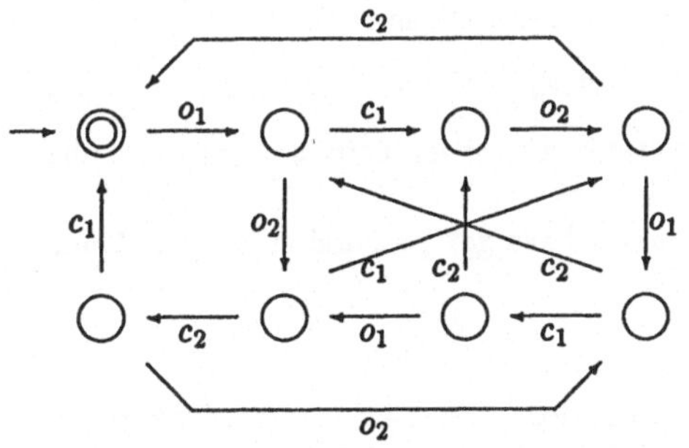

Figure 5: P b L

event	meaning
s_i	(get permission to) sit
gl_i	grab fork on the left
gr_i	grab fork on the right
e_i	eat
ll_i	lay down fork on the left
lr_i	lay down fork on the right
t_i	think

Table 2: Meaning of the events of P_i

and the total behavior:

$$t(P \text{ w } R) = (o_1; p_1; c_1; o_2; p_2; c_2)^*$$

Remark that one should not underestimate the simplicity of the above example. The computed controller could (with a little effort) as well be computed by simple intuitive reasoning. However it may be difficult to prove the correctness of the controller. If the examples become more difficult then finding a suitable controller becomes intractable by hand calculations, but using the deCODEr remains feasible.

14 The dining philosophers

Consider a number of philosophers (say k), sitting around a round table. Each of them in turn eats and thinks. To eat each philosopher needs two forks, one to the left and one to the right of his plate. Between each plate only one fork is present, so every fork has to be shared between two philosophers, but only one philosopher at the time can use it.

Each philosopher can be modeled as follows:

$$P_i =< (s_i; gl_i; gr_i; e_i; ll_i; lr_i; t_i)^*, \{gl_i, gr_i, ll_i, lr_i, e_i, t_i\}, \{s_i\} > \qquad i = 0, 1, \ldots, k-1$$

The interpretation of each event is given in table 2. To be able to model the sharing of the forks we have:

$$F_i =< ((gl_i; ll_i)|(gr_{i+1}; lr_{i+1}))^*, \emptyset, \{gl_i, ll_i, gr_{i+1}, lr_{i+1}\} > \qquad i = 0, 1, \ldots, k-1$$

Notice that fork F_i plays the role of left fork for philosopher P_i and the role of right fork for philosopher P_{i+1}. The behavior expresses, that grabbing a fork should first be followed by laying down that fork by the same philosopher before it can be grabbed again.

The total behavior we like to investigate is:

$$T = P_0 \text{ w } P_1 \text{ w } \ldots \text{ w } P_{k-1} \text{ w } F_0 \text{ w } F_1 \text{ w } \ldots \text{ w } F_{k-1}$$

Notice, that we have used the total connection here in order to be able to investigate the crucial grabbing of the forks.

Suppose, that all philosophers have got permission to eat. Then the following sequence of events is possible (take $k = 3$):

$$s_0; s_1; s_2; gl_0; gl_1; gl_2$$

Now all forks are in use, but no philosopher is able to eat. This phenomenon is called *deadlock*.

Notice, that the above trace is no trace of T. In order to use the deCODEr to prevent this process from ending in deadlock we first have to add all deadlock-ending traces to T . This problem is not as difficult as it seems to be. To do so we have to be more specific about how the weaving of two FSM's is defined.

Definition 21 *Consider two deterministic, completely defined and minimal FSM's M_P and M_R, given as*

$$M_P = (\mathbf{a}P, Q_P, \delta_P, q_P, F_P)$$
$$M_R = (\mathbf{a}R, Q_R, \delta_R, q_R, F_R)$$

then the FSM M is given as:

$$M = (\mathbf{a}P \cup \mathbf{a}R, Q_P \times Q_R, \delta, (q_P, q_R), F_P \times F_R)$$

where $\delta((p,q),a)$ is defined as:

$$\begin{array}{lll} a \in \mathbf{a}P \ \wedge\ a \notin \mathbf{a}R: & \delta((p,q),a) = (\delta(p,a),q) & \\ a \notin \mathbf{a}P \ \wedge\ a \in \mathbf{a}R: & \delta((p,q),a) = (p,\delta(q,a)) & \\ a \in \mathbf{a}P \cup \mathbf{a}R: & \delta((p,q),a) = [\emptyset] & \textit{if } \delta_P(p,a) = [\emptyset] \vee \delta(q,a) = [\emptyset] \\ & \delta((p,q),a) = (\delta_P(p,a), \delta_R(q,a)) & \textit{otherwise} \end{array}$$

The FSM representing the weaving of P and R is denoted by M_{PR} and constructed out of M by deleting all unreachable states[11].

The constructed automaton M_{PR} is again deterministic, complete but need not be minimal any more. It represents the behavior of the total connection of P and R.

Next we define a *deadlock state* in M_{PR}.

Definition 22 *A state (p,q) in the FSM M_{PR} is called a deadlock state if:*

$$(p,q) \neq [\emptyset] \ \wedge\ (p,q) \notin F_P \times F_R \ \wedge\ (\forall a : a \in \mathbf{a}P \cup \mathbf{a}R : \delta((p,q),a) = [\emptyset])$$

A path leading to a deadlock state is a trace not leading to a final state (so not part of the trace set of the total connection) but possible by weaving prefixes of traces of P and R. Such a path therefore leads to deadlock[12].

Notice, that minimizing the FSM M_{PR} results in a FSM that exactly represents the behavior of the total connection of P and R but omits the possibility of detecting deadlock.

If we construct M_{PF} it turns out, that there exists a deadlock state. To prevent the connection to end in deadlock we first have to add all deadlock-ending traces to the behavior (simply by making this deadlock state a final state) and secondly by using the deCODEr with desired behavior according to $F = F_0 \text{ w } F_1 \text{ w } \ldots \text{ w } F_{k-1}$.

Reconsider process T. Because the events e_i and t_i are not important to us at this moment we omit these from P_i. So we use:

$$P_i = < (s_i; gl_i; gr_i; ll_i; lr_i)^*, \{gl_i, gr_i, ll_i, lr_i\}, \{s_i\} > \qquad i = 0, 1, \ldots, k-1$$

and consider

$$T = P \text{ w } F$$

with

$$P = P_0 \text{ w } P_1 \text{ w } \ldots \text{ w } P_{k-1}$$
$$F = F_0 \text{ w } F_1 \text{ w } \ldots \text{ w } F_{k-1}$$

[11] A state is unreachable if no path exists form the initial state to that state.

[12] It is possible to make all this formal, i.e. give a definition of deadlock (corresponding to our intuitive ideas) and prove that deadlock in this sence is possible if and only if the FSM of the total connection has at least one deadlock state. However it is outside the scope of this article to do all this (we only like to show that the deCODEr can be used to prevent deadlock).

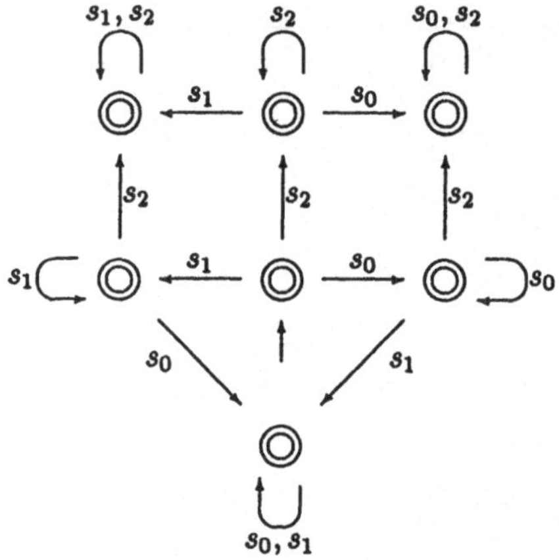

Figure 6: First butler

Now add to T all deadlock-ending traces by making the deadlock state in the corresponding FSM M_{PF} a final state and use F as desired exogenous behavior of T (notice that because of the connection all events of F are exogenous events in T).

The deCODEr gives us the controller (a *butler* in this case) (for $k = 3$ this butler is displayed in figure 6). The butler can only prevent the total process to end in deadlock by forbidding one (randomly chosen) philosopher ever to eat. This is not what we like. The resulting controller is unable to notice if a philosopher is ready with eating and therefore can not use the fact that this philosopher does not need the forks any more.

To be able to find a nicer controller we add to P_i an extra event a_i, meaning that philosopher P_i asks the butler if he may sit down. The butler then may give him permission to sit down by letting event s_i occur. So we have:

$$P_i = < (a_i; s_i; gl_i; gr_i; ll_i; lr_i)^*, \{gl_i, gr_i, ll_i, lr_i\}, \{a_i, s_i\} > \qquad i = 0, 1, \ldots, k-1$$

Using the deCODEr we find (for $k = 3$) the butler as in figure 7.

This butler behaves precisely as we should think he should. He gives permission to at most two philosophers to start eating. A third demand is retained until one of the philosophers asks again (and thereby letting the butler know that he has finished eating).

15 Conclusions

In this paper discrete event systems are defined using (an extended version of) trace theory. It has turned out, that this way of modeling gives the possibility to formulate a control problem (CODE) and construct a controller, that is defined in exactly the same way as the original plant. So it is not necessary to interpret plant and controller differently (as for example in the supervisor control theory of [RaWo]).

The philosophers-example illustrates that CODE can be used to avoid deadlock.

An advantage of this way of modeling is further, that the behavior of the process is given as a set of traces, while nothing is said about (and nothing need to be known of) how such traces are actually given. This means, that it is possible to give such a behavior by means of finite state automatons or regular expressions (as is done in this paper), by means of (possibly) infinite state automatons (as is done in [RaWo]), or by other means (although in this last case it may be impossible to compute for instance the blend of such (in a strange way defined) trace sets).

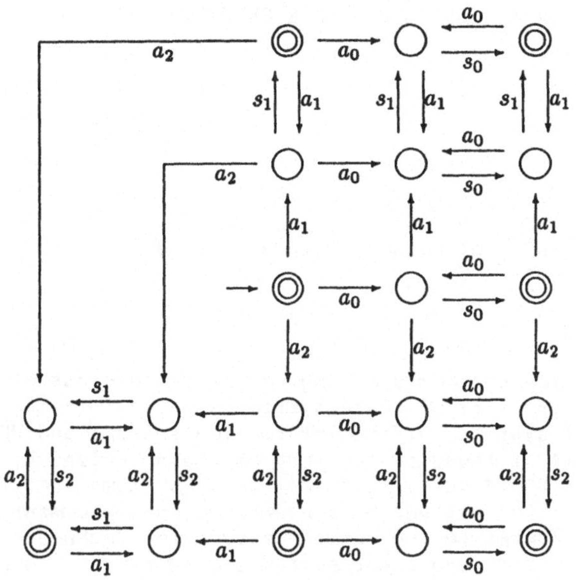

Figure 7: Second butler

References

[JvdS] J.L.A. van de Snepscheut (1985) *Trace theory and VLSI design* (Lecture notes in computer science, nr. 200), Springer Verlag

[HoUl] J.E. Hopcroft and J.D. Ullman (1979) *Introduction to automata theory, languages and computation*, Addison Wesley

[RaWo] P.J. Ramadge and W.M. Wonham (1985) *Supervisory control of a class of discrete event processes*, systems control group report 8515, Dept. of electl. engrg., univ. of Toronto

PROTOCOL VERIFICATION USING DISCRETE-EVENT MODELS

Michał Sajkowski
Technical University of Poznań, Poznań, Poland

1. INTRODUCTION

Communication protocols belong to the broad class of discrete-event systems. Other examples of these systems are flexible manufacturing systems, process control systems, office information systems and VLSI circuits. The behaviour of all of these systems can be characterized by starting and ending times of their activities. All these systems inherently involve the consideration of the notions of *concurrency*, *nondeterminism*, *time* and *communication*. Therefore the formal description technique (FDT) used to the description of a discrete-event system should be able to express these notions.

The satisfaction of these requirements, which corresponds to the modelling power of a FDT, influences the analyzability of this technique. The analyzability of a FDT means here the ability to the analysis of the specification written in this FDT. Therefore, the nature of a discrete-event system influences also the formal technique used for the verification of its specification.

In this paper we present a new approach to the verification of a communication protocol modelled as a discrete-event system. This approach is based on the analysis of a communication protocol considered as a time-driven system.

The paper is organized as follows. First, in the second section, some useful definitions dealing with protocol engineering, discrete-event systems and time-driven systems are given. Then, in the third section, the selected discrete-event models applied to protocol verification are discussed. The fourth section gives the principles of the modelling a protocol as a time-driven system. The analyzed protocol is specified by means of time augmented Petri nets. In the fifth section the idea of the verification technique is presented. The technique combines time constraints based projection and the examination of the safeness of certain places in timed Petri net model. A simple example is provided. In the last section the usefullness of discrete-event models in protocol engineering is discussed and topics for future research are suggested.

2. DEFINITIONS

Such a scope of a paper causes that some basic definitions dealing with protocol engineering, discrete-event systems and time-driven systems should be given. The first definition specifies the protocol engineering itself.

Protocol engineering is the application of scientific disciplines to the production of high-quality communication protocols on time and within budget.

These disciplines deal with the consecutive stages in the communication protocol production process. These stages include: specification verification, performance prediction, implementation, testing, certification and assessment, performance evaluation, documentation, and possibly synthesis and conversion. The major disciplines in protocol engineering have been presented by Piatkowski (1983), who is also the author of protocol engineering term. The state of art in protocol engineering has been discussed further in (Piatkowski1986).

We will define here, for the purpose of this paper, two stages of protocol production process only: specification and verification. For the definitions and scope of other stages see e.g. (Sajkowski 1985).

Protocol specification we understand the description which defines the required behaviour of a protocol.

The specification may be formal or informal. The formal specification applies formal language to the description of a protocol behaviour. The term "formal" means "expressed in a precise mathematical way using symbols with specific meaning" (Løvengreen 1985).

Protocol verification is understood as the demonstration of the correctness, completeness and consistency of the protocol design represented by its formal specification.

Now we will give the definitions of the most important notions in the area of discrete-event systems.

Event is defined as an instantaneous elementary action. The elementary action is understood as the indivisible one.

Discrete-event system is understood as a dynamic system in which events occur at discrete times. The intervals between events have not to be identical. The behaviour of the discrete-event system is then specified by a set of concurrent and/or nondeterministic time-consuming activities, which are performed according to prescribed ordering. Each event is seen as the starting or ending of the activity (Cohen 1985, Cohen 1987, Dietz 1987, Garzia 1986, Ho 1987, Lin 1986).

Discrete-event model we understand the model of discrete-event system. Therefore, the model of a discrete-event system D is another discrete-event system M which reflects some characteristics of D.

We will obtain the *mathematical model of discrete-event system* if we form it by a mathematical object or structure. Such a structure was proposed by Zeigler (Garzia 1986), where the discrete-event system is specified by 6-tuple: "

- The set of possible external events.
- The set of system states, in sequence.
- The set of output values.
- A transition function which combines the current state and external events to yield a new state.
- An output function that maps the current state to an output value.
- A time-advance function that controls how long the system remains in a given state. "

Now we will give two general definitions, based on (Løvengreen 1985), which describe the notions of concurrency and nondeterminism.

Concurrency called also *true concurrency* means that in a system the elementary actions may overlap in time.

Nondeterminism is understood as an impossibility to uniquely determine the future behaviour of the system knowing the future input event to the system.

Finally, according to (Coolahan 1983), we will define time-driven system.

Time-driven system we understand as one in which the time spent for the execution of the system functions is critical to their successful performance, and in which the global clock drives the repetitive realization of similar activities at regular intervals.

3. DISCRETE-EVENT MODELS

There are a lot of FDTs applied earlier for protocol specification (Sajkowski 1983, Sajkowski 1984). However, a limited number of them can be considered as the methods possessing the power being enough for the modeling of discrete-event systems.

To this group belong all FDTs which satisfy the crucial requirement of discrete-event systems, namely the specification of a *time-advance function*. The next two requirements which should be satisfied by a FDT are *concurrency* and *nondeterminism*. The former is obvious if we recall our definition of true concurrency, the latter results from the fact that discrete-event systems are typically nondeterministic ones.

Among the FDTs applied earlier for communication protocol specification, only time-augmented Petri nets, temporal logic and formal specification languages satisfy three requirements given above (Sajkowski 1984). We will add to them newly developed FDTs, which not necessarily have been applied to protocol specification and which satisfy the required criteria. In this way we can form the set of FDTs relevant to the specification of discrete-event systems. This set is the following:
- Coordinated Concurrent Activities.
- Time-augmented Petri nets.
- Real-time temporal logic.
- Real-time languages.
- Timed communicating sequential processes.
- Algebras for timed processes.
- Real-time attribute grammars.

We will discuss now these FDTs. We will focus on the satisfaction by a FDT of the criteria of modeling concurrency, time and nondeterminism.

3.1. Coordinated Concurrent Activities

The technique named Coordinated Concurrent Activities (CCA), introduced in (Aggarwal 1987a, Aggarwal 1987b), corresponds to the well-known model of coupled finite state machines. The CCA model is an extension to the previous Selection/Resolution model (Aggarwal 1983, Kurshan 1985). This extension covers the introduction of: continuous time, activity termination and the precise semantics of activities coordination.

In CCA, the model of distributed system consists of N Activity Machines running in parallel. These machines coordinate by exchanging the current status of the running activities. The status corresponds to the selection in S/R model. Each Activity Machine has a transition graph, called Activity Graph, which specifies the sequencing of its activities. This graph describes also the conditions for the changing of activities. These conditions result from the status of other machines during coordination.

The real-time notion is introduced into the CCA model by assuming the existence of an observer relative to which starting and ending times of the activities are synchronized, and by the imposing timing constraints on the switching of machine activities.

In this model, time is divided into successive intervals of work and synchronization. If activities in various machines remain in their active phases, then all the machines are in the work phase. However, if any of the activities comes into its termination phase, then the system of machines enters synchronization phase and a new activity is incarnated.

The nondeterminism in CCA is included in that, that the selection of an incarnated activity may be nondeterministic.

3.2. Time-Augmented Petri Nets

In Petri nets the notion of time can be associated with transitions (Ramchandani 1974, Merlin 1976, Zuberek 1980, Molloy 1982, Razouk 1985, Dugan 1984, Ajmone Marsan 1985) and places (Sifakis 1980, Coolahan 1983, Wong 1984). The timing can be deterministic (Ramchandani 1974, Merlin 1976, Zuberek 1980, Coolahan 1983, Razouk 1985) or stochastic (Molloy 1982, Dugan 1984, Ajmone Marsan 1985, Florin 1985, Lazar 1987). In the case of the deterministic timing, there may be deterministic firing time added to each transition (Ramchandani 1974) or firing interval expressed by means of two deterministic values (Merlin 1976). In the case of stochastic timing, the firing delay is a random variable with an exponential distribution for continuous time systems or with geometrical distribution for discrete time systems.

Nondeterminism is modelled by means of conflict transitions, and concurrency by the use of non-causally related transitions.

3.3. Real-Time Temporal Logic

There are known the examples of the application of temporal logic to the modelling and analysis of discrete-event systems. For instance in (Thistle 1986) linear-time temporal logic has been used for the verification of control system. However, it seems that more expressive temporal logics should be used for this purpose, for instance real-time temporal logic (Koymans 1983a, Koymans 1983b). It comes from the fact, that the linear time temporal logic considers time qualitatively, and it cannot express the time-out mechanism.

In the real-time temporal logic, for the purpose of quantitative treatment of time, two new temporal operators have been added: "before" referred to the past and "strong until in real-time t" from which other real-time operators can be derived: e.g. "eventually within real-time t from now" or "always after real-time t has elapsed from now". Therefore, in the real-time temporal logic the state possesses the component representing a kind of a global clock.

3.4. Real-Time Languages

Recently, various high-level programming languages for real-time applications in industry have been developed. The examples of them are: Ada (Ada 1983), Occam (Occam 1984), Chill (Chill 1985), LUSTRE (Caspi 1987) and ESTEREL (Berry 1985). The first three of them, i.e. Ada, Occam and Chill are asynchronous, nondeterministic and they use the notion of "absolute" time only. The last two of them, i.e. LUSTRE and ESTEREL are synchronous, deterministic and they use the "multiform" notion of time.

We will consider now the treatment of concurrency, time and nondeterminism in the representatives of these groups of languages.

For instance, in Ada the passage of absolute time is expressed by means of *delay* statement, but time-out mechanism can be described jointly by *select* and *delay* statements. The *select* statement is the basic mean for the description of nondeterminism. The parts of a program in Ada called *tasks* can run in parallel, and they communicate or synchronize by the use of rendezvous mechanism. The rendezvous can be also used in order to cause the delay up to the occurrence of a particular event.

In the second group of real-time languages, the synchronous data flow language LUSTRE provides simple way of handling time. In LUSTRE, variables may be considered to be functions of time. The variable is characterized by sequence of values and its clock. LUSTRE possesses four non-standard operators, i.e. "previous", "followed by", "when" and "current". These operators are used for the construction of nested clocks and for the operation on expressions with different clocks. LUSTRE has deterministic nature.

3.5. Timed Communicating Sequential Processes

Up to now, there are only a few approaches to add time notion to the parallel language CSP (Koymans 1985, Reed 1986, Gerth 1987).

The model of timed CSP developed by Reed and Roscoe (1986) is continuous with respect to time, assumes the existence of a conceptual global clock, a system delay constant, hiding removing the external control, and timed stability. The events in CSP are replaced by timed events. Two new processes are added, with respect to untimed CSP, namely process WAIT t ($t \geq 0$) and diverging process. The main result of the extension of CSP with time notion is the distinguishing deadlock from divergence. This model expresses non-discrete time with a least element and true concurrency.

Koymans et al. (1985) propose a real-time variant of CSP, called CSP-R. All events are related to each other by a conceptual global clock. The concurrent execution is modelled by an extension of maximal parallelism model. The main additional construct w.r.t. CSP is the real-time construct *wait d*, where *d* is a duration. CSP-R language allows to model discrete time and time-out.

Gerth and Boucher (1987) developed a model for the real-time behaviour of extended communicating sequential processes. This model is called "timed failures model". It is really a generalization of the failures model. The timed failures model allows to describe a-priori bounded delay of actions, time-out of actions, non-discrete time with a least element, true concurrency of actions, nondeterminism and abstraction.

3.6. Algebras for Timed Processes

Algebras for timed processes (Richier 1986, Quemada 1987, Nounou 1985) are related to very well known CCS (Milner 1980). These FDTs additionally incorporate the notion of time.

The Algebra for Timed Processes - ATP (Richier 1986), which gave the name to the group of similar FDTs, is an extension of process algebra specified in (Bergstra 1984). In ATP standard delay statement is expressed by means of start delay and termination delay constructs. The timed system itself is described by the use of timed state graph. Its nodes are labelled by corresponding delays. The parallel composition is asynchronous or synchronous, according to the type of actions, and is defined by operators "left merge" and "communication merge". Nondeterminism is described by means of an "alternative composition" operator.

The second example of the algebra for timed processes is Timed LOTOS (Quemada 1987). In it, a quantitative relative time notion is added to widely used ISO language LOTOS (Brinksma 1986, Lotos 1987). It is done by means of a time interval associated with an action, indicating the period of time in which this action should terminate. Time is expressed by the data type in ACT-ONE. Concurrency is modelled by parallel composition operator. The nondeterminism is described by choice operator.

The previous technique to these two above which applies algebraic specification of protocol timing behavior has been a variant of CCS proposed in (Nounou 1985). The protocol timing behaviour is modelled by marked point process, i.e. the set of events in a given time, and its attributes: time durations and probabilities of possible behaviours. The correct ordering of events is described by the use of a time constraint relation "<<". Time-out upper bound and mean-transfer time are analyzed.

3.7. Real-Time Attribute Grammars

Real-time attribute grammars are an extension of conventional attribute grammars in order to specify concurrency and real-time. The only example of this FDT is "Real-Time Asynchronous Grammar" - RTAG, proposed by Anderson and Landweber (1985).

In RTAGs, terminal symbols correspond to input events, i.e. receiving a message, or output events, i.e. sending a message. The parallel composition of events is expressed using curly brackets notation. Time constraints are described by means of a special terminal symbol /*timer*/, which has a single integer attribute *interval*. Productions in which /*timer*/ is engaged are called timed productions. The timed production is able to specify the time-out mechanism.

3.8. Discussion of Approaches

In this section we have discussed the FDTs, which in our opinion are the only ones sufficient for the modelling of discrete-event systems. FDTs, considered here, have been applied to the specification of various instances of discrete-event systems.

Communication protocols have been described by means of Coordinated Concurrent Activities (Aggarwal 1987a), time-augmented Petri nets (Merlin 1976, Razouk 1985, Ajmone Marsan 1985, Sajkowski 1986, Lazar 1987, Sajkowski 1987), real-time languages (Bochmann 1981), algebras for timed processes (Nounou 1985, Richier 1986, Quemada 1987) and real-time attribute grammars (Anderson 1985).

Flexible manufacturing systems have been specified only by means of CCA (Aggarwal 1987b) and time-augmented Petri nets (Alla 1986, Hillion 1987).

Control systems have been described by the use of CCA (Aggarwal 1987b, Katzenelson 1986), time-augmented Petri nets (Coolahan 1983) and real-time languages (Plessmann 1986, Caspi 1987).

It is seen that two FDTs: CCA and time-augmented Petri nets have been applied to all major examples of discrete-event systems. The biggest number of FDTs has been used for the description of communication protocols.

There are some similarities existing between certain FDTs. For instance CCA, certain real-time languages (Ada, Occam) and Timed CSP use the rendezvous concept, which requires synchronous communication. CCA, some timed CSP (Reed 1986, Gerth 1987) use the continuous time notion, whereas ATP, CSP-R, ESTEREL and RTAG apply discrete time. Time-augmented Petri nets, depending on the approach, apply discrete or continuous notion of time. CCA, time-augmented Petri nets, real-time temporal logic, Timed CSP (Reed 1986) and Timed LOTOS assume explicitly the existence of a global clock.Finally in all but one FDTs, discussed here, nondeterminism can be expressed.

This exception are deterministic real-time languages like LUSTRE and ESTEREL.

It should be noticed, that almost all of FDTs use the asynchronous cooperation between processes, and synchronous communication between actions in the different processes. The exceptions are languages LUSTRE and ESTEREL which use synchronous cooperation and Chill which applies asynchronous communication. For the definitions of asynchronous or synchronous cooperation and communication see (Bergstra 1985).

Let us consider now the techniques used for the verification of the discrete-event system specification written by the use of given FDT. For specifications in CCA, reachability analysis and simulation of sample trajectories of a discrete-event system are applied (Aggarwal 1985a, Aggarwal 1985b). Time-augmented Petri nets descriptions of a discrete-event system are verified by the use of reachability analysis (Merlin 1976, Razouk 1985, Sajkowski 1986, Stotts 1986, Sajkowski 1987) and invariant analysis (Alla 1986, Hillion 1987). The specifications written in real-time temporal logic, real-time languages and Timed CSP require assertion proving techniques for the verification of the properties of discrete-event systems. The systems modelled by means of algebras for timed processes are verified by the use of algebraic verification, e.g. applying observation equivalence notion of CCS (Richier 1986). Discrete-event system model applying RTAG FDT can be verified by means of any algebraic verification technique.

As it is seen form the above comparison of FDTs, the time-augmented Petri nets are one of the techniques suitable for the description of discrete-event systems. We will show, in the next sections, a new approach to the application of this FDT to communication protocol analysis.

4. PROTOCOL AS A TIME-DRIVEN SYSTEM

We have applied a description of a communication protocol using the model of a time-driven system proposed by Coolahan and Roussopoulos (1983). Therefore we describe a protocol using time-augmented Petri nets (see Appendix), and then we add a global clock construction and we distinguish the final transition in the modelled protocol.

The global clock construction (called also a driving cycle or master timing mechanism) has the significant role in the time-driven system analysis. This construction consists of a marked place p_1, called master timing process, and the transition t_1 connected to p_1 by an elementary loop. The master timing process has the execution time T_1 associated with it. In our solution the execution time T_1 has a little changed semantics. It models the delay before the expiration of which the protocol should provide the required service. In classical time-driven systems, time T_1 drives the repetitive realizations of the remainder of Petri net model.

Such an approach comes from the fact, that even in the case of so called time-independent protocol, its user will not be waiting endlessly for the service provision. Therefore a certain time limit should be imposed on the service provision. Hence, in practice, every real-life protocol is the time-dependent one. It implies that the formal protocol specification is really a formal specification of its timing behaviour. A new approach to the verification of such a specification is presented in the next section.

Our model of a protocol has the following properties:

- The firing of transitions can be simultaneous.
- All events, i.e. the starting and ending of the process execution, are related with a global clock.
- The cooperation between protocol entities is asynchronous.
- The communication between protocol actions in (different) entities is synchronous.

As an example we have used the user-server protocol (Brand 1983). The time-driven model of this protocol is given in Figure 1.

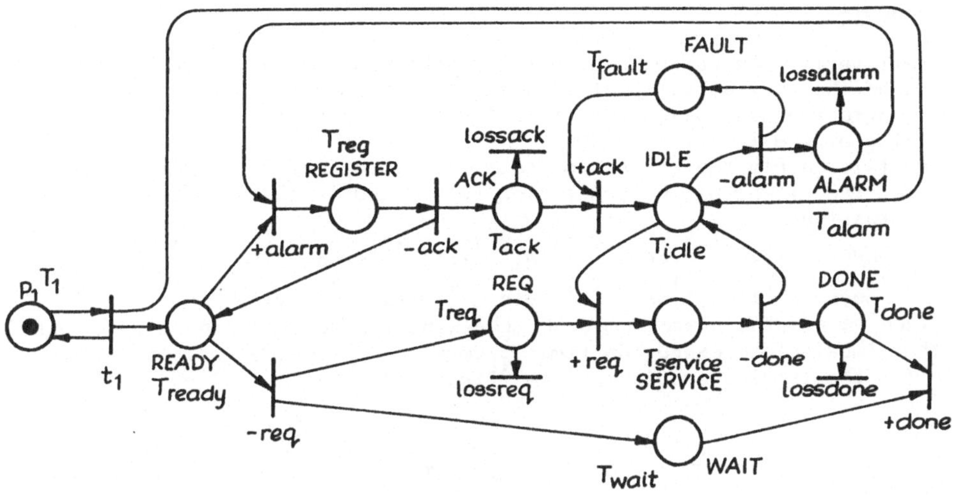

FIGURE 1 The time-driven model of the user-server protocol

5. VERIFICATION OF A PROTOCOL AS A TIME-DRIVEN SYSTEM

The verification of the communication protocol specified as a time-driven system is performed in two steps. First the time constraints based projection is applied to the time-driven model of a protocol. Then the examination of the safeness of certain places in the reduced protocol model is carried out.

5.1. Time Constraints Based Projection

The projection via time constraints is proposed as a new technique for the avoidance of state explosion. It differs from the classical projection technique (Lam 1984) , cause it reduces the set of reachable states and hence the analysis complexity by means of the use of time constraints derived from the protocol model. Therefore, the complete protocol model is projected into a plane on which certain time constraints are satisfied. Then the reduced model of a protocol is analysed only.

Hence, *the time constraints based projection* is understood as the creation of the image of the complete protocol model by the cut off these parts of the model which will never occur for given time constraints.

5.2. Examination of the Safeness of Places

For the protocol described as a time-driven system, the protocol properties can be verified by the examination of the safeness of certain places of the constructions existing in this model. The safeness is verified by means of formulae derived in (Coolahan 1983) for various constructions like: simple places, synchronized parallel paths, independent cycles and shared resources.

5.3. Example

Lets verify the semantic property of the correct execution of the purpose of the user-server protocol. It is defined here as the service provision to the user before the time limit will expire, and is modelled by the firing of the transition +done.

We first apply the projection to our protocol model. It eliminates, for instance, the firing of the transition -alarm and is based on the following time expression:

$$(T_{ready} < T_{idle}) \wedge (T_{ready} + T_{req} \leq T_{idle}) \wedge (t_{+req} < t_{-alarm}) \wedge \wedge (T_{done} < T_{idle}) \wedge (T_{wait} < 2T_{idle} - T_{ready} + T_{service}) \quad (1)$$

Then we add the place named SP (checking the service provision), and we set the global clock execution time T_1 equal to T_b, where T_b is the time limit for the service provision (see Figure 2).

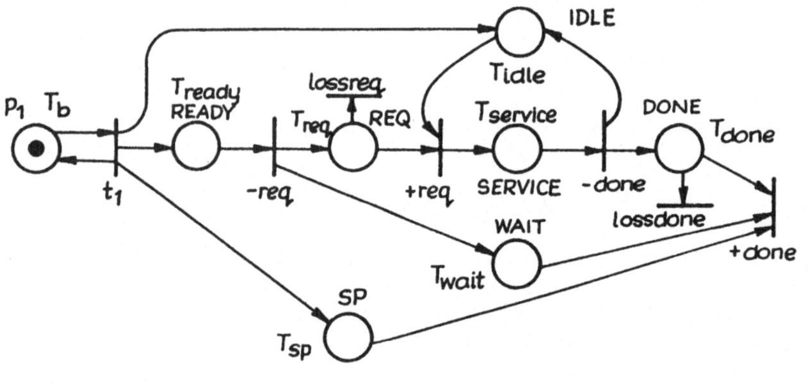

FIGURE 2 The projection of the protocol model for service provision checking

We will use now the formula for the safeness of the final place in the path of the synchronized parallel paths construction,(see Appendix). The condition for the service provision is satisfied in the case of the safeness of the place SP. Lets consider the final place SP in the path p_i: SP, and the path p_j: READY_REQ_SERVICE_DONE. Then we have:

$$(T_{ready} + T_{req} + T_{service} + T_{done}) - T_{sp} \leq (T_b/1) - T_{sp}$$

that is:

$$T_{ready} + T_{req} + T_{service} + T_{done} \leq T_b \quad (2)$$

And similarly for the path p_i and the path p_k: READY_WAIT :

$$T_{ready} + T_{wait} \leq T_b \quad (3)$$

In order to describe the lack of the loss of messages, the following time expressions should be added:

$$(T_{req} \leq T_{maxreqdelay}) \wedge (T_{done} \leq T_{maxdonedelay}) \quad (4)$$

6. CONCLUSIONS

Existing previously FDTs, without the notion of time included, have been used for any discrete-event system description. However such a description is insufficient for the purpose of sophisticated verification and performance prediction, which consider time constraints.

We have discussed here seven discrete-event models, incorporating the notion of time, which can be applied for the specification and verification of communication protocols and other discrete-event systems. It is seen that CCA, time-augmented Petri nets and algebras for timed processes have practical significance in these areas. The most promising new techniques are CCA and algebras for timed processes. It comes from the fact that these FDTs are based on a strong mathematical background.

The design and analysis of complex real-life protocols require the existence of appropriate development tools (for a comprehensive survey of the usage of protocol development tools see (Bochmann 1987)). We will indicate now the tools for specification development, constructed for the FDTs discussed here.

For CCA two tools have been developed, that is SPANNER (Aggarwal 1985b) and SIMUL (Aggarwal 1985a). For the Selection/Resolution model, preceding the CCA model, COSPAN tool has been designed (Katzenelson 1986). Time-augmented Petri nets have the biggest number of tools, both for deterministic time and stochastic Petri nets, for instance TINA (Roux 1986), GreatSPN (Chiola 1987), GTPN Analyzer (Holliday 1985), DEEP (Dugan 1985). For algebras for timed processes CUPID/Analyst tool applying algebraic verification is available (Barghouti 1987). There are other tools developed e.g. for LOTOS language (Turner 1987), however the tool for Timed LOTOS has not been announced yet. For RTAGs there exists RTAG Analyzer (Anderson 1985). The author is not aware of the tools existing for other FDTs discussed here.

Therefore, one of the topics for further research is the design and implementation of software tools for other discrete-event models like real-time temporal logic, real-time languages and timed CSP, in order to make them more attractive.

The second direction is the addition of verification tools to existing implementation oriented tools, e.g. RTAG Analyzer.

The third question which can be considered is the application of discrete-event simulators, used up to now to flexible manufacturing systems analysis - e.g. SEDRIC (Valette 1985), to the examination of communication protocols and other discrete-event systems.

The fourth problem which remains to be done is the development of FDTs themselves and the determination of the impact of a discrete-event system modelled on the FDT used.

The fifth point that should be addressed is a separate examination of protocol engineering and discrete-event systems fields in order to provide formal definitions and reasonable taxonomy of notions relevant to them.

Finally, the presented verification technique should be further developed in order to apply to the complete set of protocol properties and to make it more friendly to its users.

APPENDIX

Time-augmented Petri nets (Coolahan 1983)

In this Petri net model, places represent processes and a nonnegative execution time T_k is assigned to the each place k. The input transition to the place models the beginning of the execution of the process. A transition will be able to fire if all its input places possess tokens and for all of them the required execution times have expired. The transition will fire immediately when is enabled. The firing time of the transition is equal to zero. If more than one transition is enabled, then may happen that one of them will fire only. Then the choice of the transition is nondeterministic.

A formula for the safeness of the final place p_{if} in the path p_i of a synchronized parallel paths construction, in the presence of time, is the following (Coolahan 1983):

$$P_j - P_i \leq (T_1/F_{if}) - T_{if} \qquad \text{(A1)}$$

where:

T_1: global clock execution time

P_i (P_j, respectively): time of the traversing the path p_i (p_j, respectively)

F_{if}: maximum relative firing frequency of the input transition to the final place p_{if}

T_{if}: execution time of p_{if}.

The notation used in the protocol model

V: a process

T_V: the execution time of a process

X: transmitted message

-x: sending a message

+x: receiving a message

T_x: sojourn time of a message in the channel

lossx: the loss of message

$T_{maxxdelay}$: maximum allowed delay for a message X in a channel

t_{-x} (t_{+x}, resp.) : global time of sending (receiving) message X

$t_{-y} < t_{+x}$: nondeterministic choice of -y among transitions -y, +x.

REFERENCES

Ada. (1983). The Programming Language Ada. Reference Manual. LNCS 155. Springer-Verlag, Berlin.

Aggarwal, S., Kurshan, R.P., and Sabnani, K. (1983). A Calculus for Protocol Specification and Validation. In H. Rudin and C.H. West (Eds.), Protocol Specification, Testing, and Verification, III. North Holland, Amsterdam.

Aggarwal, S. and Har'El, Z. (1985a). SIMUL: A Tool for the Simulation and Analysis of Protocols, Tech. Memo, AT&T Bell Labs, Murray Hill, USA.

Aggarwal, S., Barbara, D., and Meth, K.Z. (1985b). SPANNER: A Tool for the Specification, Analysis, and Evaluation of Protocols. To appear in IEEE Trans. on Software Engineering.

Aggarwal, S., Barbara, D., and Courcoubetis, C. (1987a). LAN Broadcast Protocols for Implementing the CCA Model. Proc. IFIP Int'l Symp. on Protocol specification, Testing, and Verification, 7th, Zurich, May 1987.

Aggarwal, S., Barbara, D., and Courcoubetis, C. (1987b). Real-Time Coordination of Concurrent Activities. In B. Sarikaya and G.V. Bochmann (Eds.), Protocol Specification, Testing, and Verification, VI. North Holland, Amsterdam.

Ajmone Marsan, M. and Chiola, G. (1985). Modeling Discrete Event Systems With Stochastic Petri Nets. Proc. ISCAS 85 Symp., Kyoto, June 1985. IEEE Computer Society Press.

Alla, H. and Ladet, P. (1986). Coloured Petri Nets: A Tool for Modelling, Validation and Simulation of FMS. In A. Kusiak (Ed.), Flexible Manufacturing Systems: Methods and Studies. North Holland, Amsterdam.

Anderson, D.P. and Landweber, L.H. (1985). A Grammar-Based Methodology for Protocol Specification and Implementation. Proc. Data Communications Symp., 9th, Whistler Mountain, September 1985. IEEE Computer Society Press.

Barghouti, N., Nounou, N., and Yemini, Y. (1987). An Interactive Protocol Development Environment. In B. Sarikaya and G.v. Bochmann (Eds.), Protocol Specification, Testing and Verification, VI. North Holland, Amsterdam.

Bergstra, J.A. and Klop, J.W. (1984). Algebra of Communicating Processes. CS R8421. Dept. of Computer Science, Center for Mathematics and Computer Science, Amsterdam, The Netherlands.

Bergstra, J.A., Klop, J.W., and Tucker. J.V. (1985). Process Algebra with Asynchronous Communication Mechanism. In S.D. Brookes, A.W. Roscoe and G. Winskel (Eds.), Seminar on Concurrency. LNCS 197, Springer-Verlag, Berlin.

Berry, G. and Cosserat, L. (1985). The ESTEREL Synchronous Programming Language and Its Mathematical Semantics. In S.D. Brookes, A.W. Roscoe and G. Winskel (Eds.), Seminar on Concurrency. LNCS 197, Springer-Verlag, Berlin.

Bochmann, G.v. and Pickens, J.R. (1981). A Methodology for the Specification of a Message Transport System. In R.P. Uhlig (Ed.), Computer Message Systems. North Holland, Amsterdam.

Bochmann, G.v. (1987). Usage of Protocol Development Tools: The Results of a Survey. Proc. IFIP Int'l Symp. on Protocol Specification, Testing, and Verification, 7th, Zurich, May 1987.

Brand, D. and Zafiropulo, P. (1983). On Communicating Finite-State Machines. Journal of the ACM, 30(2): 323-342.

Brinksma, E. (1986). A Tutorial on LOTOS. In M. Diaz (Ed.), Protocol Specification, Testing, and Verification, V. North Holland, Amsterdam.

Caspi, P., Pilaud, D., Halbwachs, N., and Plaice, J.A. (1987). LUSTRE: A Declarative Language for Programming Synchronous Systems. Proc. Annual ACM Symp. on Principles of Programming Languages, 14th, Munich, January 1987.

Chill. (1985). The CCITT High Level Programming Language (CHILL). Recommendation Z.200, Geneva, Switzerland.

Chiola, G. (1987). A Graphical Petri Net Tool for Performance Analysis. In Proc. Int'l Workshop on Modelling Techniques and Performance Evaluation, 3rd, Paris, March 1987.

Cohen, G., Dubois, D., Quadrat, J.P., and Viot, M. (1985). A Linear-System Theoretic View of Discrete-Event Processes and Its Use for Performance Evaluation in Manufacturing. IEEE Trans. on Automatic Control, AC-30 (3): 210-220.

Cohen, G., Moller, P., Quadrat, J.P., and Viot, M. (1987). A 2-D Discrete Event Linear System Theory. In Proc. Algebres Exotiques et Systemes a Evenements Discrets Seminar, Issy-les-Moulineaux, June 1987.

Coolahan, J.E. and Roussopoulos, N. (1983). Timing Requirements for Time-Driven Systems Using Augmented Petri Nets. IEEE Trans. on Software Engineering, SE-9(5): 603-613.

Dietz, J.L.G. and van Hee, K.M. (1987). A Framework for the Conceptual Modeling of Discrete Dynamic Systems. Proc. Temporal Aspects in Information Systems Conf., 2nd, Sophia Antipolis, May 1987.

Dugan, J.B., Trivedi, K.S., Geist, R.M., and Nicola, V.F. (1984). Extended Stochastic Petri Nets: Application and Analysis. In E. Gelenbe (Ed.), Performance 84. North Holland, Amsterdam.

Dugan, J.B., Bobbio, A., Ciardo, G., and Trivedi, K. (1985). The Design of a Unified Package for the Solution of Stochastic Petri Net Models. Proc. Int'l Workshop on Timed Petri Nets, 1st, Turin, July 1985.

Florin, G. and Natkin, S. (1985). Les Reseaux de Petri Stochastiques. T.S.I., 4(1): 143-160.

Garzia, R.F., Garzia, M.R., and Zeigler, B.P. (1986). Discrete-Event Simulation. IEEE Spectrum, 23(12): 32-36.

Gerth, R. and Boucher, A. (1987). A Timed Failures Model for Extended Communicating Processes. Proc. ICALP87 Colloquium, Karlsruhe, July 1987.

Hillion, H.p. and Proth, J-M. (1987). Performance Evaluation of Job-Shop Systems Using Timed Event Graphs. Proc. Int'l Conf. on Production Systems, 2nd, Paris, April 1987.

Ho, Y-C. (1987). Performance Evaluation and Perturbation Analysis of Discrete Event Dynamic Systems. IEEE Trans. on Automatic Control, AC-32(7): 563-572.

Holliday, M.A. and Vernon, M.K. (1985). A Generalized Timed Petri Net Model for Performance Analysis. Proc. Int'l Workshop on Timed Petri Nets, 1st, Turin, July 1985.

Katzenelson, J. and Kurshan, R. (1986). S/R: A Language for Specifying Protocols and Other Coordinating Processes. Proc. Annual Int'l Phoenix Conf. on Computers and Communications, 5th, Scottsdale, March 1986, IEEE Computer Society Press.

Koymans, R. and de Roever, W.P. (1983a). Examples of a Real-Time Temporal Logic Specification. In B.T. Denvir, W.T. Harwood, M.I. Jackson, and M.J. Wray (Eds.), The Analysis of Concurrent Systems. LNCS 207. Springer-Verlag, Berlin.

Koymans, R., Vytopil, J., and de Roever, W.P. (1983b). Real-Time Programming and Asynchronous Message Passing. Proc. ACM Principles of Distributed Computing Symp., 2nd, Montreal 1983.

Koymans, R., Shyamasundar, R.K., de Roever, W.P., Gerth, R., and Arun-Kumar, S. (1985). Compositional Semantics for Real-Time Distributed Computing. No. 68. Dept. of Computer Science, University of Nijmegen, Nijmegen, The Netherlands.

Kurshan, R.P. (1985). Modelling Concurrent Processes. Proceedings of Symposia in Applied Mathematics, 31, 45-57.

Lam, S.S. and Shankar, A.U. (1984). Protocol Verification via Projections. IEEE Trans. on Software Engineering, SE-10(4): 325-342.

Lazar, A.A. and Robertazzi, T.G. (1987). Markovian Petri Net Protocols with Product Form Solution. Proc. INFOCOM'87, San Francisco, March/April 1987, IEEE Computer Society Press.

Lin, F. and Wonham, W.M. (1986). Decentralized Supervisory Control of Discrete-Event Systems. No 8612. Systems Control Group, Dept. of Electrical Engineering, University of Toronto, Toronto, Canada.

Lotos. (1987). LOTOS-A Formal Description Technique Based on the Temporal Ordering of Observational Behaviour. ISO DIS8807, July 1987.

Løvengreen, H.H. (1985). On Concurrency Formalization. Ph.D. Thesis. ID-TR 1985-3. Dept. of Computer Science, Technical University of Denmark, Lyngby, Denmark.

Merlin, P.M. and Farber, D.J. (1976). Recoverability of Communication Protocols - Implications of a Theoretical Study. IEEE Trans. on Communications, COM-24(9): 1036-1043.

Milner, R. (1980). A Calculus of Communicating Systems. LNCS 92, Springer-Verlag, Berlin.

Molloy, M.K. (1982). Performance Analysis Using Stochastic Petri Nets. IEEE Trans. on Computers, C-31(9): 913-917.

Nounou, N. and Yemini, Y. (1985). Algebraic Specification-Based Performance Analysis of Communication Protocols. In Y. Yemini, R. Strom and

S. Yemini (Eds.), Protocol Specification, Testing, and Verification, IV. North Holland, Amsterdam.

Occam. (1984). Occam Programming Manual. INMOS Ltd. Prentice Hall International.

Piatkowski, T.F. (1983). Protocol Engineering. Proc. ICC'83 Conf., Boston, June 1983.

Piatkowski, T.F. (1986). The State of Art in Protocol Engineering. Proc. SIGCOMM'86 Symp., Stowe, August 1986.

Plessmann, K.W. and Tassakos, C. (1986). Die Programmiersprache OCCAM. Angewandte Informatik, 28(9): 389-399.

Quemada, J. and Fernandez, A. (1987). Introduction of Quantitative Relative Time Into LOTOS. Proc. IFIP Int'l Symp. on Protocol Specification, Testing, and Verification, 7th, Zurich, May 1987.

Ramchandani, C. (1974). Analysis of Asynchronous Concurrent Systems by Petri Nets. Ph.D. Thesis. MAC-TR-120, MIT, Cambridge, USA.

Razouk, R.R. and Phelps, C.V. (1985). Performance Analysis Using Timed Petri Nets. In Y. Yemini, R. Strom and S. Yemini (Eds.), Protocol Specification, Testing, and Verification, IV. North Holland, Amsterdam.

Reed, G.M. and Roscoe, A.W. (1986). A Timed Model for Communicating Sequential Processes. Proc. ICALP'86 Colloquium, 13th, Rennes, July 1986. Springer-Verlag, Berlin.

Richier, J.L., Sifakis, J., and Voiron, J. (1986). ATP-An Algebra for Timed Processes. No 1. Projet CESAR, LGI, Grenoble, France.

Roux, J.L. and Berthomieu, B. (1986). Verification of a Local Area Network Protocol with TINA, a Software Package for Time Petri Nets. Proc. European Workshop on Application and Theory of Petri Nets, 7th, Oxford, June/July 1986.

Sajkowski, M. and Stroiński, M. (1983). Evaluation of Formal Methods for Communication Protocol Specification from Protocol Designer's Viewpoint. Proc. Int'l Conf. on Software Engineering for Telecommunication Switching Systems, 5th, Lund, July 1983. IEE, London.

Sajkowski, M. (1984). Evaluation of Formal Description Techniques Applied to Communication Protocols Development. Proc. Int'l Symp. on Data Communication and Computer Networks 'Networks India 84'. Madras, October 1984.

Sajkowski, M. (1985). Protocol Verification Techniques: Status Quo and Perspectives. In Y. Yemini, R. Strom and S. Yemini (Eds.), Protocol Specification, Testing, and Verification, IV. North Holland, Amsterdam.

Sajkowski, M. (1986). On Verifying Time-Dependent Protocols. Proc. Int'l. Conf. on Software Engineering for Telecommunication Switching Systems, 6th, Eindhoven, April 1986. IEE, London.

Sajkowski, M. (1987). Protocol Verification in the Presence of Time. In B. Sarikaya and G.v. Bochmann (Eds.), Protocol Specification, Testing, and Verification, VI. North Holland, Amsterdam.

Sifakis, J. (1980). Performance Evaluation Using Nets. In W. Brauer (Ed.), Net Theory and Applications. LNCS 84, Springer-Verlag, Berlin.

Stotts, P.D. and Pratt, T.W. (1986). Petri Net Reachability Trees for Concurrent Execution Rules. TR-1679. Computer Science Dept., University of Maryland, College Park, USA.

Thistle, J.G. and Wonham, W.M. (1986). Control Problems in a Temporal Logic Framework. Int. J. Control, 44(4): 943-976.

Turner, K.J. (1987). LOTOS-A Practical Formal Description Technique for OSI. Proc. Int'l Open Systems Conf., March 1987.

Valette, R., Thomas, V, and Bachmann, S. (1985). SEDRIC: un simulateur a evenements discrets base sur les reseaux de Petri. R.A.I.R.O. APII, 19(5): 423-436.

Wong, C.Y., Dillon, T.S., and Forward, K.E. (1984). Analysis of Timing

Aspects of Communicating Computer Systems Using Timed Places Petri Nets. In Proc. ICCC Conf., 7th, Sydney, October/November 1984.

Zuberek, W.M. (1980). Timed Petri Nets and Preliminary Performance Evaluation. In Proc. Int'l Symp. on Computer Architecture, 7th, May 1980.

Analysis and Control of Discrete Event Systems Represented by Petri Nets

Atsunobu Ichikawa
Department of Systems Science
Tokyo Institute of Technology
Nagatsuta, Midori-ku
Yokohama 227, Japan

Kunihiko Hiraishi
International Institute for Advanced Study of Social Information Science
Fujitsu Limited
Numazu Shizuoka, Japan

1. Introduction

When we intend to control a discrete event system (DES) we need to describe the system behavior in a mathematical form. Some of formal representations, such as, a formal language, a sequential machine or a finite automaton have been used for this purpose (Ramadge and Wonham (1982), Cohen et al (1984)). A time domain representation was also used by Ho and Cassandras (1983). A formal representation is a kind of external or a black box type of model in a sense that it does describe the system input output relation in terms of state transition but not describe the structure of the system which actually realizes state transitions. When we want to realize a control system, we need a structural model of a system. By structural model, we mean the model which describes the structure of the system and a state transition mechanism in the system. The structural model must also be capable of representing sufficiently large class of DES.

We shall utilize a Petri net as a structural model in analysis and design of a control system of DES since the Petri net has been recognized as a suitable mean of describing a DES, particularly when a system is asynchronous concurrent (Petri 1962, Peterson 1981, Reisig 1985). In contrast with a state machine, a Petri net has no explicit input and output from the outside in its definition. Presence of inputs from the outside is implicitly assumed in selecting a transition to fire among fireable transitions. Presence of outputs to the outside is also implicitly assumed to be a sequence of markings observed from the outside. This is, however, not the case in real systems. We are often prohibited from access to some of transitions, and token counts on some of places are not seen from the outside. We need, therefore, to define explicitly the external input and output ports of a Petri net.

In this paper we first describe briefly the outcome of the authors' study conducted on a Petri net with the external input and output: they are; modeling capability, observability, decision-free fireability, reachability and control system design. Then we extend some of the results, reachability in particular, to a wider class of Petri net. The major contribution of this paper is, therefore, to give a necessary and sufficient condition for reachability in a class of Petri nets and to utilize it in the control system design of DES.

The reachability problem of a Petri net, which is to decide whether or not a marking is reachable in a Petri net from its initial marking, has been drawing considerable attention since many problems defined on the Petri net are reducible to the problem. The structure of the reachability set has been extensively studied. Though it may be very complicated for a general Petri net, the reachability set is shown to be semilinear and decidable for some restricted classes of Petri nets . They are less-than-6-places Petri nets (Hopcroft and Pansiot 1979), reversible nets (Araki and Kasami 1977), persistent nets (Landweber and Robertson 1978, Grabowski 1980, Muller 1980, and Mayr 1981) and weakly persistent nets (Yamasaki 1981, Yamasaki 1984). Necessary and sufficient conditions for reachability (NSCR) have also been obtained for restricted classes of Petri nets, such as for marked graphs (Murata 1977) , for forward conflict-free (Ichikawa *et al* 1985), for constrained firing nets (Ichikawa and Hiraishi 1983) and backward conflict-free Petri nets (Hiraishi and Ichikawa 1986) .

In this paper, a NSCR is obtained for a class of trap circuit Petri nets where any fundamental circuit contained in the net is a trap. This result is extended to a class of deadlock circuit Petri net. A sufficient condition for reachability which seems to be very close to necessary and sufficient is obtained for a class of trap containing circuit Petri net where any fundamental circuit contains a trap.

The necessary and/or sufficient condirtions are then utilized for designing the control system of DES represented by these classes of Petri nets.

2. Definitions and Notation

2.1 Petri net

A *Petri net* (a net) is a five-tuple,

$$M = (P, T, B^+, B^-, m), \quad C = (P, T, B^+, B^-) \qquad (1)$$

where $P = \{p_1, p_2,..., p_m\}$ is a finite set of *places* ; $T = \{t_1, t_2,..., t_n\}$ is a finite set of *transitions*; $P \cap T= \phi$; $B^+= [b^+_{ij}]$ and $B^- = [b^-_{ij}]$ are $m\times n$ *incidence matrices* from transitions to places and from places to transitions, respectively , $B = B^+ - B^-$ is used whenever convenient; $m : P \to N$, N is the set of natural numbers, is a *marking* ; C is a *Petri net sturucture.*

Marking and firing

We write m(Q), $Q \subset P$, to indicate the token count in a set of places Q, at the marking m. If m(Q) = 0 we say that Q is *token-free* or *unmarked* at the marking m. If m(Q) = 0 we say that Q is not token-free or *marked*. Without confusion, we also write a marking in vector form. $m = (m(p_1), m(p_2),...,m(p_m))^T$ is the marking denoted as the *m*-dimentional non-negative integer vector.

Transition t_j is fireable when each input place p_i of t_j contains at least b^-_{ij} number of tokens. Firing of transition t_j results the removal of b^-_{ij} tokens from each input place p_i and the addition of b^+_{ij} tokens in each output place p_i of the transition t_j.

When we intend to control a system represented by a Petri net from the outside of the system, we need to introduce the outside time-scale into the Petri net. So we assume that transitions in the net fire at discrete time, $k = 0,1,2,\cdots$, which are not necessary at regular intervals. We allow more than one transitions to fire at a time when they are simultaneously fireable. We do not allow, however, a transition to fire more than once at a time even if it is possible by token counts in the input places. A *firing vector* at time k, u(k), is a *m*-dimentional {0,1} vector. Its j-th conponent is one if transition t_j fires at time k and zero if not. Thus, we have the marking transition equation and the firing inequality.

$$m(k+1) = m(k) + B\,u(k), \quad m(0) = m^0, \quad k = 0,1,... \tag{2}$$

$$m(k) \geqq B^- u(k), \quad k = 0,1,... \tag{3}$$

We can extend the definitions and notation to the sequence of transition firings. Let a firing sequence be $\sigma = u(0)u(1)\cdots u(k)$. A *firng count vector* (firing count) $\overline{\sigma}$ is a *m*-dimentional positive integer vector whose j-th component is the number of firing occurence of transition t_j in a firing sequence σ. A sequnce β is said to be included in a firing sequnce α if $\overline{\beta} \leqq \overline{\alpha}$. A firing count x is said to be fireable if it has a firing sequence $\sigma = u(0)u(1)\cdots u(k)$ where each u(k) satisfies Inequality (3) at each time k. When a firing count x fires in a Petri net M = (C, m) , the resulting marking m' is given by

$$m' = m + B\,x \tag{4}$$

We call Equation (4) the matrix equation of Petri net M.

We also write $m\,[\alpha > m'$, where α is any of a firing count, a firing sequence, a firing vector or a set of transitions, to indicate that α is fireable at marking m and that the firing of α yields marking m'. When we do not need to specify the rsulting marking we simply write m $[\alpha>$ to indicate α is fireable at marking m. For example, Equation (2) together with Inequality (3) are writen as m(k) [u(k)> m(k+1).

Given a Petri net M and a firing count x, a *firing count subnet* is a subnet of the Petri net, $M_x = (P_x, T_x, B^+_x, B^-_x, m_x)$, which consists of a set of transitions included in the firing count x and their incident input and output places, where B^+_x, B^-_x, and m_x are, respectively, the projection of B^+, B^- and m on x. If a firing count x has to fire, each transition in the firing count subnet M_x must fire at least once.

The set of reachable markings or the *reachability set* $R(C, m^0)$ of a Petri net $M = (C, m^0)$ is $\{ m \mid m^0 [x> m$, for some $x \}$. If $m \in R(C, m^0)$ we say that the marking m is *reachable* in M.

Substructure

Let t_j be a transition. Then ${}^{\bullet}t_j = \{ p_i \mid b^-_{ij} \geqq 1 \}$ and $t_j{}^{\bullet} = \{ p_i \mid b^+_{ij} \geqq 1 \}$ are the set of input places and the set of output places, respectively, of transition t_j. The set of input transitions, ${}^{\bullet}p_i = \{ t_j \mid b^+_{ij} \geqq 1 \}$, and the set of output transitions, $p_i{}^{\bullet} = \{ t_j \mid b^-_{ij} \geqq 1 \}$ of a place p_i are similarly defined. We extend this notation to a set of transitions or places, such as, ${}^{\bullet}S$, $S^{\bullet}$ for $S \subset T$ and ${}^{\bullet}Q$, $Q^{\bullet}$ for $Q \subset P$, where, for example, ${}^{\bullet}S = \{ p \mid p \in {}^{\bullet}t, t \in S \subset T \}$.

A Petri net is a *single arc* if each b^+_{ij} and b^-_{ij} is equal to either zero or one.

A *deadlock* D of a Petri net M is the set of places defined by

$$D = \{ p \mid {}^{\bullet}D \subset D^{\bullet} \}$$

that is, any transition which has at least one output place in a deadlock has at least one input place in the deadlock. When a deadlock once becomes token-free, it remains token-free by firing of any transition. A place which has no input transition is the smallest deadlock. We call this a *single place deadlock.*

A *trap* Tr is the set of places defined by

$$Tr = \{ p \mid Tr^{\bullet} \subset {}^{\bullet}Tr \}$$

that is, any transition which has at least one input place in a trap has at least one output place in the trap. When a trap has at least a token, then it never becomes token-free by firing of any transition. A place which has no output transition is the smallest trap. We call this a *single place trap.*

A *directed path* (path) in a Petri net is a chain $p_1t_1p_2t_2 \cdots t_sp_{s+1}$, where $p_i \in {}^{\bullet}t_i$ and $p_{i+1} \in t_i{}^{\bullet}$, $i = 1,2,...s$. A directed path is a *directed circuit* (circuit) if $p_i = p_{s+1}$. A circuit is *fundamental* if it is not a sum of other circuits.A set of places included in a path or a circuit is simply called, without confusion, a path or a circuit, respectively.

Restriction of Petri net

A Petri net is a *marked graph* if each place has at most one input transition and at most one output transition.

A Petri net is *structurely forward (backward) conflict-free* if each place has at most one output (input) transition.

A Petri net M is a *trap circuit Petri net* (tc-net) if each fundamental circuit in M is a trap. Similarly a Petri net M is a *deadlock circuit net* (dc-net) if each fundamental circuit in M is a deadlock. A tc-net is equivalent to a normal net defined by Yamasaki(1984). The autors use the name tc- rather than normal since the name tc-net indicates the structural nature of the net and suggests the significance of a dc-net.

Given a Petri net $M = (P, T, B^+, B^-, m)$, a *reverse net* is a Petri net $M^{-1} = (P, T, B^{+'}, B^{-'}, m)$ where $B^{+'} = B^-$ and $B^{-'} = B^+$. The reverse net is the Petri net obtained by reversing the direction of all the arcs of the original Petri net.

A Petri net is *persistent* if for all $t_1, t_2 \in T$, $t_1 = t_2$ and any reachable marking m, $m\,[t_1>$ and $m\,[t_2>$ imply $m\,[t_1t_2>$; that is, if any two transitions are fireable at a reachable marking the firing of one transition does not make the other unfireable (Landweber and Robertson 1978). A Petri net is *weakly persistent* if for any two sequences $\alpha, \beta, \alpha = \beta$ and any reachable marking m, $m\,[\alpha>$ and $m\,[\beta>$ imply $m\,[\alpha\gamma>$ for some γ which $\overline{\gamma} = \overline{\beta}$ (Yamasaki 1981).

Additive independance of integer vector

A set of integer vectors x_i, $i=1,2,\ldots,r$, is said to be *additively independent* if an equation $\sum_i \alpha_i x_i = 0$, $\alpha_i \in \{-1,0,1\}$ holds if and only if all α_i are zero.

1.2 Petri Net with External Input and Output

When we intend to control a system represented by a Petri net from the outside we must have external input ports in the Petri net. The input ports are provided by adding an auxiliary input place, called an external input place, to a transition we want to control its firing. We may not be able to control firings of all transitions from the outside. Some of transitions may not be controlled and are left to fire spontaneously when they become fireable without intervention from the outside. Similarly, we may not be able to observe token counts in all the places. Token counts in some of places are not seen from the outside. Thus we have the following structure of a Petri net with external input and output.

A *Petri net with external input and output* (PNIO) is a four-tuple;

$$(C, Q, R, m) \tag{5}$$

where C is a Petri net structure; $Q = \{\, q(t_j) \mid t_j \in U \subset T \,\}$ is a *set of external input places*, and $R \subset T$ is a *set of external output places.*

A *control input* from the outside to a PNIO is a sequence of marking on

Q, v(k), k = 0,1,2,•••, which we are able to set at any sequence of non-negative |Q|-dimentional integer vectors.

After a control input is given, a PNIO has no intervention from the outside with regard to the firing of fireable transitions. Transitons which are fireable are left to fire spontaneously. The following notion of decision-free firing (df-firing) thus becomes necessary.

A Petri net is *decision-free at marking m* if all the transitions fireable at the marking m can fire simultaneouly, that is, the net is behaviorally conflict-free at the marking m. *Decision-free firing* (df-firing) is the spontaneous firing of transitions without further intervention from the outside in a decision-free Petri net at marking m. A Petri net is *decision-free* (df-net) if it is decision-free at any marking reachable from the initial marking through a sequence of df-firing. We use the name decision-free instead of conflict-free, since conflict-free has long been used to indicate structual conflict-free.

3. Df-firing Petri Net

3.1 Df-firing Petri net is equivalent to Turing Machine

In this section we shall discuss the modeling capability (computational power) of a df-net.

It is well known that;

-the modeling capability of a Petri net is more than that of a finite state machine since a Petri net can simulate the machine ,

-the modeling capability of a Petri net is less than that of a Turing machine since a Petri net has no ability of detecting zero token of a place when it is unbounded, and

-any extention of the Petri net which actually adds the ability of detecting zero token to a Petri net makes a net equivalent to a Truing machine.

In a first look a df-net seems to be a restriction of a Petri net since the net is constrained such that it is behaviorally conflict-free at any marking reachable by a sequence of df-firing. It is quite interesting to know, however, that the modeling capability of a df- net is equivalent to a Turing machine.

This is proven by showing that a df-firing Petri net can simulate a register machine which in turn is equivalent to a Turing machine.

A register machine is a two-tuple (Re, Pr) where Re is a finite set of registers which can store any large non-negative integers; Pr is a program which is a finite sequence of instructions. Shepardson and Sturgeis (1963) has shown that a register machine with a program consists of instructions of the following three kinds can simulate a Turing machine:

(1) I(n): increase register n by 1,

(2) D(n): decrease register n by 1 if it is not zero,
(3) J(n)[s]: Jump to statement s if register n is zero.

To simulate a register machine by a df-net, we construct a df-net consists of three kinds of elementary df-nets each of which can simulate each of the above three kinds of instructions (Sasaki 1986).

Let a given register machine has r number of registers and a program consists of s number of instructions. Let $p_{r1}, p_{r2}, \ldots, p_{rr}$, be places which represent the registers, and let $p_{c0}, p_{c1}, p_{c2}, \ldots, p_{cs}$, be places which represent program counters. For each instruction contained in the program, we use one of three elementary df-nets shown in FIg.1 a), b) and c) corresponding to I(n), D(n) and J(n)[s] of the instruction, respectively. Note that the obtained net is df-fireable starting any token count in p_{ri} , i=1,2,...,r and a token in p_{c0}, since the elementary nets are all df-fireable and the program of the register machine must be consecutively executable. Thus we have the following theorem.

Theorem 1 A df-net is equivalent to a Turing machine.

This property of the df-net is quite desireable in one hand since it assures that the df-net can represents any of real life DES as far as they are rigorously described. This property is not at all desireable, on the other hand, in the sense that it prohibits us from analyzing most of significant problems concerned with the property of the df-net. Typical example of these is the df-fireability itself. The df-fireability problem, which is defined as to decide whether or not a Petri net M = (C, m) is df-fireable, is undecidable since the problem is easily shown to be equivalent to the halting problem of the Turing machine. This property requires us to restrict the df-net within a such class that some of significant problems become decidable.

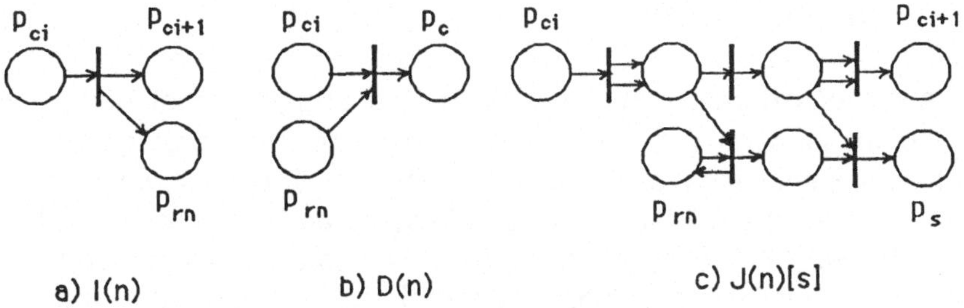

Figure 1 Petri nets representing instructions

3.2 Making a Petri Net df-fireable

Even if the df-fireability problem is undecidable, It may be possible to make a given Petri net df-fireable by chosing properly a set of external input places, an initial marking and/or a control input sequence. From this point of view, the following classification of the df-fireability becomes significant.

Level 0: a Petri net is df-fireable at level 0 if it is df-fireable for any initial marking, any choice of external input places and any control input sequence.

Level 1: a Petri net is df-fireable at level 1 if there exists a set of external input places such that it is df-fireable for any initial marking and any control input sequence.

Level 2: a Petri net is df-fireable at level 2 if there exist a set of external input places and a control input sequence such that it is df-fireable for any initial marking.

Level 3: a Petri net is df-fireable at level 3 if there exist a set of external input places, an initial marking and an control input sequence such that it is df-fireable.

The following theorems which have been obtained by the previous research (Ichikawa, Yokoyama and Kurogi 1985) will be usefull;

Theorem 2 Level 0 and Level 1 are equivalent.

Theorem 3 A Petri net is df-fireable at level 1 if and only if it is structurely forward conflict-free.

Theorem 4 A Petri net is df-fireable at Level 2.

Theorem 4 is obvious, since any Petri net can be made df-fireable at Level 2 by adding an external input place to each transition which has input places in common with other transitions. Level 3 is, therefore, of no significance in determining a set of external input places in order to make a Petri net df-fireable.

4. Observability of Marking

When a set of external output places R is a proper subset of the set of places P, the marking on P is not directly observed and must be estimated from the observed sequence of the marking on R. The observation problem thus arises. Once the initial marking is estimated and the control input sequence is known, the current marking is immediately computable. The observation problem is, therefore, essentially to estimate the initial marking of the net. Without loss of generality we shall limit our study within a class of Petri nets with the external output (PNO).

Observation Problem

Given a PNO, estimate the initial marking from the observed sequence of the marking on the set of external output places; is the observation problem.

The problem can be devided into two problems.

Firing sequence problem : Given a PNO and an observed output sequence, estimate the firing sequence of transitions.

Initial marking problem : Given a PNO and a firing sequence σ, estimate initial markings that give the firing sequence.

The necessary and sufficient condition for the solution of the firing sequence problem to be unique is given by the previous research (Ichikawa and Ogasawara 1986).

Theorem 5 Given a PNO and an observed output sequence, let B_R be the projection of the incidence matrix B onto only the set of external output places R. The firing sequence that gives the observed output sequence can be uniquely determined if and only if the set of column vectors of the matrix B_R does not contain zero vector and is additively independent.

The initial markings which give the specified sequnece of df-firing can be obtained by the following equation. This is derived by successive substitution of Equations (2) and Inequality (3):

$$m^0 \in \underline{m}^0 \cap m^0 \tag{6}$$

$$\underline{m}^0 = \cap_k \{ m \mid m \geqq \underline{m}^0{}_k \}$$

$$\underline{m}^0{}_k = \sum_{t_j \in \sigma(k)} b^-_{\cdot j} - \sum_{n=1}^{k-1} \sum_{t_j \in \sigma(n)} t_j$$

$$m^0 = \cap_k \{ m \mid m \ngeqq \overline{m}^0{}_{kj} \}$$

$$\overline{m}^0{}_{kj} = \underline{m}^0{}_{kj} + b^-_{\cdot i}$$

Equation (6) constitues an algorithm to estimate the initial marking from the observed sequence of the output marking. The algorithm is effective for a large-scale df-net, since the computation is straightforward and contains no iterative procedure.

5. Necessary and Sufficient Condition for Reachability

In this section we shall give the necessary and sufficient condition for reachability in a trap circuit Petri net(tc-net).

As described in Section 3, a df-net is equivalent to a Turing machine.

The reachability problem of the df-net is, therefore, undecidable. To control a DES represented by a Petri net, we have at least to know whether the target marking is reachable or not. At this point, we shall limit our concern within a such class of Petri net that the reachability problem is at least decidable.

We shall deal with a single arc Petri net. A single arc Petri net is equivalent to a Petri net in its modeling capability when we consider a conventional Petri net, that is, the Petri net where the firing of any fireable transition is subject to the control from the outside. They are, however, not equivalent each other in the class of df-nets. A df-net employs mutiple arc in the simulation of a register machine in order to realize a type of instruction J(n)[s] or, in other words, in order to have a capability of detecting zero token. This capability can not be achieved by a single arc df-net. Thus a single arc df-net is not equivalent to a Turing machine. So we can expect that the reachability problem is decidable for a class of single arc df- nets and this is true as shown in succeeding sections.

Decidability is not only our concern for an application to real systems since even if it is decidable the amount of computation may exceed our capasity. It is quite desireable to have a necessary and sufficient condition for reachability (NSCR) expressed in terms of an initial marking, a target marking and a structure of a Petri net.

NSCRs have been obtained so far for some restricted classes of Petri nets. A marked graph (Murata 1977), a structually conflict-free Petri net (Ichikawa, Yokoyama and Kurogi 1985, Hiraishi and Ichikawa 1986) are these classes. If we take a close look at these conditions so far obtained, we find that the specified structure, such as marked graph and structurely conflict-free, is only sufficient condition for the obtained NSCRs to be true. This finding suggest us that it may be worthwhile to try to answer the qestion under what structural constraint a NSCR is obtainable. This is done in the following.

To obtain a NSCR expressed in terms of an initial marking, a target marking and a Petri net structure means that we must be able to estimate a sequence of markings, without computation, to such an extent that Inequality (3) can be verified at each time k = 1,2,.... We shall seek a structure that enables us this verification.

5.1 Preparation

A Petri net considered in this section is a single arc Petri net.

Lemma 1 Let a Petri net be $M = (P, T, B^+, B^-, m^0)$. If a firing count vector x is fireable at m^0, then the following two conditions hold:

1) all deadlock in the firing count subnet M_x are marked at the marking m^0, and

2) all traps are marked in M_x at $m^f = m^0 + B\,x$.

proof : Condition 1) : If there is a token-free deadlock in M_x at m^0, then a transition having an input place in the deadlock can not fire. This violates that any transition in M_x must fire at least once.
Condition 2) : If there is a token-free trap in M_x at m^0, then a transition which has an output place in the trap do not fire in the firing sequence. This violates that any transition in M_x must fire at least once.
Lemma 2 Let a Petri net be M = (P, T, B^+, B^-, m^0). If no transition can fire at a marking m, then there exists at least a token-free deadlock.
Proof : Assume that a transition, say t_j, can not fire at m. It has at least one token-free input place p_k. Let a transition which has p_k as one of output places be t_s. The t_s cannot fire at m. Repeat this procedure, then we reach either of the following two cases since the Petri net is finite:
Case 1 : there is a token-free single place deadlock.
Case 2 : there is a token-free circuit. In this case, if there is no path which is to deposit a token into the circuit then the circuit is token-free deadlock.
If there is a path which is to deposit a token into the circuite, we find the path is token-free by applying the above procedure. Repeating this procedure, we finally have a token-free deadlock.
Lemma 3 Given a Petri net M = (P, T, B^+, B^-, m^0) and a firing count vector x. If $m^f = m^0 + B x \geqq 0$, then all single place deadlocks in M_x are marked at m^0.
Proof : If there is a token-free single place deadlock at m^0, then token count of the place at m^f must be negative after firing of its output transitions.
Lemma 4 If a Petri net M = (P, T, B^+, B^-, m^0) has no circuit, then the marking m^f is reachable in M if and only if the equation $m^f = m^0 + B x$ has a non-negative integer vector solution.
Proof : Only-if-part is imeadiate from Equation (2).
If-part: Assume that M_x has no fireable transition, then from Lemma 2 there exists a token-free deadlock. Since M has no circuit, this means there is at least one token-free single place deadlock. Since this violates Lemma 3, there must exists at least one transition fireable. Consider the situation that the transition fired. Let the remainning firing count vector be x'. Then $M_{x'}$ must have at least one transition fireable from the same reason described above. Repeating this procedure, we can generate the firing sequence which satifies a specified firing count vector.
Lemma 5 Let a Petri net be M = (P, T, B^+, B^-, m^0). A marking m^f is reachable through a firing sequence σ in M if and only if the marking m^0 is reachable through the reverse sequence σ^{-1} in the reverse net M^{-1} = (P, T, B^-, B^+, m^f).
Proof : Proof is immediate if we subsitute m^0 by m^f, m^f by m^0, B^- by B^+ and B by -B in Equations (2) and Inequality (3).
Lemma 6 Let a Petri net M = (P, T, B^+, B^-, m^0) be a tc-net. Then any marked circuit in M does not become token-free by firing of any transition in M.

Proof : Obvious from the definition of tc-net.

5.2 Trap Circiut Petri Net

Lemma 2 and 4 indicate importance of the presence of token-free deadlock of circuit type in the firing sequence. In this respect, a Petri net that any marked deadlock does not become unmarked during a firing sequence is of significance. And from Lemma 6 the tc-net has such a property.

Theorem 6 Let a Petri net $M = (P, T, B^+, B^-, m^0)$ be a tc-net. Then a marking m^f is reachable in M if and only if there exists a non-negative integer vector x which satisies the following two conditions,

1) x is the minimal solution of $m^f = m^0 + Bx$, and

2) all deadlocks in M_x are marked at m^0.

Proof : Only-if-part : Condition 1) is immediate from Equation (4). Condition 2) is also immediate from Lemma 2. If x is not the minimal solution of $m^f = m^0 + Bx$ then there exists the minimal solution $x' \leqq x$. Since $M_{x'} \subset M_x$ for $x' \leqq x$, there is no token-free deadlock in $M_{x'}$ if there is no token-free deadlock in M_x.

If-part : From Lemma 2 and Condition 2) there exists at least a transition that can fire. Consider the situation after the transition fires. Let the remainning firing count vector be x'. There is no token-free deadlock of circuit type in $M_{x'}$ from Lemma 6 , and there is no token-free single place deadlock from Lemma 3. There exists, therefore, no token-free deadlock in $M_{x'}$ and thus there is at least one transition fireable. Repeating this procedure we can generate firing sequence from m^0 which satisfies the firing count vector x.

While a number of solutions of the matrix equation of a Petri net may be infinite, a number of the minimal solutions is finite. So the reachability in a tc-net is decidable within finite amount of computation.

Theorem 6 includes as its sepecial case the NSCR for a marked graph by Murata(1977) and the NSCR for a structually forward conflict-free Petri net (Ichikawa, Yokoyama and Kurogi 1985).

5.3 Deadlock circuit Petri net

Using Lemma 4 we have the following NSCR in a dc- net.

Corollary 1 Let a Petri net $M = (P, T, B^+, B^-, m^0)$ be a dc-net. Then a marking m^f is reachable in M if and only if there exists a non-negative integer vector x such that the following two conditions hold;

1) x is the minimal solution of the equation $m^f = m^0 + B\,x$, and

2) all traps in M_x are marked at the marking m^f.

5.4 Sufficient Condition for Reachability

If we carefully examine the if-part of the proof of Theorem 6, we find it most essential that a marked deadlock in the firing subnet at the initial marking does not become unmarked by any transition firing. The tc-net is sufficient but not necessary to assure this property. We can weaken a little bit the restriction imposed on the structure of Petri nets.

A Petri net is a *trap containning circuit Petri net* (tcc-net) if any circuit in the net contains a trap. Then we have next.

Theorem 7 Let a Petri net $M = (P, T, B^+, B^-, m^0)$ be a tcc-net. Then a marking m^f is reachable in M if there exists a non-negative integer vector x which satisies the following two conditions,

1) x is the minimal solution of $m^f = m^0 + B\,x$, and

2) all deadlock in M_x conntain marked trap at m^0.

Proof : If-part of the proof for Theorem 3 is also true for a tcc-net.

The reason why the condition 2) in Theorem 7 is not necessary but sufficient is the following. As we see in Lemma 2, the necessary condition requires token(s) be in a deadlock, not necessarily token(s) be in a trap within a deadlock. When token(s) exists in a deadlock and not in a contained trap at the initial marking it may be deposited into the trap during the firing sequence before it is taken out of the deadlock.

This gives us the feeling that this sufficiency is very close to the necessity. From practical point of view, it is worthwhile to examine a given net to be a tcc-net when it is not a tc-net.

A *deadlock containning circuit Petri net* (dcc-net) is similarly defined as a ttc-net and we have;

Corollary 2 Let a Petri net $M = (P, T, B^+, B^-, m^0)$ be a dcc-net. Then a marking m^f is reachable in M if there exists a non-negative integer vector x which satisies the following two conditions,

1) x is the minimal solution of $m^f = m^0 + B\,x$, and

2) all traps in M_x are marked at m^f.

6. Control System Design

There are two types of control in a DES represented by a Petri net. One is to control a firing sequence such that the sequence follows a prescribed firing sequence. We call this type of control a *firing control.* The other is to bring a Petri net to the specified marking. We call this type of control a *marking control.* For each type control, we have to carry out two tasks. One is to provide a set of the external input places, and the other is to determine a control input sequence on the set of external input places.

We shall limit our study within the class of tc- and dc-net so that the NSCR obtained in the last section are utilized in the control system design.

6.1 Firing Control

The firing control may be devided into two classes. One is the *firing sequence control* where the firing sequence has to follow the prescribed sequence. The other is the *firing count control* where the specified firing count has to be realized. We shall discuss here only the firing sequence control, since the firing count control is similar with the marking control in its concept and in available theorems.

The initial marking which enables a specified firing sequence is already obtained in Section 3, Equation (6) for a Petri net without particular restriction.

If the set $\underline{m}^0 \cap m^0$ obtained by Equation (6) for a prescribed firing sequence σ is non-empty, then σ is realizable as a df-firing sequence starting from an initial marking $m^0 \in \underline{m}^0 \cap m^0$. We say this case the initial marking control for a prescribed firing sequence. This means that we need only to set the initial marking and need not to apply a control input sequence.

If the set $\underline{m}^0 \cap m^0$ for the prescribed firing sequence σ is empty then we can not realize σ by an initial marking control and we need a control input sequence.

The following algorithm will give the firing sequence control.

Algorithm 1

1) Compute $\underline{m}^0$ and m^0 by Equation (6) for a prescribed sequence of firing σ.

2) If the set $\underline{m}^0 \cap m^0$ is non-empty then choose a $m^0 \in \underline{m}^0 \cap m^0$. Deposit the m^0 over the places and let the net start the df-firing. The df-firing will realize the prescribed sequence of firing. If $\underline{m}^0 \cap m^0$ is empty, then go to Step 3).

3) Choose a $m^0 \in \underline{m}^0$. Compute $m^f = m^0 + B\,\sigma$, where σ is the firing count of σ. Let S be the set of transitions which are not fireable in m^f. Add an external input place q(t) to each transition $t \in X + T - S$, where X is the set of transitions included in the firing sequence σ. $Q = \{q(t)\}$ thus formed is the set of external input places.

4) Deposit tokens sequentially into coresponding external input places according to the prescribed firing sequence.

6.2 Marking Control

We shall again limit our study within the classes of tc-, dc- , ttc- and dcc-nets, in order to utilize the necessary and/or sufficient conditions for reachability in these classes of Petri nets.

First of all, we shall show that a tc-net which satifies the NSCR is weakly persistent.

Theorem 8 A Petri net M = (C, m) is weakly persistent if and only if it is a tc-net and all deadlocks are marked at the initial marking m.

Proof : Only-if-part : Consider firing sequences α, β and $\alpha\gamma$ with $\overline{\gamma} = \overline{\beta}$. Since the sequences α and β are fireable, a firing count subnet $M_{\alpha\beta} = M_{\alpha} \cup M_{\beta} = M\alpha\gamma$ has no token-free deadlock. Therefore, the sequence $\alpha\gamma$ is fireable.

If-part : Assume the net is not a tc-net, then there is a circuit c which is not a trap. There are at least two transitions which have their input places in c. One is transition t that has no output place in c. The other is transition t' that has an output place in c. Clearly a sequence t' first then t is always fireable, but a sequence t first and then t' is not allways fireable. The net is not, therefore, weakly persistent.

Firing count bounded df-firing

Given a Petri net M = (C, m) which is a tc-net and a firing count vector x, a firing sequence defined by the following procedure is a *x-bounded df-firing sequence.*

1) Set k = 0, $m(0) = m^0$ and x(0) = x.
2) Let u(k) be the maximal df-firing vector at m(k).
3) If u(k) = 0 or x(k) – u(k) = 0, then stop. Otherwise, let x(k+1) = x(k) - u(k), m(k+1) = m(k) + B u(k) and k=k+1, then back to Step 2).

A x-bounded df-firing sequence is said to be df-fireable if there exists an integer K such that $x = \sum^{K}_{k=0} u(k)$.

Then we have the following.

Theorem 9 If a Petri net M = (C, m) is a tc-net, then x-bounded df-firing sequence is fireable if x is fireable.

Proof : For any two firing count vector x_1 and x_2 , $x_2 \leqq x_1$ which are fireable, $x_3 = x_1 - x_2$ has firing sequence, since the tc- net is persistent from Theorem 8. Assume that a df-firing sequence come to time k. x(k) = x - (x-x(k)) has a firing sequence since x-x(k) has a firing sequence u(0)u(1)u(2)•••u(k-1), and $x(k) \leqq x$. Therefore, there is at least one fireable transition at m(k) which is contained in x(k). u(k) becomes zero if and only if x(k) becomes zero.

Theorem 10 Let a Petri net M = (C, m) be a tc-net. Assume that the target marking m^f is reachable in M. Let s be a minimal solution of $m^f = m^0 + B s$ and has a firing sequence. Let S be a set of transitions which are fireable at the marking m^f. Let $Q = \{ q(t_k) \mid t_k \in S \}$ be a set of external input places. Deposit s_k number of tokens in each external input place $q(t_k)$. Then the marking m^f is reached by df-firing in M.

Proof : It is obvious from Theorem 9 that each t_k in S fires s_k times. We only need to show that a transition t_j in T – S which is not fireable at m^f actually fires s_j times. It is clear that t_j can fire at least s_j times since s has df-firing sequence. We need, therefore, to show that the firing of t_j ends after s_j times firings. Assume that t_j can fires s_{j+1} times.

Since s is the solution of $m^f = m^0 + B\,s$, and has a firing sequence, there exists a firing sequence which leads to m^f without firing $t_j\ s_{j+i}$ times. This means that t_j can fire at m^f, since the net is weakly persistent from Theorem 8.

Control Design Procedure for Marking Control

Given a tc-net $M = (C, m^0)$ and the target marking m^f, the procedure to find the set of external input places and the control input is given by the following.

Algorithm 2

1) Solve Equation $m^f = m^0 + B\,x$ and find the minimal solutions, $s_a, s_b, \ldots$
If no solution of the equation exists m^f is not reachable.

2) Among the minimal solutions find one that satisfies the conditions of Theorem 6. If no such solution exists m^f is not reachable.

3) Let S be the set of transitions which are fireable at m^f. Add an external input place $q(t_s)$ to each $t_s \in S$, and let the set $Q(S) = \{ q(t_s) \mid t_s \in S \}$ be the set of external input places.

4) Deposit coresponding number of tokens s_s to each $q(t_s)$ at time k=0. Then let M be df-firing.

Theorem 6, 9 and 10 assure that the net reaches to the target marking m^f. Token number s is not necessarily deposited all at time k=0. They can be deposited consecutively into Q(S) at any time k such that $\Sigma_k v(k) = s$.

Similar design method is developed for a dc-net, utilizing Collorary 1 instead of Teorem 5.

6.3 Firing Count Control

The difference between the firng count control and the marking control is that in the former the firing count vector x is prescribed and in the latter the initial and the target markings m^0 and m^f, respectively, are specified. The matrix equation $m^f = m^0 + B\,x$ connects these two. The design procedure for the firing count control is therefore obvious.

Algorithm 3

1) Find $m^o \geqq 0$ and $m^f \geqq 0$ such that $m^f = m^0 + B\,x$ for a given firing count vector x.

2) If there is a token-free deadlock in M_x at m^0, add tokens until all the deadlock are marked.

3) Follow Algorithm 2 from Step 3).

In this case any non-negative integer vector x is made fireable by properly setting the initial marking and manipulating the control input sequence.

7. Illustrative Example

A tcc-net shown in Figure 2 is used to illustrate how the necessary and/or sufficient conditions work in the control system design. The net is tcc- but not tc- since a circuit $\{p_1, p_2, p_3\}$ is not trap but contains a trap $\{p_1, p_2\}$. Note that the Petri net is not a marked graph, a structurelly conflict-free, nor a free choice.

We shall examine whether a target marking $m^f = (0\ 1\ 0\ 2\ 1\ 0)^T$ is reachable from the initial marking $m^0 = (0\ 0\ 0\ 1\ 1\ 0\ )^T$. The matrix equation $m^f = m^0 + B\,x$, where the incidence matrix

$$B = \begin{pmatrix} -1 & 1 & 1 & 0 & 0 & 0 \\ 1 & -1 & -1 & 1 & 0 & 0 \\ 1 & 0 & 0 & -1 & -1 & 0 \\ 0 & 0 & -1 & 1 & 0 & 0 \\ 0 & 0 & 0 & 0 & -1 & 1 \\ 0 & 0 & 0 & 0 & 1 & -1 \end{pmatrix}$$

has the minimal non-negative solution $x = (2\ 1\ 1\ 1\ 1\ 1)^T$. The firing count subnet Mx is Mx = M. There exists a token-free deadlock $\{p_1, p_2, p_3\}$ in Mx at m^0 as shown in Figure 3 a), the target marking m^f is not reachable in M.

Instead, $m^f = (1\ 1\ 0\ 1\ 1\ 0)^T$ is reachable from $m^0 = (1\ 0\ 0\ 0\ 1\ 0)^T$, since any deadlock in Mx, $x = (2\ 1\ 1\ 1\ 1\ 1)^T$, contains a marked trap at m^0 as shown in Figure 3 b). The scheme of the control system obtained by Algoritm 2 is shown in Figure 4. The firing sequence is in this case either $t_1\, t_5\, t_6\, t_2\, t_1\, t_3\, t_4$ or $t_1\, t_4\, t_3\, t_1\, t_5\, t_6\, t_2,$ either sequence can bring the net to $(1\ 1\ 0\ 1\ 1\ 0)^T$.

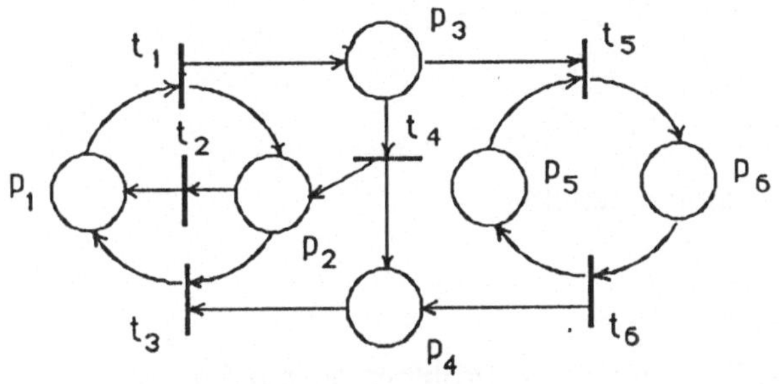

Figure 2 A structure of a tcc-net

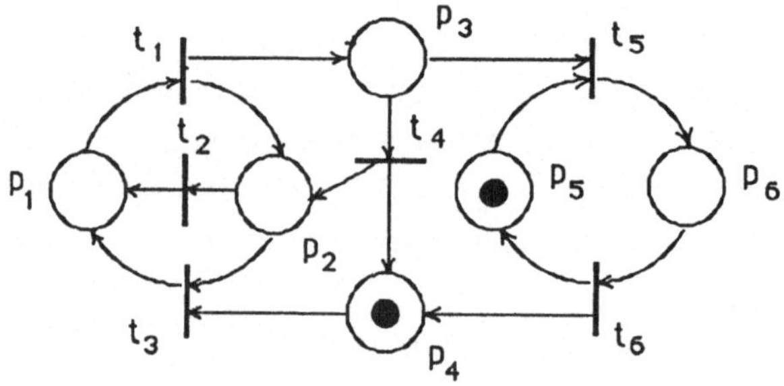

a) Token-free deadlock $\{p_1, p_2, p_3\}$

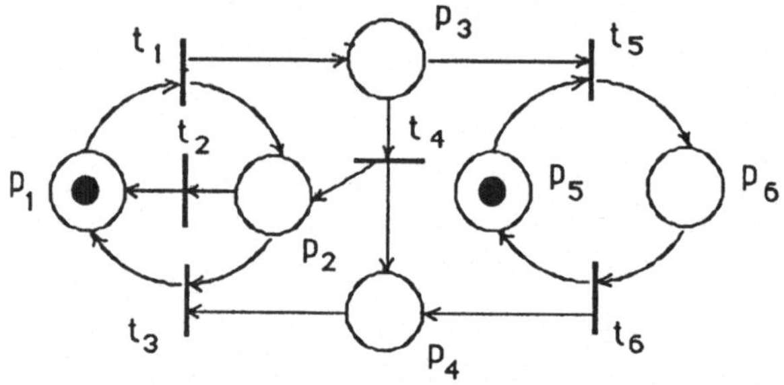

b) No token-free deadlock

Figure 3 Deadlocks at the initial marking

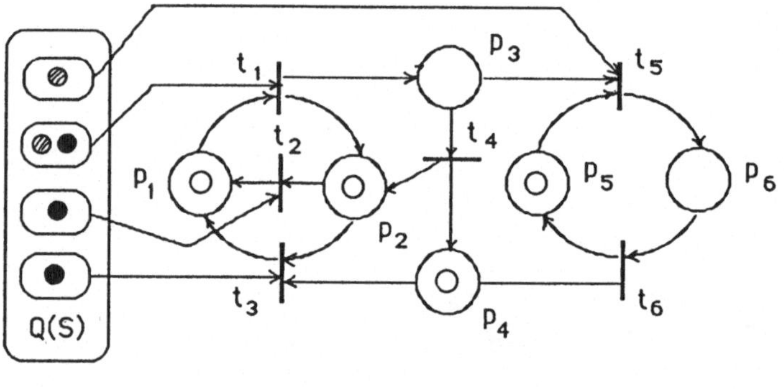

Figure 4 A control scheme

References

Araki, T. and Kasami, T.(1977). Decidable Problem on the Strong Connectivity of Petri Net Reachability Sets. Theor. Comput. Science 4 (1) ,97-119

Cohen, G., Moller, P. Quadrat, J.P. and Viot, M. (1984). Linear System Theory for Discrete Event Systems. Proc. 23-rd Int. Conf. Decision and Control. 1, 539-544

Gravowski, J. (1980) . The decidability of Persistence for Vector Addition Systems. Inf. Proc. Letters 11(1) 20-23

Hiraishi, K. and Ichikawa, A. (1986). Conflict-free Places and Fireability of a Solution of Matrix Equation in Petri Nets. Trans. Soc. Instr. Contr. Engr.. 22(7), 750-755 (in Japanese)

Ho, Y.C. and Cassandras, C. (1983). A New Approach to the Analysis of Discrete Event Systems. Automatica, 19 (2) 149-167

Hopcroft, J. and Pansiot, J.J. (1979). On the Reachability Problem for 5-dimentional Vector Addition Systems. Theor. Comput. Science 8, 135-159

Ichikawa, A. and Hiraishi, K. (1984) . Necessary and Sufficient Condition for a class of Reachability in Petri nets. Trans. Soc. Instr. Contr. Engr.. 20(8), 762-764 (in Japanese)

Ichikawa, A.and Ogasawara, K. (1986). Observation of markings in Petri Nets. Trans. Soc. Instr. Contr. Engr. 21(4), 324-330 (in Japanese)

Ichikawa, A., Yokoyama, K. and Kurogi S. (1985). Control of Event-driven Systems - Reachability and Control of Conflict-free Petri nets -. Trans. Soc. Instr. Contr. Engr. 21(4), 324-330 (in Japanese)

Ichikawa, A., Yokoyama, K. and Kurogi S. (1985). Reachability and Control of Discrete Event Systems Represented by Petri Nets. Proc. Int. Symp. Circuits and Systems 1985 , 2, 487-490

Landweber, L.H. and Robertson, E.L. (1978) . Properties of Conflict-free and Persistent Petri nets. Journal of the ACM. 25, 352-364 .

Mayr, E. W. (1981). Persistence of Vector Replacement System is Decidable. Acta Informatica 15, 309-318

Mayr, E. W. (1984). An Algorithm for the General Petri Net Reachability Problem. SIAM J. Comput. 13(3) 441-460

Muller, H. (1980). Decidability of Reachbility in Persistent Vector Replacement Systems. Lecture Notes in Computer Science, 88, 426-438

Murata, T. (1977). Circuit Theoretic Analysis and Synthesis of Marked Graph. IEEE Trans. Circuits and Systems. CAS-24(7), 400-405

Peterson J.L.(1981) . Petri Net Theory and The Modeling of Systems, Prentice-Hall

Petri C.A.(1962). Fundamentals of a Theory of Asynchronous Information Flow,. Proc. IFIP Congress 1962. 386-390

Ramadge, P.J. and Wonham, W.M. (1982). Algebraic Decomposition of Controlled Sequential Machines. Proc. 8-th World Congress IFAC, 1, 313-318

Ramadge, P.J. and Wonham, W.M.(1982) Supervision of Discret Event Processes. Proc. 21st IEEE Conf. Decision and Control. 3, 1228-1229

Reisig, W. (1985) Petri Nets , An Introduction. Springer-Verlag

Sasaki, I. (1986). Modeling and Language Acception Power of Decision-free Petri nets. BS thesis, Department of Systems Science, Tokyo Institute Technology.

Shepardson, J. and Sturgis, H. (1963) . Computability of Recursive Functions, Journal of the ACM, 10(2), 217-255

Yamasaki H. (1981) . On Weak Persistency of Petri nets. Inf. Proc. Letters, 13(1) : 94-97

Yamasaki H. (1984). Normal Petri Nets. Theor, Comput. Science, 31 (3) 307-315

DATA FLOW PROGRAMMING FOR PARALLEL IMPLEMENTATION OF

DIGITAL SIGNAL PROCESSING SYSTEMS[1]

Edward A. Lee

Dept. of Electrical Engineering and Computer Science
U.C. Berkeley, Cory Hall, Berkeley CA 94720, USA

Presented at: IIASA workshop on
Discrete Event Systems: Models and Applications
3-7 August, 1987, Sopron, Hungary

ABSTRACT

Digital signal processing systems constitute a special class of discrete event systems where events correspond to samples of signals. A data flow description of such systems can capture much of the information required for high performance, cost-effective, parallel implementation. A formal model called *synchronous data flow* (SDF) is a useful special case of data flow and subclass of discrete event systems where the events are deterministic and periodic. An SDF description of an algorithm can be first analyzed for implementability, then an implementation can be synthesized. In the analysis phase we can check for (1) stability of the buffers, (2) freedom from deadlocks, and (3) adequate concurrency to meet a given performance specification. The synthesis phase consists primarily of constructing a periodic schedule and mapping the algorithm onto parallel processors. The resulting schedule is said to be *static* and is far less costly to implement than *dynamic*, or run-time scheduling.

Although many digital signal processing systems can be accurately described within the SDF model, the model needs to be generalized to be broadly applicable. In particular, the expanded model should accommodate asynchronous systems and systems with data dependent computations. To some degree, dynamic scheduling becomes essential. However, in order to achieve high performance and low cost, fully dynamic scheduling should be avoided. Limited extensions to the SDF model are described which are inexpensive to implement and can be used to describe a variety of systems with asynchronous events.

1. Digital Signal Processing and Discrete Event Systems

Digital signals are sequences of numbers called samples. Usually signals are processed to produce new signals at the same sample rate, as shown in figure 1a, or at a sample rate related by a rational factor, as shown in figure 1b. Both are discrete-time systems, although the multi-rate system is a little different from the usual conception of a discrete-time system. Time is discretized differently in different parts of the system. Such systems are said to be synchronous, and are a special

[1] *This research was sponsored by an IBM faculty development grant and National Science Foundation Presidential Young Investigator award.*

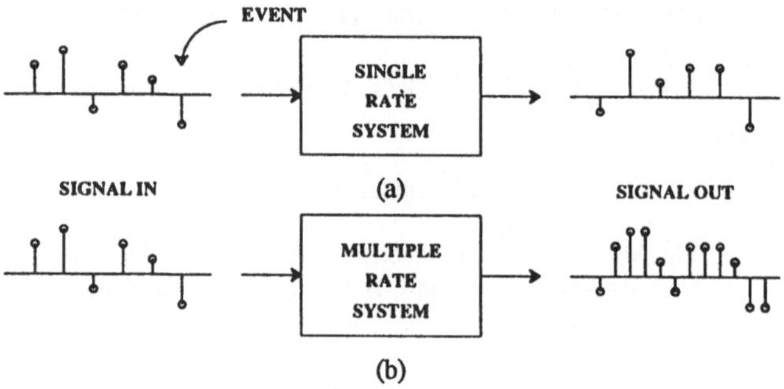

Figure 1. Single sample rate (a) and multi-rate (b) digital signal processing systems.

subset of discrete event systems. The arrival of a sample is an event, and typically the events are periodic and deterministic.

A complete signal processing system may be a complex interconnection of synchronous subsystems. Such an interconnection is conveniently described by a block diagram, or a *data flow* graph, an example of which is shown in figure 2a. Blocks in the block diagram (nodes in the data flow graph) represent computations performed on signals, and the paths between blocks (branches or arcs in the data flow graph) represent the routing of signal samples (called *tokens*). The fundamental premise of data flow program graphs [1] is that nodes can fire whenever there is sufficient data on their input branches. Since nodes can fire simultaneously, the concurrency available in an algorithm is exhibited in a data flow description.

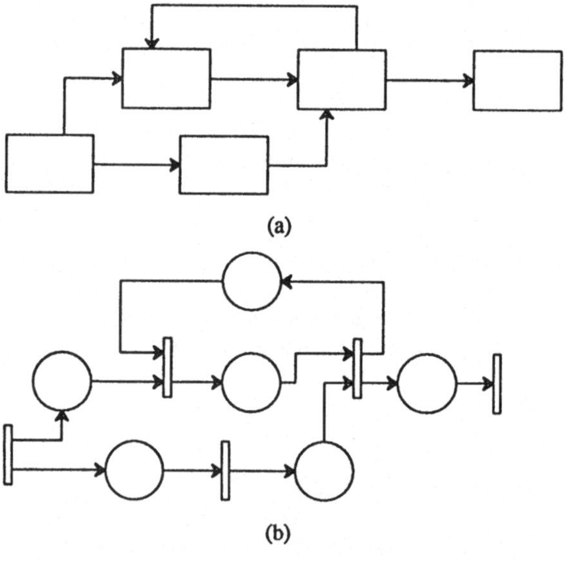

Figure 2. (a) A signal processing system is built by connecting subsystems. The boxes represent computations and the arcs represent signal paths. (b) The petri net equivalent of (a) has transitions instead of boxes and places instead of arcs. The assumption is that transitions take some time to fire (the computation time of the boxes).

The connection between data flow graphs and Petri nets [2] should be obvious to those readers familiar with Petri nets. The graph in figure 2a is modeled as a Petri net in figure 2b. The nodes in the data flow graph correspond to transitions in the Petri net, and the branches in the data flow graph correspond to the places in the Petri net. In the Petri net model we assume that the transitions take some time to fire (the execution time of a computation). Also, tokens can accumulate in the places, but in a signal processing system they should be processed in the same order of their arrival.

We make the important distinction between data flow program graphs, data flow languages, and data flow machines. The term "data flow language" is often used to refer to languages that are easily translated into data flow program graphs, even if the syntax of the language bears no resemblance to a graph. Such languages are usually functional languages [3], and are often intended for use with data flow machines. Data flow machines evaluate data flow graphs. Such machines use costly hardware or software to determine when to fire nodes in the graph. Since the scheduling is done at run time, it is said to *dynamic*. A good survey of such machines is given by Srini [4]. Here we view data flow program graphs as the input from the programmer (preferably graphical input), and the target architecture is a conventional multiprocessor, not a data flow machine.

Our overall aim is the automatic synthesis of cost-effective implementations of signal processing systems from a high level description of the algorithm. The objective is to map the data flow graph onto processors such that we either maximize the throughput subject to a constraint on the number of processors, or minimize the number of processors subject to a throughput constraint. In essence, what is required is a scheduling strategy. In addition, we will need a way to check the correctness of a graph.

For signal processing, data flow descriptions of algorithms have important advantages.

Appropriateness:

Signal processing algorithms are often described as data flow graphs by choice when there is no compelling reason to do so. In fact, this appropriateness has led dozens of researchers to develop so called "block diagram languages" for signal processing.

Parallelizability:

As mentioned above, the concurrency in an algorithm is evident in a data flow description. Signal processing applications are frequently computation intensive, so parallel processing is essential.

Modularity:

Subsystems that have been implemented in the past and are well understood can be easily reused by splicing them into a new data flow graph.

Synchronous data flow [5,6] is a special case of data flow that has properties that are particularly convenient for implementation. A synchronous data flow node is defined to be one where the same number of tokens are consumed on each input branch and produced on each output branch each time the node fires. A synchronous data flow node is shown in figure 3. The numbers adjacent to each branch indicate the number of tokens produced and consumed when the node fires. An SDF graph is an interconnection of synchronous data flow nodes, and is clearly capable of representing synchronous multi-rate signal processing systems. The main advantage of specializing to SDF is the ability to generate static schedules at compile time and ensure correctness of the schedule.

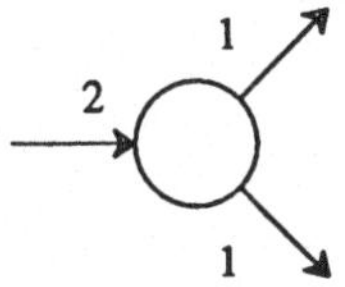

Figure 3. An SDF node. The numbers adjacent to each branch indicate the number of tokens produced and consumed each time the node fires.

An example of an SDF description of signal processing algorithm (a second order recursive digital filter) is shown in figure 4. The triangle at the left is a data flow node that can fire at any time, because it has no inputs, each time producing one token. This models the signal source. Hence the assumption is that the computation can run forever, and must run forever without deadlock. This assumption is peculiar to signal processing applications, and is one of the important differences between signal processing and general purpose computations.

All of the nodes in figure 4 consume and produce a single token on each input and output branch. A graph consisting exclusively of such nodes is called a *homogeneous* SDF graph, and corresponds to a single-sample-rate signal processing system. Also, the nodes in the graph represent elementary computations (additions, multiplications, forks), so the graph is said to be *atomic*. A larger granularity may be desirable. The graph in figure 5 represents a voiceband data modem, and each node is a complicated computation that may be specified hierarchically as an SDF graph. Descriptions at this level of granularity are called large grain data flow [7]. The techniques discussed in this paper apply equally well to all levels of granularity and any mixture of levels.

The labels "D" on some of the branches in figure 4 refer to *delays*. The term "delay" is used in the signal processing sense to mean a sample offset between the input and the output (a z^{-1} operator). We define a unit delay on an arc from node A to node B to mean that the nth sample consumed by B will be the $(n-1)^{th}$ sample produced by A. This implies that the first sample that B consumes is not produced by A at all, but is part of the initial state of the buffer connecting the two. Indeed a delay of

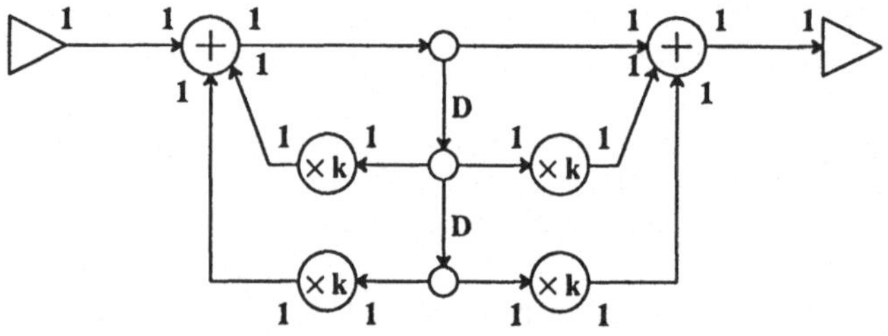

Figure 4. A data flow graph for a second-order recursive digital filter. The empty circles are "fork" nodes, which simply replicate each input token on all output paths. The "1" adjacent to each node input or output indicates that a single token is consumed or produced when the node fires.

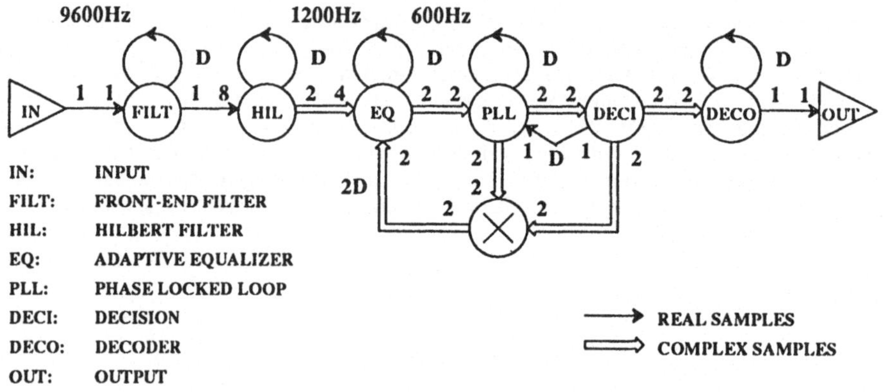

Figure 5. A SDF graph showing a voiceband data modem. Note the multiplicity of sample rates. For emphasis, signal paths that carry complex signals are shown with double lines, although these arcs are no different from the arcs carrying real signals except that the rate of flow of tokens is twice the sample rate of the complex signal.

d samples on a branch is implemented in SDF simply by initializing the arc buffer with *d* zero samples. The inscription *dD* will be placed near the arc to illustrate the delay. In the Petri net model, delays correspond simply to an initial marking.

Notably absent from the SDF model is data dependent routing of tokens. Since tokens drive the firing of nodes, data dependent firing is also absent. The corresponding feature of the Petri net model that is missing is conflicts, which would introduce an indeterminacy. Although most DSP systems can modeled without this indeterminacy, some practical systems are excluded by this restriction. An example of a computation that we may wish to perform is shown in figure 6. It is equivalent to the functional statement

$$z = \text{if } (x) \text{ then } f(y) \text{ else } g(y).$$

The nodes labeled *switch* and *select* are asynchronous because it is not possible to specify *a-priori* how many tokens will be produced on their outputs or consumed on their inputs. A range (0,1) is specified instead. Since the firing of nodes in this graph is dependent on the data (the value of x), it is not possible to construct a static schedule (a schedule constructed at compile time that is valid throughout the computation). Our approach is to introduce limited dynamic scheduling to handle only those situations that absolutely require dynamic scheduling. Hence the price (in overhead) of dynamic scheduling is not paid when not necessary.

2. OBJECTIVES OF THIS PAPER

Given an SDF graph describing a digital signal processing algorithm, we wish to analyze it for implementability (the analysis phase) and synthesize a cost effective implementation (the synthesis phase). In the analysis phase we check for (1) stable buffers, (2) freedom from deadlocks, and (3) adequate concurrency to meet the performance specification. All three can be done in polynomial time. The relevant techniques have been described in two closely related papers [6,5], so we skip some details in order to highlight the intuition and point out the relationship to other models used for discrete event systems.

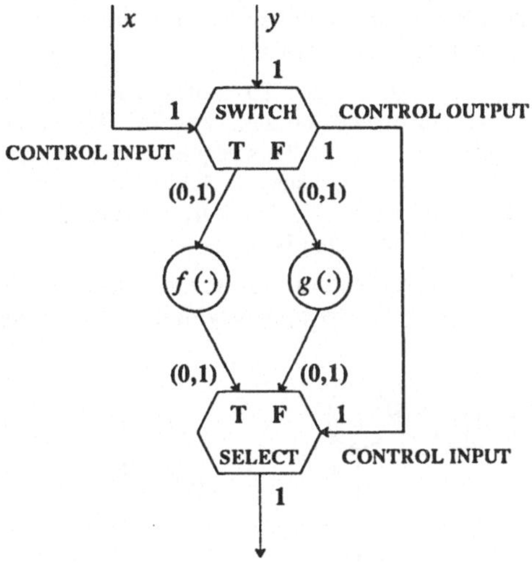

Figure 6. Two asynchronous nodes are shown used in a conditional construct implementing an *if-then-else*. The *switch* routes tokens to one of two output paths depending on a control input and the *select* selects tokens from one of two inputs depending on the control token. The notation "(0,1)" indicates that when the node fires, either zero or one sample will be produced or consumed.

For cost reasons, the number of processors available for the implementation of the SDF graph may be limited. Hence we cannot simply assign one processor to each node. Even if we could, this would probably result in a gross under-utilization of the hardware resources. The problem that we consider in this paper is to schedule the SDF graph onto one or more parallel processors such that we maximize the throughput (or minimize the period of a periodic schedule). Usually for a signal processing system the number of processors is selected so that a required throughput is just barely met. To simplify the discussion, we assume that the parallel processors share memory without contention, so there are no communication delays between nodes mapped onto different processors. Practical parallel architectures with this feature have in fact been proposed [8,9]. A useful (and interesting) special case is a single processor. In the single processor case it is clear that we are discussing compilation techniques.

The history of models related to SDF is extensively reviewed in [5] so the interested reader is referred to that paper. Nonetheless, one prior paper is closely enough related to require mentioning. In 1966 Karp and Miller introduced *computation graphs*, which are essentially equivalent to SDF graphs but are intended to describe general computations [10]. In particular, Karp and Miller discuss graphs that *terminate*, or deadlock after some time. They concentrate on fundamental theoretical considerations, for example proving that computation graphs are *determinate*, meaning that any admissible execution yields the same results. Such a theorem of course underlies the validity of data flow.

Also described in [5] is a software implementation of an SDF programming system called Gabriel. We again refer the interested reader to that paper.

3. THE ANALYSIS PHASE

In the analysis phase we check the implementability of the system by checking (1) stability of the buffers, (2) freedom from deadlocks, and (3) adequate concurrency.

3.1. Stability of the Buffers

We assume that an SDF graph describes a repetitive computation to be performed on an infinite stream of input data, so the desired schedule is periodic. It is not always possible to construct a practical periodic schedule for an SDF graph, however. In particular, for some graphs the buffers used to implement the arcs of the data flow graph may have to be of infinite size. This indicates an error in the construction of the graph, and must be identified.

Consider the SDF graph of figure 7(a). To start the computation, node 1 can be invoked because it has no input arcs and hence needs no data samples. After invoking node 1, node 2 can be invoked, after which node 3 can be invoked. This sequence can be repeated. But node 1 produces twice as many samples on arc 2 as node 3 consumes. An infinite repetition of this schedule therefore causes an infinite accumulation of samples in the buffer associated with arc 3. This implies an unbounded memory requirement, which is clearly not practical.

In a DSP sense, the SDF graph has *inconsistent sample rates*. Node 3 expects as inputs two signals with the same sample rate but gets two signals with different sample rates. The SDF graph of

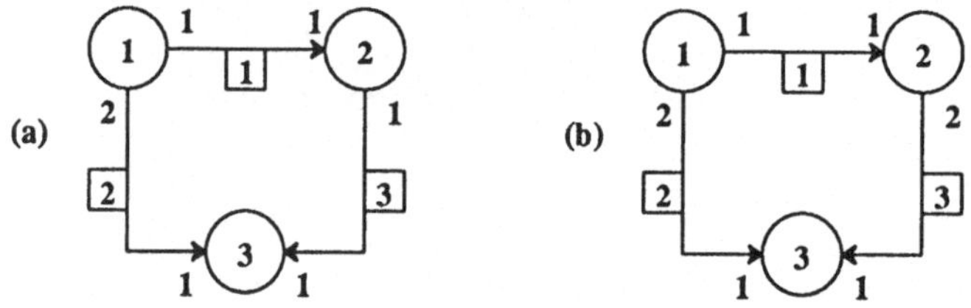

Figure 7. (a) An example of a defective SDF graph with sample rate inconsistencies. The flags on the arcs simply identify them with a number for reference in the text. (b) A corrected SDF graph with consistent sample rates. The flags attached to the arcs simply identify them with a number.

figure 7(b) does not have this problem. A *periodic admissible sequential schedule* repeats the invocations {1,2,3,3}. Node 3 is invoked twice as often as the other two. It is possible to automatically check for consistent sample rates and simultaneously determine the relative frequency with which each node must be invoked. To do this, we need a little formalism.

A SDF graph can be characterized by a matrix similar to the incidence matrix associated with directed graphs in graph theory. It is constructed by first numbering each node and arc, as done in figure 7, and assigning a column to each node and a row to each arc. The $(i,j)^{th}$ entry in the matrix is the amount of data produced by node j on arc i each time it is invoked. If node j consumes data from arc i, the number is negative, and if it is not connected to arc i, then the number is zero. For the graphs in figure 7 we get

$$\Gamma_a = \begin{bmatrix} 1 & -1 & 0 \\ 2 & 0 & -1 \\ 0 & 1 & -1 \end{bmatrix} \qquad \Gamma_b = \begin{bmatrix} 1 & -1 & 0 \\ 2 & 0 & -1 \\ 0 & 2 & -1 \end{bmatrix} \tag{1}$$

This matrix is called a *topology matrix*, and need not be square, in general.

If a node has a connection to itself (a *self loop*), then only one entry in Γ describes this link. This entry gives the net difference between the amount of data produced on this link and the amount consumed each time the node is invoked. This difference should clearly be zero for a correctly constructed graph, so the Γ entry describing a self loop should be a zero row.

We can conceptually replace each arc with a FIFO queue (buffer) to pass data from one node to another. The size of the queue will vary at different times in the execution. Define the vector $\mathbf{b}(n)$ to contain the number of tokens in each queue at time n. The vector $\mathbf{b}(n)$ thus specifies the marking at time n in the equivalent Petri net model.

For the sequential (single processor) schedule, only one node can be invoked at a time, and for the purposes of scheduling it does not matter how long each node runs. Thus, the time index n can simply be incremented each time a node finishes and a new node is begun. We specify the node invoked at time n with a vector $\mathbf{v}(n)$, which has a one in the position corresponding to the number of the node that is invoked at time n and zeros for each node that is not invoked. For the systems in figure 7, $\mathbf{v}(n)$ can take one of three values for a sequential schedule,

$$\mathbf{v}(n) = \begin{bmatrix} 1 \\ 0 \\ 0 \end{bmatrix} \; OR \; \begin{bmatrix} 0 \\ 1 \\ 0 \end{bmatrix} \; OR \; \begin{bmatrix} 0 \\ 0 \\ 1 \end{bmatrix} \tag{2}$$

depending on which of the three nodes is invoked. Each time a node is invoked, it will consume data from zero or more input arcs and produce data on zero or more output arcs. The change in the size of the queues caused by invoking a node is given by

$$\mathbf{b}(n+1) = \mathbf{b}(n) + \Gamma\mathbf{v}(n) \tag{3}$$

The topology matrix Γ characterizes the effect on the buffers of invoking a node.

To initialize the recursion (3) we set $\mathbf{b}(0)$ to reflect the number of delays on each arc. The initial condition for the queues in figure 7 is thus

$$\mathbf{b}(0) = \begin{bmatrix} 0 \\ 0 \\ 0 \end{bmatrix},$$

and in figure 8 is

$$\mathbf{b}(0) = \begin{bmatrix} 1 \\ 2 \end{bmatrix}. \tag{4}$$

Recall that this corresponds to an initial marking of a Petri net. Because of these initial conditions, node 2 can be invoked once and node 3 twice before node 1 is invoked at all. Delays, therefore, affect the way the system starts up. Clearly, every directed loop must have at least one delay, or the system cannot be started. Automatic identification of this condition is discussed in the next

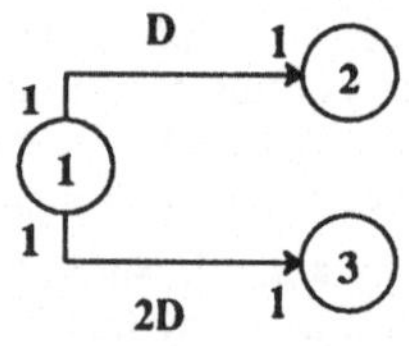

Figure 8. An example of an SDF graph with delays on the arcs.

subsection.

Inconsistent sample rates preclude construction of a periodic sequential schedule with bounded memory requirements. A necessary condition for the existence of such a schedule is that $rank(\Gamma) = s-1$, where s is the number of nodes. This can be seen by observing that stable periodic execution requires that

$$\mathbf{b}(n+K) = \mathbf{b}(n) \tag{5}$$

where K is the number of node firings in one cycle of the periodic schedule. From (3) we also see that

$$\mathbf{b}(n+K) = \mathbf{b}(n) + \Gamma\mathbf{q} \tag{6}$$

where

$$\mathbf{q} = \mathbf{v}(n) + \cdots + \mathbf{v}(n+K-1). \tag{7}$$

Equations (5) and (6) can only both be true if

$$\Gamma\mathbf{q} = \mathbf{0}$$

or $\mathbf{q}$ is in the right nullspace of Γ. For $\mathbf{q}$ to be non-trivial, Γ cannot have full rank. It is proven in [6] that it must have rank $s-1$.

The topology matrix Γ_a for the graph in figure 7a has rank three, so no periodic admissible sequential schedule can be constructed. The topology matrix Γ_b for the graph in figure 7b has rank two, so a schedule may be possible. It is also proven in [6] that a topology matrix with the proper rank has a strictly positive (element-wise) integer vector $\mathbf{q}$ in its right nullspace. For figure 7b, a set of such vectors is

$$\mathbf{q} = J \begin{bmatrix} 1 \\ 1 \\ 2 \end{bmatrix},$$

for any positive integer J. Notice that the dimension of $\mathbf{q}$ is s, the number of nodes. Notice further that $\mathbf{q}$ specifies the number of times we should invoke each node in one cycle of a periodic schedule, as can be seen from (7). Node 3 gets invoked twice as often as the other two nodes, for any positive integer J.

Valuable information is obtained from the topology matrix. Its rank can be used to verify consistent sample rates, which is necessary for stable buffers, and its nullspace gives the relative frequency with which nodes must be invoked.

3.2. Freedom From Deadlocks

Even with consistent sample rates, it may not be possible to construct a periodic admissible sequential schedule. Two examples of SDF graphs with consistent sample rates but no such schedules are shown in figure 9. Directed loops with insufficient delays are an error in the construction of the SDF graph and must be identified to the user. It is shown in [6] that a large class of scheduling algorithms will always run to completion if a periodic admissible sequential schedule exists, and will fail otherwise. Running such an algorithm is a simple way of verifying the

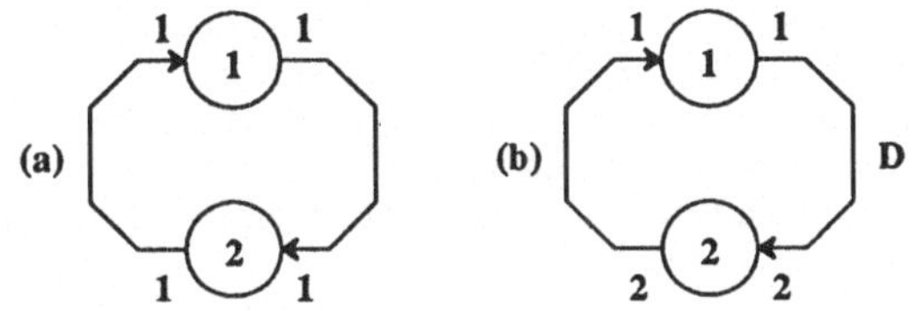

Figure 9. Two SDF graphs with consistent sample rates but no admissible schedule.

correctness of the SDF graph. The class of algorithms is described in the section below on the synthesis phase.

3.3. Checking for Adequate Concurrency

The *iteration period* is the length of one cycle of a periodic schedule. Throughput is inversely proportional to the iteration period. The iteration period is bounded from below in graphs with directed loops. If the iteration period bound exceeds the iteration period required for a particular signal processing algorithm, then the algorithm has to be modified. No amount of parallel hardware will overcome the problem. Hence the iteration period bound would be useful information for the designer of the algorithm.

For homogeneous SDF graphs (all nodes produce or consume one token on each input and output) the iteration period bound has been shown to be the worst case (over all directed loops) of the total computation time in the loop divided by the number of delays [11, 12]. An alternative point of view is that the iteration period bound is the unique eigenvalue of a matrix used to describe the graph in a max-algebra [13, 14]. The iteration period bound can be found in polynomial time [13], but existing techniques only apply to homogeneous SDF graphs. An algorithm for translating general SDF graphs into homogeneous SDF graphs is given in [15].

4. THE SYNTHESIS PHASE

The synthesis phase consists of constructing schedules for single or multiple processors. We begin with the single processor problem.

4.1. Scheduling for a Single Processor

Given a positive integer vector **q** in the nullspace of Γ, one cycle of a periodic schedule invokes each node the number of times specified by **q**. A sequential schedule can be constructed by selecting a *runnable* node, using (3) to determine its effect on the buffer sizes, and continuing until all nodes have been invoked the number of times given by **q**. We define a class of algorithms.

DEFINITION *(CLASS S ALGORITHMS):* Given a positive integer vector **q** s.t. $\Gamma \mathbf{q} = \mathbf{0}$ and an initial state for the buffers **b**(0), the ith node is said to be *runnable* at a given time if it has not been run q_i times and running it will not cause a buffer size to become negative. A *class S algorithm* ("S" for Sequential) is any algorithm that schedules a node if it is runnable, updates $\mathbf{b}(n)$ and stops (*terminates*) only when no more nodes are runnable. If a class S algorithm terminates before it has scheduled each node the number of times specified in the **q** vector, then it is said to be *deadlocked*.

Class S algorithms construct static schedules by simulating the effects on the buffers of an actual run for one cycle of a periodic schedule. That is, the nodes need not actually run. Any dynamic (run time) scheduling algorithm becomes a class S algorithm simply by specifying a stopping condition, which depends on the vector **q**. It is proven in [6] that any class S algorithm will run to completion if a periodic admissible sequential schedule exists for a given SDF graph. Hence, successful completion of the algorithm guarantees that there are no directed loops with insufficient delay. A suitable class S algorithm for sequential scheduling is

1. Solve for the smallest positive integer vector **q** in the right nullspace of Γ.
2. Form an arbitrarily ordered list L of all nodes in the system.
3. For each $\alpha \varepsilon L$, schedule α if it is runnable, trying each node once.
4. If each node α has been scheduled q_α times, STOP.
5. If no node in L can be scheduled, indicate a deadlock (an error in the graph).
6. Else, go to 3 and repeat.

The only question remaining for single processor schedules is the complexity of the first step above. Our technique is simple. We begin with any node A in the graph and assume it will be run once in one cycle of the periodic schedule (i.e. let $q_A = 1$). Assume node B is connected to node A. We can find q_B with a simple division, possibly getting a fraction, but always getting a rational number. A node cannot be invoked a fractional number of times, so we will have to correct for this later. We do the same for any node C adjacent to B. A simple recursive algorithm computes these rational numbers in linear time (a linear function of the number of arcs, not the number of nodes). The resulting vector **q** has rational entries and is in the nullspace of Γ. To get the smallest *integer* vector in the nullspace of Γ we use Euclid's algorithm to find the least common multiple of all the denominators. Actually, three simultaneous objectives are accomplished with one pass through the graph. Sample rate consistency is checked, a vector (with rational entries) in the nullspace of Γ is found, and Euclid's algorithm is used to find the least common multiple of all the denominators.

SDF offers concrete advantages for single processor implementations. The ability to interconnect modular blocks of code (nodes) in a natural way could considerably ease the task of programming high performance signal processors, even if the blocks of code themselves are programmed in assembly language. But a single processor implementation cannot take advantage of the explicit concurrency in an SDF description. The next section is dedicated to explaining how the concurrency in the description can be used to improve the throughput of a multiprocessor implementation.

4.2. Scheduling for Parallel Processors

Clearly, if a workable schedule for a single processor can be generated, then a workable schedule for a multiprocessor system can also be generated. Trivially, all the computation could be scheduled onto only one of the processors. Usually, however, the throughput can be increased substantially by distributing the load more evenly. It is shown in [6] that the multiprocessor scheduling problem can be reduced to a familiar problem in operations research for which good heuristic methods are available. We again give the intuition without the details. We assume for now homogeneous parallel processors sharing memory without contention, and consider only *blocked schedules*. A blocked schedule is one where one cycle of the schedule must finish on all processors before the next cycle can begin on any (cf. Schwartz [12]).

A blocked periodic admissible *parallel* schedule is a set of lists $\{\psi_i \, ; \, i = 1, \cdots, M\}$ where M is the number of processors, and ψ_i specifies a periodic schedule for processor i. If **p** is the *smallest* positive integer vector in the nullspace of Γ then a cycle of a schedule must invoke each node the number of times given by $\mathbf{q} = J\mathbf{p}$ for some positive integer J. J is called the *blocking factor*, and for blocked schedules, there is sometimes a speed advantage to using J greater than unity. If the "best" blocking factor is known, then construction of a good parallel schedule is not hard.

The task of the scheduler is to construct a schedule that avoids deadlocks and minimizes the *iteration period*, defined more generally to be the run time for one cycle of the schedule divided by J. The first step is to construct a graph describing the precedences in $\mathbf{q} = J\mathbf{p}$ invocations of each node. The graph will be acyclic. A precise class S algorithm accomplishing this construction is given in [6] so we merely illustrate it with the example in figure 10a. Node 1 should be invoked twice as often as the other two nodes, so $\mathbf{p} = [2\ 1\ 1]^T$. Further, given the delays on the arcs, we note that there are three periodic admissible sequential schedules with unity blocking factor, $\phi_1 = \{1,3,1,2\}$, $\phi_2 = \{3,1,1,2\}$, or $\phi_1 = \{1,1,3,2\}$. A schedule that is not admissible is $\phi_1 = \{2,1,3,1\}$, because node 2 is not immediately runnable. Figure 10b shows the precedences involved in all three schedules. Figure 10c shows the precedences using a blocking factor of two (J=2).

The self-loops in figure 10a imply that successive invocations of the same node cannot overlap in time. Some practical SDF implementations have such precedences in order to preserve the

(a)

(b)

(c)

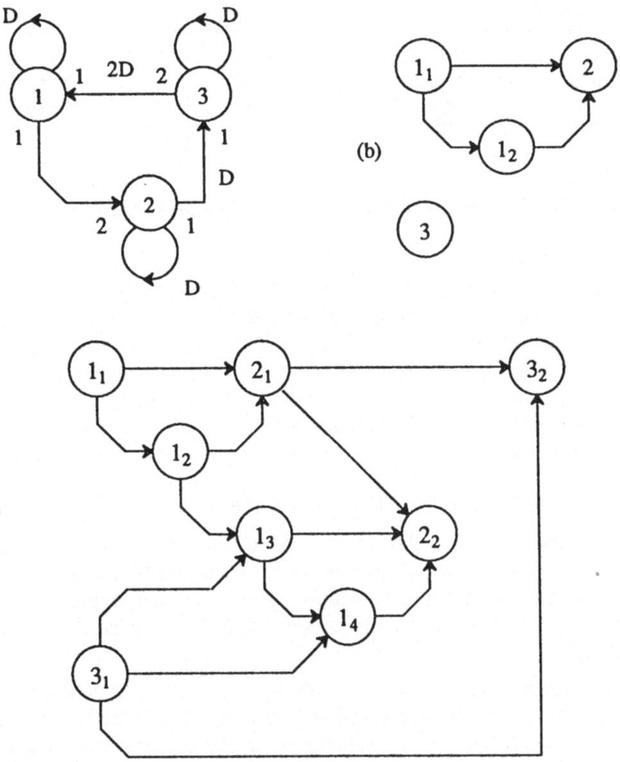

Figure 10. (a) An SDF graph with self-loops. (b) An acyclic precedence graph for unity blocking factor, $J=1$. (c) An acyclic precedence graph for $J=2$.

integrity of the buffers between nodes. In other words, two processors accessing the same buffer at the same time may not be tolerable, depending on how the buffers are implemented. The self-loops are also required, of course, if the node has a *state* that is updated when it is invoked. We will henceforth assume that all nodes have self loops, thus avoiding the potential implementation difficulties. Note that this may increase the iteration bound.

If we have two processors available, a schedule for $J=1$ is

$$\psi_1 = \{1,1,2\}$$

$$\psi_2 = \{3\} .$$

When this system starts up, nodes 3 and 1 will run concurrently. The precise timing of the run depends on the run time of the nodes. If we assume that the run time of node 1 is a single time unit, the run time of node 2 is two time units, and the run time of node 3 is three time units, then the timing is shown in figure 11a. The shaded region represents idle time. A schedule constructed for $J=2$, using the precedence graph of figure 10c will perform better. An example is

$$\psi_1 = \{1,1,2,1,2\}$$

$$\psi_2 = \{3,1,3\}$$

and its timing is shown in figure 11b. There is no idle time, so no faster schedule exists.

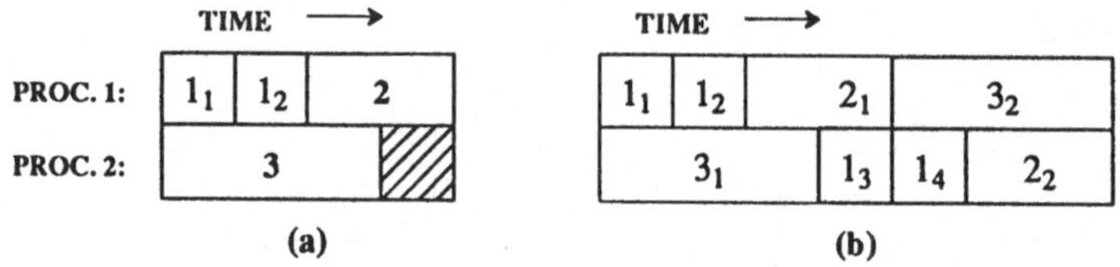

Figure 11. One period of each of two periodic schedules for the SDF graph of figure 11. In (a) J=1 while in (b) J=2.

The construction of the acyclic precedence graph is handled by the class S algorithm given in [6]. The remaining problem of constructing a parallel schedule given an acyclic precedence graph is a familiar one. It is identical with assembly line problems in operations research, and can be solved for the optimal schedule, but the problem is NP complete [16]. This may not be a problem for small SDF graphs, and for large ones we can use well studied heuristic methods, the best being members of a family of "critical path" methods [17]. An early example, known as the Hu-level-scheduling algorithm [18], closely approximates an optimal solution for most graphs [19, 17], and is simple.

5. EXTENSIONS TO HANDLE LIMITED ASYNCHRONY

It has been mentioned that the SDF model cannot accommodate some useful constructs that involve data dependent routing. An example is shown in figure 6. Because of the data dependent routing of tokens, some form of dynamic scheduling is in order. Our approach is to introduce limited dynamic scheduling only where it is actually needed, rather than generalizing the system to do all scheduling dynamically.

Observe that the graph in figure 6 can be divided into three synchronous subgraphs, (1) the graph represented by $f(\cdot)$, (2) the graph represented by $g(\cdot)$, and (3) the rest of the system. Consider a single processor implementation. Static schedules can be constructed for each of the three subgraphs. The system is started such that the schedule for subgraph (3) is invoked first. When the switch node fires, schedule (3) is suspended and either schedule (1) or (2) is invoked to run through one cycle. The code to do this in effect is implementing a dynamic scheduler, but decisions are only being made at run time if they have to be.

Consider a multiprocessor schedule. The problem is a little more complicated for several reasons. For one, if $f(\cdot)$ and $g(\cdot)$ are complicated subsystems then it is desirable to distribute them over several processors. For another, if the execution time of $f(\cdot)$ and $g(\cdot)$ are different then a worst case execution time must be assumed. Our approach is to put dynamic scheduling code into all processors, making all processors available for the subsystem $g(\cdot)$ or $f(\cdot)$. This is easily explained by example.

In figure 12 we have shown how one such if-then-else might be scheduled. Beginning at the left, subgraph (3) (the synchronous subgraph containing the switch and select) are scheduled like any ordinary SDF graph, until the switch node is invoked. The boxes indicate time occupied by nodes in the subgraph on each of three processors. When the switch node is invoked, special code is inserted (shown as cross-hatched boxes) to implement a conditional branch in each of the processors. Then schedules for each of subgraphs (1) (for $f(\cdot)$) and (2) (for $g(\cdot)$) are created assuming that all three processors become available immediately after the switch code is executed. These two schedules are shown at the top and the bottom of the figure. After this, we wish to return to scheduling subgraph (3), but we need to know when the three processors are available. The code for subgraphs (1) and (2) is padded with no-ops (boxes with horizontal lines) so that both subgraphs finish at the same time on each processor. Then we can resume scheduling subgraph (3), as shown on the right.

6. CONCLUSION

We have outlined a paradigm called synchronous data flow for the description of digital signal processing algorithms. The description permits interpolation and decimation, and restricts neither the

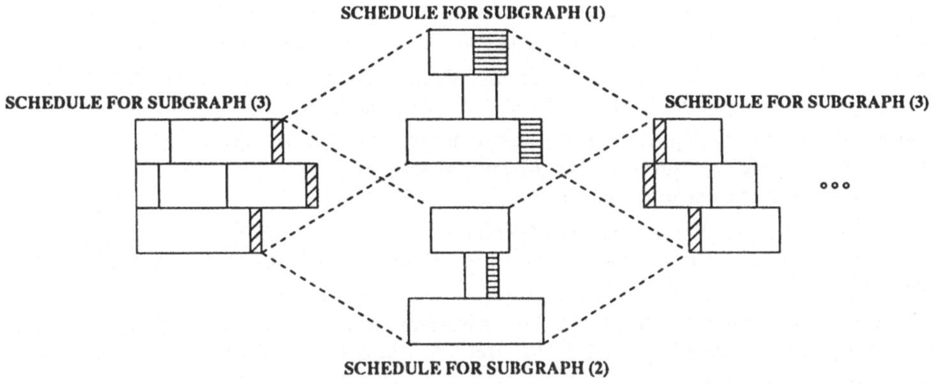

Figure 12. An illustration of scheduling for an if-then-else construct (figure 6) on multiple processors. The blank boxes indicate time taken by nodes on each of three processors. The cross-hatched boxes indicate code associated with the switch and select nodes. This code implements the dynamic scheduling. The top and bottom subschedules correspond to the two possible outcomes of the if-then-else. The boxes with horizontal lines indicate no-ops inserted so that the two possible outcomes finish at the same time on each of the three processors.

granularity (complexity) of a node nor the language in which it is programmed. It is hierarchical and encourages a structured methodology for building a system. Most importantly, SDF graphs explicitly display concurrency and permit automatic scheduling onto parallel processors. We illustrated how the SDF paradigm can be used to generate code for DSP microcomputers, including the management of limited forms of asynchrony that support conditionals. We also introduced the notion of static buffering. Using these techniques, we believe that compilers can be constructed which efficiently map SDF descriptions onto a wide variety of hardware architectures, thereby eliminating many of the costly translations from one description to another that are necessary under current methodologies.

References

1. Davis, A. L. and Keller, R. M., "Data Flow Program Graphs," *Computer* **15**(2)(February 1982).
2. Peterson, J. L., *Petri Net Theory and the Modeling of Systems*, Prentice-Hall Inc., Englewood Cliffs, NJ (1981).
3. Ackerman, W. B., "Data Flow Languages," *Computer* **15**(2)(Feb., 1982).
4. Srini, V., "An Architectural Comparison of Dataflow Systems," *Computer* **19**(3)(March 1986).
5. Lee, E. A. and Messerschmitt, D. G., "Synchronous Data Flow," *IEEE Proceedings*, (1987). To appear
6. Lee, E. A. and Messerschmitt, D. G., "Static Scheduling of Synchronous Data Flow Programs for Digital Signal Processing," *IEEE Trans. on Computers* **C-36**(2)(January 1987).
7. Babb, R. G., "Parallel Processing with Large Grain Data Flow Techniques," *Computer* **17**(7)(July, 1984).
8. Lee, E. A. and Messerschmitt, D. G., "Pipeline Interleaved Programmable DSPs: Architecture," *IEEE Trans. on ASSP*, (1987). To appear
9. Lee, E. A. and Messerschmitt, D. G., "Pipeline Interleaved Programmable DSPs: Synchronous Data Flow Programming," *IEEE Trans. on ASSP*, (1987). To appear
10. Karp, R. M. and Miller, R. E., "Properties of a Model for Parallel Computations: Determinacy, Termination, Queueing," *SIAM Journal* **14** pp. 1390-1411 (November, 1966).

11. Renfors, M. and Neuvo, Y., "The Maximum Sampling Rate of Digital Filters Under Hardware Speed Constraints," *IEEE Trans. on Circuits and Systems* **CAS-28**(3)(March 1981).

12. Schwartz, D. A., "Synchronous Multiprocessor Realizations of Shift-Invariant Flow Graphs," *Georgia Institute of Technology Technical Report DSPL-85-2*, (July 1985). PhD Dissertation

13. Cohen, G., Dubois, D., Quadrat, J. P., and Viot, M., "A Linear-System-Theoretic View of Discrete-Event Processes and its Use for Performance evaluation in Manufacturing," *IEEE Trans. on Automatic Control* **AC-30** pp. 210-220 (1985).

14. Olsder, G. J., "Some Results on the Minimal Realization of Discrete-Event Dynamic Systems," *Report 85-35*, Department of Mathematics, Delft University of Technology, The Netherlands, (1985).

15. Lee, E. A., "A Coupled Hardware and Software Architecture for Programmable Digital Signal Processors," *Memorandum No. UCB/ERL M86/54, EECS Dept., UC Berkeley*, (1986). PhD Dissertation

16. Coffman, E. G. Jr., *Computer and Job Scheduling Theory*, Wiley, New York (1976).

17. Adam, T. L., Chandy, K. M., and Dickson, J. R., "A Comparison of List Schedules for Parallel Processing Systems," *Comm. ACM* **17**(12) pp. 685-690 (Dec., 1974).

18. Hu, T. C., "Parallel Sequencing and Assembly Line Problems," *Operations Research* **9**(6) pp. 841-848 (1961).

19. Kohler, W. H., "A Preliminary Evaluation of the Critical Path Method for Scheduling Tasks on Multiprocessor Systems," *IEEE Trans. on Computers*, pp. 1235-1238 (Dec., 1975).

ON AN ANALOGY OF MINIMAL REALIZATIONS IN CONVENTIONAL AND DISCRETE-EVENT DYNAMIC SYSTEMS

Geert J. Olsder, Remco E. de Vries
Department of Mathematics and Informatics
Delft University of Technology
P.O. Box 356, 2600 AJ DELFT
The Netherlands

ABSTRACT

Recently an analogy between the conventional linear system theory and the relatively new theory on discrete-event dynamic systems has been shown to exist. A mapping which relates these two theories will be investigated, specifically with respect to the theory of minimal realizations.

1. INTRODUCTION

A very large class of dynamic systems, such as the material flow in production or assembly lines, the message flow in communication networks and jobs in multi-programmed computer systems comes under the heading of Discrete-Event Dynamic Systems (DEDS). In contrast to conventional dynamic systems described by difference equations, where the underlying time set is independent of the evolution of the system, the evolution of DEDS is described by sudden changes of state (hereafter referred to as discrete events) which occur at instants determined by the dynamics of the system. Examples of points of time at which discrete events occur are the beginning and the completion of a task or the arrival of a message or a customer. Unfortunately, for several DEDS mathematical models, evaluation of their performance or studies pertaining to optimal decisions are lacking (with the noticeable exception of many problems in queueing theory).

Recently, at INRIA, France, a new approach [1] has been promoted, showing similarities with conventional system theory [2] and for which the underlying algebra is the so-called max-algebra [3], an absorbing semi-ring. The elements in this max-algebra are the real numbers (and minus infinity) and the admissible operations are maximization and addition. One of the basic difficulties in proving statements in the max-algebra is that the inverse of the maximum operator does not exist. In spite of this, it is surprising to learn, that many analogies exist with respect to the conventional linear algebra, such as Cramer's rule and the theorem of Cayley-Hamilton [4].

The current paper is a continuation of [5], in which it is investigated whether the conventional minimal realization theory [2] does have an analogy in the theory of DEDS. Minimal realizations are important for compact mathematical descriptions without dimensional redundancies and also for design purposes. In [5] a procedure was given (repeated in this paper) which constructs a state space description of a DEDS if the impulse response (or, equivalently, the Markov-parameters) is given. In this paper we will prove, at least for some cases, that the procedure given leads to a correct state space description. We restrict ourselves to single input/single output systems.

In section 2 the precise problem statement is given, as well as the standard theorem in conventional system theory about minimal realizations. The basic idea for the proofs of theorems to follow is also given. The characteristic equation plays a crucial role and it is given, in max-algebra-setting, in section 3. Section 4 gives the procedure indicated above. Section 5 gives an example as to elucidate how the procedure works and finally, in section 6, a proof is given for the correctness of the procedure (in some specific cases; a general proof seems to be hard). The paper concludes with the references.

2. PROBLEM STATEMENT

In conventional linear system theory one considers models described by

$$x(k+1) = Ax(k) + bu(k), \qquad y(k) = c'x(k) \tag{1}$$

or written componentwise,

$$x_i(k+1) = \sum_{j=1}^{n} a_{ij}x_j(k) + b_i u(k), \quad i = 1,\dots,n;$$

$$y(k) = \sum_{j=1}^{n} c_j x_j(k). \tag{2}$$

The state vector x is n-dimensional, the input u and the output y are scalars. We will only consider single-input/single-output systems. Matrix A has size n×n and the vectors b and c have n components. The symbol ' denotes transpose. In (1) and (2) two operations are used, viz. addition and multiplication. We will speak of a linear system as a discrete-event dynamic system (DEDS) if the system can still be written in the form of (1) or (2) with the only difference that the operations of addition and multiplication are replaced by maximization and addition respectively. These operations will be denoted by $\oplus$ and $\otimes$. Thus a linear discrete-event dynamic system can be written as

$$x_i(k+1) = \sum_{j=1}^{n}{}_{\oplus} (a_{ij} \otimes x_j(k)) \oplus b_i \otimes u(k), \ i = 1,\dots,n;$$

$$y(k) = \sum_{j=1}^{n}{}_{\oplus} (c_j \otimes x_j(k)). \tag{3}$$

A possible interpretation of such a system is that of a network with n nodes in which $x_i(k)$ is the earliest time instant at which node i is activated for the k-th time; u(k) is the time instant at which the resource becomes available for the k-th time; y(k) is the time instant at which the output (a finished product) becomes available for the k-th time; a_{ij}, b_i and c_i are processing times at nodes and transportation times between the nodes. As the * symbol is quite often omitted in conventional analysis, this will also be done with respect to the $\otimes$ symbol if no confusion is possible. The summation symbol $\sum$ provided with the index $_\oplus$ refers to the maximization of the elements concerned.

In this paper we will be concerned whether a DEDS analogy exists of a well-known theorem on minimal realizations in conventional system theory. We will now first formulate this theorem in the conventional setting.

The external description of (1) is given by

$$y(k) = \sum_{j=1}^{k-1} g_{k-j} u(j), \tag{4}$$

where $\{g_j\}$ is the impulse response; the quantities g_j are sometimes called Markov-parameters and they satisfy

$$g_j = c'A^{j-1}b, \quad j = 1,2,\dots \,. \tag{5}$$

In realization theory one starts from the sequence $\{g_j\}$ and tries to construct A, b, c, such that (5) is satisfied. In order to find A, b, c, one constructs the semi-infinite Hankel matrix H;

$$H = \begin{pmatrix} g_1 & g_2 & g_3 & \cdots\cdots \\ g_2 & g_3 & g_4 & \cdots\cdots \\ g_3 & g_4 & g_5 & \cdots\cdots \\ \vdots & \vdots & \vdots & \\ \vdots & \vdots & \vdots & \end{pmatrix} \tag{6}$$

The truncated Hankel matrix $H_{k,\ell}$ is obtained from H by deleting the rows k+1,k+2,... and the columns $\ell+1,\ell+2,\dots$. The proof of the following theorem can for instance be found in [2].

• *Theorem 1:* To the sequence $\{g_i\}_{i=1}^{\infty}$ corresponds a finite dimensional realization of dimension n if and only if rank $H_{\ell,\ell} = n$ for $\ell = n+1,n+2,\dots$, equivalently det $H_{\ell,\ell} = 0$ for $\ell = n+1,n+2,\dots$. If moreover det $H_{n,n} \neq 0$, then this realization is minimal. If the (n+1) vector $\gamma = (a_n,a_{n-1},\dots,a_1,1)'$ satisfies $H_{n+1,n+1}\gamma = 0$, then

$$A = \begin{pmatrix} 0 & 1 & 0 & \cdots & 0 \\ \vdots & \ddots & \ddots & \ddots & \vdots \\ \vdots & & \ddots & \ddots & 0 \\ 0 & \cdots & \cdots & 0 & 1 \\ -a_n & \cdots & \cdots & -a_2 & -a_1 \end{pmatrix}, \quad b = \begin{pmatrix} g_1 \\ \vdots \\ \vdots \\ g_n \end{pmatrix}, \quad c = \begin{pmatrix} 1 \\ 0 \\ \vdots \\ 0 \end{pmatrix} \tag{7}$$

is such a minimal realization. □

For a possible formulation of Theorem 1 in DEDS-setting, the starting point will be the conventional analysis. Instead of considering quantities (scalars) α and β, we will consider exp (αs) and exp (βs) and then apply conventional analysis. The quantity s is a (real) parameter. Since

$$e^{s\alpha} * e^{s\beta} = e^{s(\alpha+\beta)} = e^{s(\alpha\otimes\beta)}, \tag{8}$$

$$\lim_{s\to\infty} \frac{1}{s} \log (e^{s\alpha} + e^{s\beta}) = \lim_{s\to\infty} \frac{1}{s} \log e^{s\max(\alpha,\beta)} = \lim_{s\to\infty} \frac{1}{s} \log e^{s(\alpha\oplus\beta)}, \tag{9}$$

we hope to obtain results in the max-algebra setting by considering the exponential behaviour of the quantities introduced.

Instead of considering the mapping $\alpha \to \exp(\alpha s)$ one could equally well use the mapping $\alpha \to z^{\alpha}$ and then let $z \to \infty$. We will, however, stick to the first mapping. If instead of (1) we consider

$$e^{sx(k+1)} = e^{sA}e^{sx(k)} + e^{sb}e^{su(k)}, \qquad (10)$$

where the exponential of matrices and vectors is defined componentwise, e.g. $(\exp(sA))_{ij} = \exp(sa_{ij})$, and if we now consider the limit as $s \to \infty$, then the exponential growth of (10) is exactly given by (3)!

3. THE CHARACTERISTIC EQUATION

Since the characteristic equation of a matrix in max-algebra setting will be used frequently in the sections to come, it will be introduced briefly. For a more detailed analysis see [4], [5].

An equivalence of the determinant in max-algebra is the dominant, which is defined as

$$\operatorname{dom}(A) = \lim_{s\to\infty} \frac{1}{s} \log |\det(\exp(sA))|. \qquad (11)$$

An eigenvalue λ is defined by means of the equation

$$Ax = \lambda x$$

in max-algebra sense. If such a λ and x ($\neq \varepsilon$) exist, then λ is called an eigenvalue and x an eigenvector. The element ε introduced is the neutral element with respect to $\oplus$, i.e. $a \oplus \varepsilon = a$ for all $a \in \mathbb{R}$. In fact, $\varepsilon = -\infty$. In order to define the characteristic equation, we start with the following equation in conventional analysis sense;

$$\hat{A}(s)\hat{x}(s) = \hat{\lambda}(s)\hat{x}(s) \qquad (12)$$

where

$$\hat{x}_i(s) = \exp\{x_i s\}, \quad i = 1,\ldots,n; \quad \hat{\lambda}(s) = \exp\{\lambda s\}, \ \{\hat{A}(s)\}_{ij} = e^{a_{ij}s}.$$

Since $\hat{A}(s)$ is a positive matrix, the Perron-Frobenius theorem teaches us that at least one such $\hat{\lambda}(s)$ and $\hat{x}(s)$ exist. For finite s, $\hat{\lambda}(s)$ satisfies the characteristic equation

$$\hat{\lambda}^n(s) + \hat{c}_{n-1}(s)\hat{\lambda}^{n-1}(s) + \ldots + \hat{c}_1(s)\hat{\lambda}(s) + \hat{c}_0 = 0 \qquad (13)$$

where

$$\hat{c}_{n-k}(s) = (-1)^k \sum_{i_1<i_2<\ldots<i_k} \det \begin{pmatrix} \hat{a}_{i_1 i_1}(s) & \ldots & \hat{a}_{i_1 i_k}(s) \\ \vdots & & \vdots \\ \hat{a}_{i_k i_1}(s) & \ldots & \hat{a}_{i_k i_k}(s) \end{pmatrix}, \quad k = 1,\ldots,n.$$

For $s \to \infty$ we get

$$\lim_{s\to\infty} \hat{c}_{n-k}(s) = \lim_{s\to\infty} (-1)^k \tilde{d}_{n-k} \exp(sc^*_{n-k}),$$

where

$$c^*_{n-k} = \sum_{\oplus\, i_1<i_2<\dots<i_k} \operatorname{dom} \begin{pmatrix} a_{i_1 i_1} & \dots & a_{i_1 i_k} \\ \vdots & & \\ a_{i_k i_1} & \dots & a_{i_k i_k} \end{pmatrix} \tag{14}$$

and where $\tilde{d}_{n-k}$ is determined by a counting procedure. This quantity equals the number of even minus the number of odd permutations which determine the value of c^*_{n-k}. Thus $\tilde{d}_{n-k}$ can be negative, zero or positive. The terms in (13) will now be rearranged in the following way; if $(-1)^k\tilde{d}_{n-k}$ is positive, then the corresponding term, i.e. the k-th term in (13), will remain at the left-hand side; if it is negative then the corresponding term will be moved to the right-hand side of the equation. If $(-1)^k\tilde{d}_{n-k}$ is zero, the corresponding term is deleted. Thus an equation arises for which all coefficients at left- and right-hand side are positive for s sufficiently large. If we now take logarithms of both sides of this equation - the logarithm of a matrix is taken componentwise -, divide the result by s and take the limit as $s \to \infty$, then we get

$$\lambda^n \oplus \sum_{\oplus\, k\in N} c^*_{n-k}\lambda^{n-k} = c^*_{n-1}\lambda^{n-1} \oplus \sum_{\oplus\, k\in \bar{N}} c^*_{n-k}\lambda^{n-k}, \tag{15}$$

where N and $\bar{N}$ are nonoverlapping subsets of $\{2,\dots,n\}$. Equation (15) is called the characteristic equation in max-algebra sense.

• *Theorem 2:* The characteristic equation (15) has at least one (real) solution. If λ is replaced by A in (15), we get an identity. This latter result is called the Cayley-Hamilton theorem in the max-algebra. □

This theorem has been proved in [4]. A different proof can be found in [6]. Unlike the situation in ordinary calculus, in the max-algebra not every polynomial is a characteristic polynomial. This will be elucidated by means of polynomials of degree two, for which there are two forms;

$$\lambda^2 = c_1\lambda \oplus c_2, \tag{16}$$

$$\lambda^2 \oplus c_2 = c_1\lambda. \tag{17}$$

The form $\lambda^2 \oplus c_1\lambda = c_2$ is not possible as a characteristic equation as follows easily from the derivation of this equation.

• *Lemma 1:* For any values of c_1 and c_2 (16) is a characteristic equation, for (17) to be a characteristic equation it is necessary and sufficient that $c_2 \le c_1^2$ $(= c_1 \otimes c_1)$.

Proof: The matrix $\begin{pmatrix} \varepsilon & 0 \\ c_2 & c_1 \end{pmatrix}$ has as characteristic equation formula (16) from which the first assertion of the lemma follows. To prove the second assertion we start with

$$A = \begin{pmatrix} a_{11} & a_{12} \\ a_{21} & a_{22} \end{pmatrix} .$$

The coefficients c_1 and c_2 satisfy $c_1 = a_{11} \oplus a_{22}$, $c_2 = \text{dom}\,(A)$.
Because c_2 belongs to the left-hand side of (17), it is determined by an even permutation, hence $c_2 = a_{11} \otimes a_{22}$;

$$c_2 = a_{11} \otimes a_{22} \leq a_{11} \otimes a_{22} \oplus a_{11}^2 \oplus a_{22}^2 = (a_{11} \oplus a_{22})^2 = c_1^2,$$

which proves the necessity. For the sufficiency-part of the proof we construct a matrix A which has (17) as characteristic equation. Choose $a_{11} = c_1$, $a_{22} = c_2 - c_1$ and choose a_{12} and a_{21} such that $a_{12} \otimes a_{21} < a_{11} \otimes a_{22}$. □

4. PROCEDURE FOR MINIMAL REALIZATIONS

The starting point is the sequence $\{g_i\}_1^\infty$, and a procedure will be given which yields A, b and c such that

$$g_j = c' \otimes A^{j-1} \otimes b, \quad j = 1,2,\ldots . \tag{18}$$

The procedure consists of the following seven steps.
1. Construct the Hankel matrix H as in (6).
2. Find a linear dependency among the least possible number of successive columns of H. The coefficients describing this dependency are constant (i.e. irrespective of which sequence of successive columns of H is taken, provided the order is the same). These coefficients, called $\{c_i\}_{i=1}^{n+1}$, with $c_{n+1} = 0$, determine a polynomial equation. (A requirement is that this must be a characteristic equation; see discussion in section 6.) Vectors $a_i \in (\mathbb{R} \cup \{\varepsilon\})^n$, $i = 1,\ldots,k$, are called linearly dependent if the index set $\{1,\ldots,k\}$ can be divided into two disjunct parts S and $\overline{S}$, and scalars $\lambda_i \in (\mathbb{R} \cup \{\varepsilon\})$, $i = 1,\ldots,k$, exist such that

$$\sum_{\oplus\, i \in S} \lambda_i a_i = \sum_{\oplus\, i \in \overline{S}} \lambda_i a_i .$$

   An equivalent statement is that $\text{dom}(a_1,\ldots,a_k,a_{k+1},\ldots,a_n) = \varepsilon$, where $a_{k+1},\ldots,a_n$ are arbitrary n-vectors.
3. Apply the operation exp $\{s.\}$ to the elements of H, after the transformation called $\hat{H}(s)$, and to the (characteristic) equation.
4. Extend $\hat{H}(s)$ to $\hat{H}_e(s)$ by adding lower-order exponentials to the elements such that

$$\hat{H}_{e_{n+1,n+1}}(s)\begin{pmatrix} \underline{+}e^{c_n s} \\ \vdots \\ \vdots \\ \underline{+}e^{c_1 s} \\ 1 \end{pmatrix} = 0, \qquad \hat{H}_{e_{n+1,n+1}} = \hat{H}'_{e_{n+1,n+1}},$$

where the $\underline{+}$ signs depend on the place of the corresponding term in the characteristic equation; left-hand or right-hand side.

5. Apply Theorem 1 and construct $\bar{A}(s)$, $\bar{b}(s)$ and $\bar{c}(s)$.
6. Make a coordinate transformation such that the dominant elements of the transformed $\bar{A}(s)$, $\bar{b}(s)$ and $\bar{c}(s)$ are nonnegative.
7. Study the experimental behaviour of the transformed $\bar{A}(s)$, $\bar{b}(s)$ and $\bar{c}(s)$ as $s \to \infty$, i.e. apply operation $\lim_{s\to\infty} (1/s)\log(.)$ to all elements.

   This yields the minimal realization in max-algebra sense of the given impulse response.

Section 5 will give an example to illustrate this procedure and section 6 gives a proof of the correctness of this procedure for some specific cases. A general proof is currently not available unfortunately.

5. EXAMPLE

In this section an example is given which shows how the procedure of section 4 can be applied.

Suppose we are given the Markov-parameters

$$g_1 = 3,\ g_2 = 5,\ g_3 = 8\tfrac{1}{2},\ g_4 = 12\tfrac{1}{2},\ g_5 = 16\tfrac{1}{2},\ \ldots\ . \tag{19}$$

The columns of the Hankel matrix H,

$$H = \begin{pmatrix} 3 & 5 & 8\tfrac{1}{2} & 12\tfrac{1}{2} & 16\tfrac{1}{2} & \ldots\ldots \\ 5 & 8\tfrac{1}{2} & 12\tfrac{1}{2} & \ldots\ldots\ldots\ldots & & \\ 8\tfrac{1}{2} & 12\tfrac{1}{2} & \ldots\ldots\ldots\ldots\ldots & & & \\ \vdots & \vdots & & & & \\ \vdots & \vdots & & & & \end{pmatrix}$$

satisfy

$$0 \otimes \text{i-th column} \oplus 6 \otimes \text{(i-2)-nd column} = 4 \otimes \text{(i-1)-st column},$$

$$i = 3,4,\ldots\ .$$

The corresponding polynomial equation is

$$\lambda^2 \oplus 6 = 4\lambda.$$

Matrix $\hat{H}(s)$ can be constructed and subsequently $\hat{H}_e(s)$, which satisfies

$$e^{6s} * \text{(i-2)-nd column} - e^{4s} * \text{(i-1)-st column} + e^{0s} * \text{i-th column} = 0.$$

Matrix $\hat{H}_{e_{3,3}}(s)$ is

$$\hat{H}_{e_{3,3}}(s) = \begin{pmatrix} e^{3s} & e^{5s}+e^{4\frac{1}{2}s} & e^{8\frac{1}{2}s} \\ e^{5s}+e^{4\frac{1}{2}s} & e^{8\frac{1}{2}s} & e^{12\frac{1}{2}s}-e^{11s}-e^{10\frac{1}{2}s} \\ e^{8\frac{1}{2}s} & e^{12\frac{1}{2}s}-e^{11s}-e^{10\frac{1}{2}s} & e^{16\frac{1}{2}s}-e^{15s}-2e^{14\frac{1}{2}s} \end{pmatrix},$$

from which

$$\bar{A}(s) = \begin{pmatrix} 0 & 1 \\ -e^{6s} & e^{4s} \end{pmatrix}, \quad \bar{b}(s) = \begin{pmatrix} e^{3s} \\ e^{5s} + e^{4\frac{1}{2}s} \end{pmatrix}, \quad \bar{c}(s) = \begin{pmatrix} 1 \\ 0 \end{pmatrix}.$$

Application of the transformation matrix

$$P = \begin{pmatrix} 1 & 1 \\ e^{2s}+e^{1\frac{1}{2}s} & e^{5s} \end{pmatrix}, \quad \det P = e^{5s} - e^{2s} - e^{1\frac{1}{2}s} \tag{20}$$

yields

$$\bar{\bar{A}} = P^{-1}\bar{A}P = \frac{1}{\det P} \begin{pmatrix} e^{7s}+e^{6\frac{1}{2}s}-e^{5\frac{1}{2}s} & e^{10s}-e^{9s}+e^{6s} \\ e^{5\frac{1}{2}s}-e^{4s}-2e^{3\frac{1}{2}s}-e^{3s} & e^{9s}-e^{7s}-e^{6\frac{1}{2}s}-e^{6s} \end{pmatrix}$$

$$\bar{\bar{b}} = P^{-1}\bar{b} = \frac{1}{\det P} \begin{pmatrix} e^{8s}-e^{5s}-e^{4\frac{1}{2}s} \\ 0 \end{pmatrix}, \quad \bar{\bar{c}} = \begin{pmatrix} 1 \\ 1 \end{pmatrix}.$$

The DEDS which leads to the Markov-parameters we started with, can now easily be found;

$$x(k+1) = \begin{pmatrix} 2 & 5 \\ \frac{1}{2} & 4 \end{pmatrix} x(k) \oplus \begin{pmatrix} 3 \\ \varepsilon \end{pmatrix} u(k), \quad y(k) = (0 \quad 0)\, x(k). \tag{21}$$

If instead of the transformation matrix P in (20) we would choose

$$P = \begin{pmatrix} 1 & 1 \\ e^{2s}+e^{1\frac{1}{2}s} & e^{4s} \end{pmatrix},$$

then the following DEDS would result;

$$x(k+1) = \begin{pmatrix} 2 & 2 \\ 1\frac{1}{2} & 4 \end{pmatrix} x(k) \oplus \begin{pmatrix} 3 \\ \varepsilon \end{pmatrix} u(k), \quad y(k) = (0 \quad 0)\, x(k). \tag{22}$$

The series $\{g_i\}$ we started with in this example was not chosen ad random; it is the impulse response of

$$x(k+1) = \begin{pmatrix} 2 & 3 \\ 1 & 4 \end{pmatrix} x(k) \oplus \begin{pmatrix} 1\frac{1}{2} \\ 0 \end{pmatrix} u(k), \quad y(k) = (1\frac{1}{2} \quad 0)\, x(k) \qquad (23)$$

such that, starting with the series $\{g_i\}$, it would be known that at least one solution, viz. (23), would exist. One easily convinces himself that each of the systems (21), (22) and (23) yields the same Markov-parameters as given in (19).

6. TWO CLASSES OF IMPULSE RESPONSES FOR WHICH THE PROCEDURE IS CORRECT

As already said, a general proof of the fact that the procedure described in section 4 indeed yields the minimal realization is not currently available. In [5] the proof was given for a specific class of impulse responses (for sake of completeness the proof is repeated here);

• *Theorem 3:* Given an impulse response $\{g_i\}_{i=1}^{\infty}$ such that for the corresponding Hankel matrix

$$0 \otimes \text{i-th column} = c_1 \otimes \text{(i-1)-st column} \oplus \dots \oplus c_n \otimes \text{(i-n)-th column},$$
$$i = n+1, n+2, \dots$$

and n is the smallest integer for which this or another dependency is possible, then the discrete-event system characterized by

$$A = \begin{pmatrix} \varepsilon & 0 & \varepsilon & \dots & \varepsilon \\ \vdots & \ddots & \ddots & \ddots & \varepsilon \\ \vdots & & \ddots & \ddots & 0 \\ \varepsilon & \dots & \dots & \varepsilon & 0 \\ c_n & \dots & \dots & \dots & c_1 \end{pmatrix}, \quad b = \begin{pmatrix} g_1 \\ \vdots \\ \vdots \\ g_n \end{pmatrix}, \quad c = \begin{pmatrix} 0 \\ \varepsilon \\ \vdots \\ \varepsilon \end{pmatrix} \qquad (24)$$

is a minimal realization.

Proof: Direct calculation yields that the impulse response corresponding to the system (24) equals the sequence $\{g_i\}_{i=1}^{\infty}$. Realization (24) is minimal, since if there would exist a lower dimensional realization, this would result in a lower-order characteristic polynomial and then, due to Theorem 2, there would be a smaller number (smaller than n+1) of successive columns of H which would be linear dependent which is contradicted by the statement of the theorem. □

We will now consider series $\{g_i\}$ for which three successive columns in the corresponding Hankel matrix are linearly dependent. Three different kinds of dependency relations exist;

$$0 \otimes \text{column } i = c_i \otimes \text{column } (i-1) \oplus c_2 \otimes \text{column } (i-2), \qquad (25)$$

$$0 \otimes \text{column } i \oplus c_2 \otimes \text{column } (i-2) = c_1 \otimes \text{column } (i-1), \qquad (26)$$

$$0 \otimes \text{column } i \oplus c_1 \otimes \text{column } (i-1) = c_2 \otimes \text{column } (i-2), \qquad (27)$$

$i = 3,4,5,\dots$. Relation (25) is a special case of the series $\{g_i\}$ considered in Theorem 3 and we know how to construct a minimal realization.

Next we consider relation (26). Without any loss of generality we can confine ourselves to $c_1 \neq \varepsilon$ and $c_2 \neq \varepsilon$.

o *Lemma 2:* If relation (26) holds, then $c_2 \leq c_1^2$.

Proof: From (26), i = 3, it follows that

$$\text{column } 3 \leq c_1 \otimes \text{column } 2,$$

and from (26), i = 4, it follows that

$$c_2 \otimes \text{column } 2 \leq c_1 \otimes \text{column } 3.$$

Combining these two results we get

$$c_2 \otimes \text{column } 2 \leq c_1 \otimes \text{column } 3 \leq c_1^2 \otimes \text{column } 2$$

and therefore $c_2 \leq c_1^2$.

We now consider three subcases; a) $c_2 = c_1^2$; b) $c_2 < c_1^2$ and $g_3 = c_1 g_2$ (and therefore $c_2 g_1 \leq c_1 g_2$); c) $c_2 < c_1^2$ and $c_2 g_1 = c_1 g_2$ (and therefore $g_3 \leq c_1 g_2$), which will be treated separately.

Ad a. Because $c_2 = c_1^2$, (26) yields

$$0 \otimes g_i \oplus c_1^2 \otimes g_{i-2} = c_1 \otimes g_{i-1}, \quad i = 3,4,\ldots \; . \tag{28}$$

Substitution of i = 3 gives $g_2 \geq c_1 \otimes g_1$ and of i = 4: $g_3 \quad c_1 \otimes g_2$ and hence $g_3 \geq c_1^2 \otimes g_1$. In general $g_i \geq c_1^2 \otimes g_{i-2}$ and (28) can be rewritten as $0 \otimes g_i = c_i \otimes g_{i-1}$ and (26) as $0 \otimes \text{column } i = c_1 \otimes \text{column } (i-1)$, $i = 3,4,\ldots$. Since this latter equation may not be true for i = 2, it cannot be concluded here that Theorem 3 can be applied such that a one-dimensional system results. Equation (28) can also be written as $0 \otimes g_i = c_1 \otimes g_{i-1} \oplus a \otimes g_{i-2}$, where a is a constant $\leq c_1^2$ and therefore

$$0 \otimes \text{column } i = c_1 \otimes \text{column } (i-1) \oplus a \otimes \text{column } (i-2),$$
$$i = 3,4,\ldots, \; a \leq c_1^2. \tag{29}$$

The conclusion is that if (26) holds with $c_2 = c_1^2$, then also (29) holds and now Theorem 3 can be applied.

Ad b. If (26) is considered componentwise, then

$$0 \otimes g_i \oplus c_2 \otimes g_{i-2} = c_1 \otimes g_{i-1}, \quad i = 3,4,\ldots \; .$$

For i = 3: $g_3 \oplus c_2 g_1 = c_1 g_2$. Since $g_3 = c_1 g_2$ by assumption, $c_2 g_1 \leq c_1 g_2$ and also $c_2 g_1 \leq g_3$. Therefore we can write $g_3 = c_1 g_2$. For i = 4 we obtain $g_4 \oplus c_2 g_2 = c_1 g_3$. Since $c_2 < c_1^2$ by assumption, $c_2 g_2 < c_1^2 g_2 = c_1 g_3$, and we can write $g_4 = c_1 g_3$. In general, $g_i = c_1 g_{i-1}$, $i = 3,4,\ldots$. As in the previous case Ad a), the first element g_1 is missing in these equalities. In order to have g_1 included, we can write

$$0 \otimes \text{column } i = c_1 \otimes \text{column } (i-1) \oplus a \otimes \text{column } (i-2), \; i = 3,4,\ldots, \tag{30}$$

where a is a constant which is $\le c_2$ as is easily shown. Theorem 3 can be applied to equations (30).

Ad c. Here we consider (26) with $c_2 < c_1^2$ and $c_2 g_1 = c_1 g_2$. Application of the procedure of section 4 yields, after the fifth step,

$$\bar{A}(s) = \begin{pmatrix} 0 & 1 \\ -e^{c_2 s} & e^{c_1 s} \end{pmatrix}, \quad \bar{b}(s) = \begin{pmatrix} b_1 \\ b_2 \end{pmatrix}, \quad \bar{c} = \begin{pmatrix} 1 \\ 0 \end{pmatrix},$$

where $b_1 = e^{g_1 s}$ and $b_2 = e^{g_2 s}$ + lower-order exponentials. A matrix P which transforms $\bar{A}$, $\bar{b}$ and $\bar{c}$ to $\bar{\bar{A}}$, $\bar{\bar{b}}$ and $\bar{\bar{c}}$ respectively with nonnegative elements is

$$P = \begin{pmatrix} 1 & 1 \\ \frac{b_2}{b_1} & e^{c_1 s} \end{pmatrix}. \tag{31}$$

It is straightforward to show that the elements of $\bar{\bar{A}}$, $\bar{\bar{b}}$ and $\bar{\bar{c}}$ are nonnegative for s sufficiently large and provided the assumptions of case c) hold. Ultimately, step 7 of the procedure leads to the DEDS

$$x(k+1) = \begin{pmatrix} c_2-c_1 & c_2-c_1 \\ * & c_1 \end{pmatrix} x(k) \oplus \begin{pmatrix} g_1 \\ \varepsilon \end{pmatrix} u(k), \; y(k) = (0 \quad 0)\, x(k), \tag{32}$$

where the element indicated by * depends on the lower-order exponentials in b_2. It is easy to show that realization (32) is minimal. If this were not true, a one-dimensional realization would exist and for the Markov-parameters we would have that $g_i = k \otimes c_1^{i-1}$, $i = 1,2,\ldots$, where k is some constant. Other choices of P exist which also lead to (other) two-dimensional realizations. If a column of P in (31) is multiplied by a positive constant for instance, then this new matrix also leads to such a realization. Another possibility is to replace the (2,2) element of P in (31) by e^{ks} where k is a constant $\ge c_1$.

Dependency-relations (25) and (26) have been considered in detail now and we next consider (27). Two subcases will be considered;
a) $g_3 < c_1 g_2$; b) $g_3 \ge c_1 g_2$.

Ad a. Consider the first scalar equation of the vector equation (27). For i = 3 we get $g_3 \oplus c_1 g_2 = c_2 g_1$. Because of the assumption $g_3 < c_1 g_2$, necessarily $c_1 g_2 = c_2 g_1$.
For i = 4 we get $g_4 \oplus c_1 g_3 = c_2 g_2$. Then either $c_1 g_3 = c_2 g_2$ $(<c_1^2 g_2)$ or $g_4 = c_2 g_2$ (and $c_1 g_3 \le g_4$). From the first possibility it follows that $c_2 < c_1^2$.
For i = 5 we get $g_5 \oplus c_1 g_4 = c_2 g_3$. Now $c_1 g_3 \le g_4$ is ruled out since

$c_1g_4 \geq c_1c_2g_2 = c_2^2g_1 > c_2g_3$ which is impossible. Therefore $c_1g_3 = c_2g_2$, $g_4 < c_1g_3$ and $c_2 < c_1$. From the equality $g_5 \oplus c_1g_4 = c_2g_3$ it follows that either $c_1g_4 = c_2g_3$ or $g_5 = c_2g_3$. By studying what happens in (27) with i = 6, it readily follows that $g_5 < c_2g_3$ and therefore $c_1g_4 = c_2g_3$.
With induction it can now be shown that

$$c_1g_{i-1} = c_2g_{i-2}, \quad i = 3,4,5,\ldots$$

from which we conclude that a linear dependence of two successive columns of H exists. By means of Theorem 3 a one-dimensional system can be constructed now.

ad b. Now equation (27) holds with $g_3 \geq c_1g_2$. By considering $g_i \oplus c_1g_{i-1} = c_2g_{i-2}$ for i = 3,4, etc., it readily follows that

$$g_i \geq c_1g_{i-1}, \; g_i = c_2g_{i-2}, \quad i = 3,4,\ldots \; .$$

Apart from (27), now also the following dependence among the columns of H exists:

$$0 \otimes \text{column } i = a \otimes \text{column } (i-1) \oplus c_2 \otimes \text{column } (i-2), \quad i = 3,4,\ldots,$$

and now Theorem 3 can be applied again (a is a constant which is $\leq c_1$).
As an example, consider

$$g_1 = 0, \; g_2 = 3, \; g_3 = 5, \; g_4 = 8, \ldots$$

then for the corresponding matrix H,

$$0 \otimes \text{column } i \oplus 2 \otimes \text{column } (i-1) = 5 \otimes \text{column } (i-2), \quad i = 3,4,\ldots,$$

and also

$$0 \otimes \text{column } i = a \otimes \text{column } (i-1) \oplus 5 \otimes \text{column } (i-2), \quad a \leq 2, \quad i = 3,4,5,\ldots \; .$$

Summarizing, we have obtained the following. If any three successive columns of a Hankel matrix are linearly dependent, then at least one of the following assertions is true:

1. any two successive columns are linearly dependent;
2. the linear dependence is of the form of (25);
3. if the linear dependence is not of the form of (25), then another linear dependence exists which does have this form.
4. if none of the above assertions holds, then the procedure (also) gives a minimal realization by construction.

Thus we have proved:

Theorem 4: Given a series $\{g_i\}_{i=1}^{\infty}$ such that for the corresponding Hankel matrix any three successive columns are linearly dependent, then a DEDS of at most state dimension two exists for which the given series is the impulse response.

□

Remark: With some obvious changes in notation and interpretation, the paper can be repeated in terms of the min-algebra instead of the max-algebra. In the min-algebra the operations are addition and minimization.

□

REFERENCES

[1] Cohen, G., D. Dubois, J.P. Quadrat and M. Viot, A Linear-System-Theoretic View of Discrete-Event Processes and its Use for Performance Evaluation in Manufacturing, IEEE Transactions on Automatic Control, Vol. AC-30, 1985, pp. 210-220.

[2] Chen, C.T., Linear System Theory and Design, Holt, Rinehart and Winston, 1984.

[3] Cuninghame Green, R., Minimax Algebra, Lecture Note No. 166 in Economics and Mathematical Systems, Springer-Verlag, 1979.

[4] Olsder, G.J., C. Roos, Cramer and Cayley-Hamilton in the Max-Algebra, to appear in Linear Algebra and its Applications. Also Report No. 85-30 of the Dept. of Mathematics and Informatics, Delft University of Technology.

[5] Olsder, G.J., Some Results on the Minimal Realization of Discrete-Event Dynamic Systems, Proc. 7th International Conference on Analysis and Optimization of Systems, Antibes (France), Lecture Note No. 83 in Control and Information Sciences, Springer-Verlag, 1986.

[6] Moller, P., Théorème de Cayley-Hamilton dans les Dioides et Application à l'Etude des Systèmes à Evénement Discrets, Ibidem.

REPRESENTATION, ANALYSIS AND SIMULATION OF MANUFACTURING SYSTEMS BY PETRI NET BASED MODELS

Francesco Archetti, Anna Sciomachen

Università di Milano, Via L. Cicognara 7, 20129 Milano, Italy

1. INTRODUCTION

The aim of this paper is to give, in a simple framework, some basic indications on the methodological tools required in order to understand, develope and analyze Petri net based models of manufacturing systems.

Petri nets have been developed in the last decade into a powerful tool to model discrete event systems. Their growing relevance is witnessed by a host of theoretical studies, the wide ranging application areas and a number of software tools for the analysis of Petri net based models (Ajmone Marsan et al. 1984a, Chiola 1985, Cumani 1985, Dugan et al. 1985, Molloy and Riddle 1986, Sciomachen 1986, Billington 1987).

Recently manufacturing systems have emerged as a most important application area. Petri nets are well suited to model the complex interactions between the elements of a manufacturing system, in particular those features of synchronization and concurrency of different activities which exert a critical influence on the overall performance of a system (Dubois 1983, Alla et al. 1985, Martinez et al. 1986).

Petri nets have been particularly successful in the specification and validation of control procedures at the workcell level (Valette et al. 1983b, Thuriot and Courvoisier 1983, Murata 1984a, Komoda et al. 1984).

There are many advantages in modelling a system using Petri nets. First of all, the system is described in a graphical form and hence it is possible to visualize the interactions among different components of complex systems. Petri nets also indicate explicity those points in the system where the control can be exercised (Valette et al. 1983a, Marabet 1986). Second, the system can be modelled hierarchically and represented in a top-down fashion at different levels of abstraction. Moreover, a systematic and complete qualitative analysis of the system is allowed by Petri net analysis techniques. Finally, performance evaluation of the system is possible by Markov techniques, when this is stochastically and computationally feasible, and by direct simulation of the net.

The above points will be dealt with in this paper and exemplified considering the representation, analysis and simulation of a machine for electronic assembly.

The basic operations of the machine are in this paper only sketched without going into unnecessary technicalities. A complete description of the operations of a specific machine of this type is given in Ahmadi et al. (1986).

For each workboard to be assembled by the machine a given number of electronic components must be picked up from a component magazine and fitted to the workboard in a specified position. There are two component magazines located in the north and south side of the machine, which can independently move to align the slot containing the specified component with the pick up position. At the center of the machine there is an arm which has one head at each of its ends. Each cycle of the machine is defined by the following operations. The arm moves first to the north magazine while the north head of the arm is preparing for picking the component and the magazine is moving to the proper position. Concurrently, the other magazine is moving and the workboard is being positioned for the component to be fitted by the south head. Only when the magazine is aligned and the arm is idle the component is picked up. As soon as the arm is idle the south head can start the preparation activities and subsequently fit the component. When both the picking and the fitting are completed, the arm moves to the south magazine and the same operations are performed in the south side of the machine reversing the roles of the heads of the arm. The very features of concurrency and synchronization which can produce, in ideal conditions, the peak performance make the machine subject to severe degradation when the concurrency is not properly exploited.

In order to plan the capacity of the assembly line and to schedule the production, an accurate estimate of the machine cycle time for a given product is required before the start of the production.

Creating a setup and sequencing for a new workboard requires extensive and time consuming preparation of the machine.

Replacing this trial and error method on the machine is the main reason behind the development of simulation models.

Petri nets have proved particularly suited to this task for a class of Computer Numerically Controlled (CNC) machines (Archetti et al. 1986, Grotzinger and Sciomachen 1987, Sciomachen et al. 1987).

In this paper the steps required for the representation, analysis and simulation of these machines will be described.

Section 2 is devoted to the basic definitions and properties of Petri nets and to the model of the machine sketched above.

In Section 3 the main technique for the logical verification of the model computing the invariants of the net (Reisig 1982) is presented. These invariants are shown to have a meaningful interpretation for the operations of the machine.

Section 4 outlines the main techniques, based on Markovian analysis and direct simulation of the net, to compute the steady-state probabilities from which the performance indices of the machine can be derived.

The numerical results of the Markovian analysis and direct simulation for the machine considered in this paper are given and compared. Their analysis and that of the invariants allow to draw useful considerations on the behavior of the machine.

2. CREATING A PETRI NET MODEL

In this Section the basic components of a Petri net model are introduced. The presentation, far from being complete, is aimed at making the paper self-contained. For a comprehensive coverage of the subject the reader is referred to the volumes Peterson (1981) and Reisig (1982), while two survey papers of a general and introductory nature are Agerwala (1979) and Murata (1984b).

As far as manufacturing systems are concerned, two recent introductory references are Kamath and Viswanadham (1986) and Beck and Krogh (1986).

A Petri net is a bipartite directed graph PN = (P,T,A) whose nodes are divided into a set of places $P = \{p_1, p_2, \ldots, p_n\}$ and a set of transitions $T = \{t_1, t_2, \ldots, t_m\}$.

$A = (P \times T) \cup (T \times P)$ is a set of directed arcs which link places to transitions and transitions to places.

Places are used to represent resources and logical conditions of the modelled system.

Places may contain tokens. The state of the net, or its marking, is given by an integer vector M(i) whose k-th component M(i;k) is the number of tokens in place p_k.

The specification of the initial marking M(0) is required in order for the Petri net to be completely defined.

For each transition a set of input places I(t) = {p: (p,t) A} and a set of output places O(t) = {p: (t,p) A} are given.

In the classical Petri net theory, a transition is enabled when there is at least 1 token in each of its input places. A transition, enabled in a marking M(i), can fire removing 1 token from each input place and placing 1 token in each output place.

Several extensions, e.g. weighted arcs and places with capacity, have been added to the classical theory to reflect the modelling requirements of specific application areas. The firing rules have been modified accordingly.

The firing of a transition t moves the system into a new marking M(i+1) = M(i) + F(t) where the k-th component of F(t) is given by F(t;k) = 1 if $p_k \in O(t) - I(t)$, F(t;k) = -1 if $p_k \in I(t) - O(t)$ and F(t;k) = 0 otherwise.

Transitions can be immediate, i.e. they fire as soon as they are enabled, or timed: in this case there is a delay, random or deterministic, between enabling and firing.

Immediate transitions are utilized in order to synchronize multiple flows before enabling a timed transition (Dubois and Stecke 1983).

Timed transitions are used to represent operations and activities of the system (Ramchadami 1974, Sifakis 1977). A general reference to the research activity in the area of timed nets is in Torino (1985).

Graphically places are represented by circles, immediate transitions by bars, timed transitions by rectangles and tokens by dots inside places.

Enabled transitions which share input places cannot fire at the same time. Indeed the firing of one transition disables the others.

In this case the transitions are not independent and have a joint firing probability. This situation is called a conflict.

Two or more transitions are concurrent when they can independently fire. Concurrent transitions represent activities which take place simultaneously.

The set of markings generated by the firing of all the enabled transitions is called reachability graph and denoted by R(PN,M(0)).

For instance, the net of Figure 2.1 (Molloy 1982) displays transitions 2 and 3 which are concurrent and transitions 4 and 5 which are in conflict.

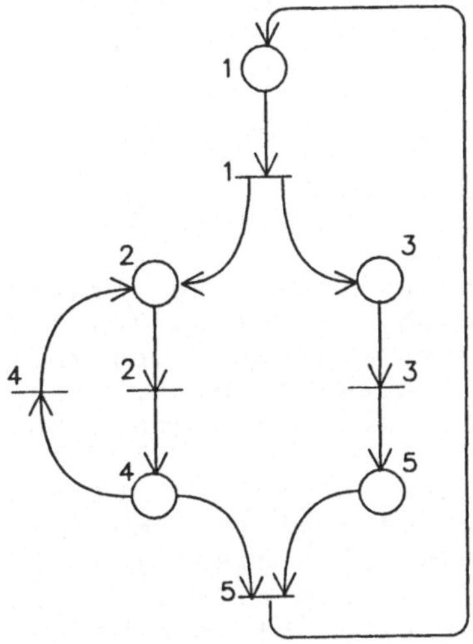

Figure 2.1

Given the initial marking M(0) = (1,0,0,0,0) the reachability graph contains the vectors (0,1,1,0,0), (0,0,1,1,0), (0,1,0,0,1) and (0,0,0,1,1).

An important feature of Petri nets is the use of inhibitor arcs. A token in a place linked by an inhibitor arc to a transition prevents the transition from firing.

Inhibitor arcs are particularly useful to model failures

of components of the system (Archetti et al. 1985, Archetti et al. 1987) and priority rules (Abraham and Sciomachen 1986).

Another useful feature of Petri nets are the double arcs. For instance, in the net of Figure 2.2 place 2 is both input and output for transition 2. Places 1, 2 and 3 can respectively represent the loading station, the output buffer of a machine and the wait station of an Automated Guided Vehicle (AGV). Places 4 and 5 represent respectively the AGV loaded and the AGV in travel mode to load a workpiece from place 1.

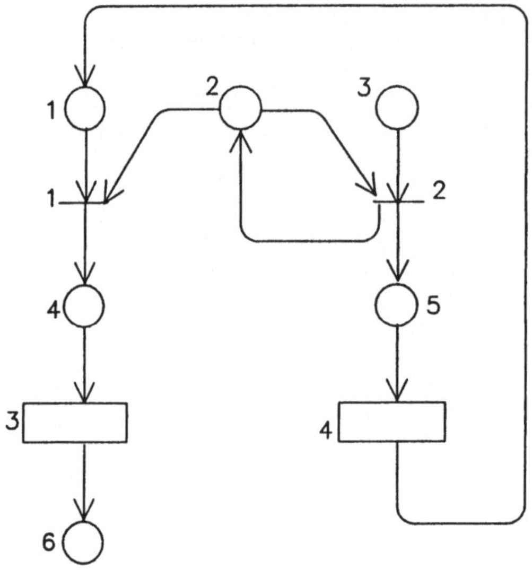

Figure 2.2

Transitions 3 and 4 represent respectively the travel time of the AGV loaded and empty. The double arc from place 2 to transition 2 represents a transportation request. Transition 2 fires but the token never leaves place 2. This property is required in order to compute statistics about the queue lenght in the output buffer.

After this simple example, we'll use the basic notions introduced above to build the Petri net model of the machine described in the Introduction. This model is displayed in Figure 2.3. The net has 26 places and 19 transitions, 13 of which are timed transitions representing all the operations and activities performed by the machine.

The basic movements of the 4 main subsystems of the machine namely arm, board, north and right magazines, are represented by the timed transitions AMS and AMN (arm moving south and arm moving north), BM (board moving), NFM (north magazine moving) and SFM (south magazine moving).

The two timed transitions NHM and SHM (north and south head moving) represent the movement of the head which has to be aligned with the picking location when the arm is moving to north/south. The two timed transitions SHT and NHT (south/north head preparing in the north/south side) represent the preparation activities required to the south/north head in order to be ready for the fitting when the arm is north/south aligned.

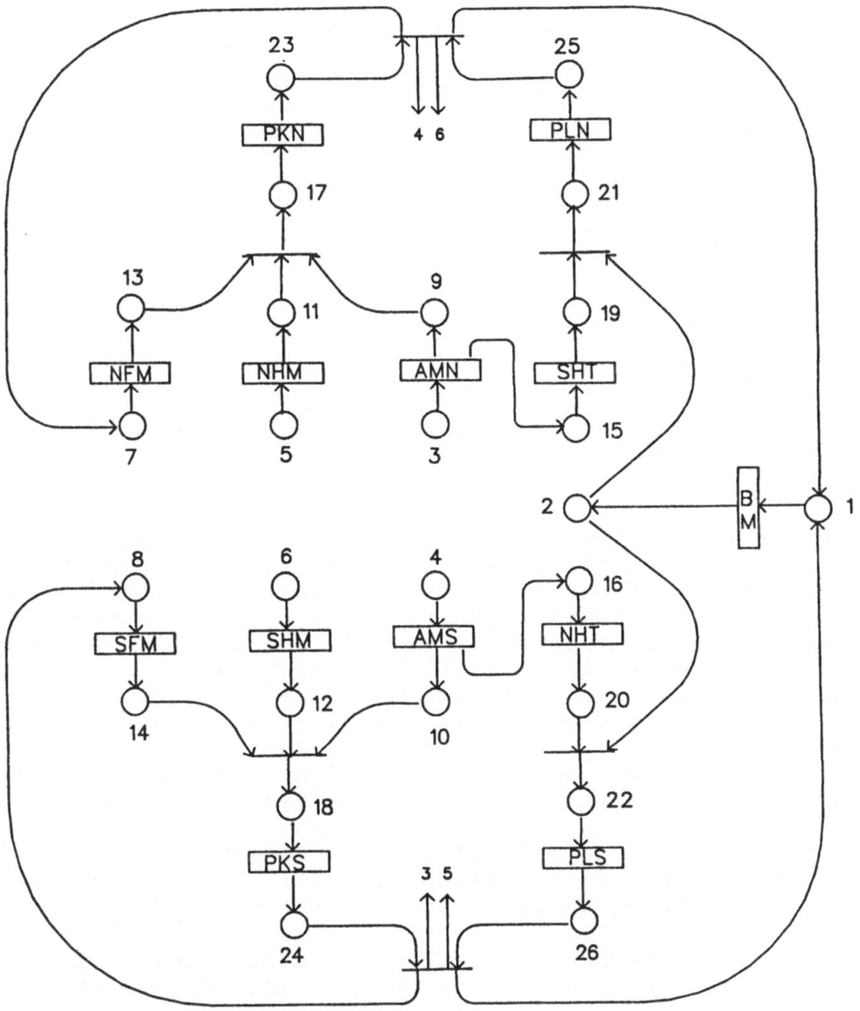

Figure 2.3

The picking and the fitting operations are represented by transitions PKN and PLN in the north side of the machine and by transitions PKS and PLS in the south side of the machine.

In this example, the time associated to each of the above transitions is expressed by an exponential random variable whose mean value is given in Table 2.1.

Table 2.1

Transition	Mean value (sec.)
BM	0.222
AMN	0.265
AMS	0.265
NHM	0.088
SHM	0.088
NFM	0.646
SFM	0.801
SHT	0.260
NHT	0.260
PKN	0.340
PKS	0.340
PLN	0.340
PKS	0.340

It can be noted that there are 6 immediate transitions in the net. These transitions are introduced in order to synchronize different concurrent operations before the beginning of the activity corresponding to the subsequent timed transition. For instance, picking can start when there is a token in place 17 in the north side or in place 18 in the south side of the machine, and the magazine involved, the arm and the corresponding head are ready. Analogously, an electronic component can be fitted only when the board is in the proper position and the corresponding head is ready. The completion of both picking and fitting represents the end of a cycle.

At the beginning an initial marking is given in which 5 tokens are in the net, in the places 1, 3, 5, 7 and 8. In this case the first picking is performed in the north side of the machine.

It can be remarked that only the main cycle of the machine operation is modelled in the net and no load/unload procedures are considered. In the initial marking, indeed, the component to be fitted in the first cycle is supposed to be ready for the head at the south end of the arm.

We observe that all the transitions enabled in the initial marking are concurrent, that is each of these transitions can be independently choosen to fire first according to the realization of the random variable expressing its firing delay. All the activities represented by these transitions are to be completed in order to start the subsequent picking and fitting operations. In particular, a picking is enabled after a time T(pick) from the initial marking, where T(pick) is given by T(pick) = Max{Time(LCM), Time(NHM), Time(AMN)}. In the same way the time T(fit) can be computed as the time elapsed from the initial marking before the start of fitting a component. This time is given by T(fit) = Max{Time(AMN) + Time(SHT), Time(BM)}.

3.SYSTEM VERIFICATION BY PETRI NETS

In this Section we shall be concerned with the qualitative aspects of the behavior of the system and of its Petri net model.

Let us first introduce the basic properties of a manufacturing system which can be analyzed using Petri nets (Naharari and Viswanadham 1984).

Conservativeness. It is related to the invariance of the number of resources and jobs in the system. A net is conservative if the number of tokens in the net is constant, that is if $\sum_{k=1}^{n} M(j;k) = \sum_{k=1}^{n} M(i;k) \quad \forall (i,j)$.

Boundedness. It is a very important property when modelling manufacturing systems. In fact it is related to the absence of overflows. A place is bounded if there exists an integer value k such that no more than k tokens can be in that place in any marking of the net. A net is bounded if all its places are bounded.

Liveness. This property means absence of deadocks. A transition is said to be live if and only if for all markings of the net there is a firing sequence which takes the net in a marking in which that transition is enabled. A net is live if all its transitions are live. If the Petri net representing the manufacturing system is live and the model is correct then no deadlocks will happen in the operation of the system.

Properness. It means that the system can be reinitialized. A net is said to be proper if the initial marking is reachable from all the markings in the reachability graph. In the case of manufacturing systems, if the corresponding net is proper then no manual intervention is required to restore the system to its initial state from any state.

In order to analyze these properties, the net is represented thru its incidence matrix C = c(i,j), i=1,...,n, j=1,...,m, where c(i,j) = -1 if $p_i \in I(t_j)$, c(i,j) = 1 if $p_i \in O(t_j)$ and c(i,j) = 0 otherwise.

Given the markings M(i) and M(j), if M(j) is reachable from M(i) then the following relation holds: M(j) = M(i) + CY where each component of the m-dimensional column vector Y denotes the number of times the corresponding transition has fired.

In particular, if the net is to return to any given state, then the existence of a positive integer vector Y such that CY = 0 is a necessary condition. Such a vector Y is called a transition invariant.

If a positive integer n-dimensional row vector X exists such that XC = 0 then XM(i) = XM(j). Such a vector is called a place invariant. This relation implies, for each invariant, that the number of tokens is constant in the places corresponding to positive components of the same invariant.

If and only if X = 1, the unit vector, the net is conservative. A net is bounded if and only if there exists a place invariant whose components are positive.

A Petri net is not proper if no transition invariant exists.

Unfortunately, at least to the authors' knowledge, there is no further condition in order to test the liveness and properness properties of a net. However it is possible in most cases, as we will show below in the case considered, to check these properties from all the invariants of the net and from some informations and considerations about the system.

The computation of the invariants for the machine considered in this paper allows to verify the behavior of the system and the correctness of the model. Indeed for the machine to get back to a state, each operation must be performed once, with the exception of the board which must execute two movements for the two fittings performed in the north and the south sides of the machine.

This requirement is reflected by the only transition invariant Y obtained from the Petri net of Figure 2.3. In this invariant all the components are equal to 1 except the component associated with transition BM (board moving) which is 2.

It can be observed, from the Petri net model and the description of the machine, that in this case this result implies that the net is proper.

Moreover, since all the transitions, for the net to get back to a marking, fire, according to the transition invariant Y, the net in this case is also live.

As far as the place invariants are concerned, the solutions of the system XC = 0 span an 8-dimensional space. A base of this space is given by the vectors V1, V2, V3, V4, V5, V6, V7 and V8 listed in Table 3.1

Linear combinations of these vectors give the invariants of interest.

In particular the invariant X1 = V1 + V2 is associated to the places representing the flow of the board in the net, i.e. all its possible states. From the definition of place invariant, the number of tokens in places 1, 2, 22, 26, 21 and 25 is constant; since this number in the initial marking is 1, this implies that 1 and only 1 token is, in any marking, in any of these places. This condition reflects the possible states of the board: moving for fitting (place 1), ready for fitting (place 2), during the fitting in the north or south side of the machine (places 21 and 22) and after the completion of the fitting operation (places 25 and 26).

Table 3.1

Place	V1	V2	V3	V4	V5	V6	V7	V8
1	1	0	0	0	0	0	-1	0
2	1	0	0	0	0	0	-1	0
3	0	0	1	0	-1	-1	1	0
4	0	0	1	0	-1	0	1	-1
5	0	0	0	0	0	1	0	0
6	0	0	0	0	0	0	0	1
7	-1	1	-1	1	1	0	0	0
8	0	0	0	0	1	0	0	0
9	1	-1	1	0	-1	1	0	0
10	0	0	1	0	-1	0	0	-1
11	0	0	0	0	0	1	0	0
12	0	0	0	0	0	0	0	1
13	-1	1	-1	1	1	0	0	0
14	0	0	0	0	1	0	0	0
15	-1	1	0	0	0	0	1	0
16	0	0	0	0	0	0	1	0
17	0	0	0	1	0	0	0	0
18	0	0	1	0	0	0	0	0
19	-1	1	0	0	0	0	1	0
20	0	0	0	0	0	0	1	0
21	0	1	0	0	0	0	0	0
22	1	0	0	0	0	0	0	0
23	0	0	0	1	0	0	0	0
24	0	0	1	0	0	0	0	0
25	0	1	0	0	0	0	0	0
26	1	0	0	0	0	0	0	0

The invariants associated to the other components of the machine, namely the arm, the north and south magazines, the picking head and the fitting head, have been found in the same way. The invariant X2 = V3 + V4 is related to the states of the arm. Indeed the corresponding non zero components represent all the places in which a token can be found. It can be easily verified that the arm can be moving to the north magazine (place 3) or to the south magazine (place 4), ready for the north and south side respectively (place 9 or 10), or can be aligned with the north/south magazine before and after the completion of the picking (places 17 and 23/18 and 24).

As far as the two magazines are concerned, the corresponding invariants X3 and X4 are given by X3 = V4 for the north magazine and by X4 = V3 + V5 for the south one. In fact, considering for instance the north magazine, it can be in four different states: moving to align the slot containing the required component (place 7), ready for picking (place 13), during the picking (place 17) and after the completion of the picking (place 23). Analogous considerations hold for the south magazine whose possible states are represented by places 4, 14, 18 and 24.

In order to perform a pick, it is required that the involved head is properly oriented with the component to be picked up. The flow of the picking head is reflected in the invariant X5 given by X5 = V3 + V4 + V6 + V8. The positive components of this invariant, in fact, are related to the positions in which the head can be. This head can be moving (place 5), ready (place 11), picking (place 17) and idle (place 23) in the north side and in the south side (places 6, 12, 18 and 24) of the machine.

The preparation activities performed by the other head, which has to fit the electronic component, as it is also shown by the invariant X6 = V1 + V2 + V7, are associated to the arm movements (places 3 and 4) which are to be completed before the head start (places 15 and 16). Then the head can be ready (places 19 and 20), fitting (places 21 and 22) or idle after the completion of the operation (places 25 and 26).

4.PERFORMANCE EVALUATION

The use of Petri nets for performance evaluation has been mostly in the areas of multiprocessing, communications and more recently manufacturing systems.

In this paper we shall not consider specific methods suggested for a particular performance evaluation problem (Sifakis 1980, Magott 1987), but only the main techniques.

Two basic approaches can be used to derive performance measures: analytical-numerical techniques and direct simulation.

The first can be applied to Stochastic Petri Nets (SPN). By this term we denote timed nets in which the firing delay of the transitions is an exponentially distributed random variable.

SPN are isomorphic to continuous time Markov chains: the states of the chain correspond to the markings of the net and the sojourn times are the same. This result, due to Molloy (1981) and extended by other authors (Ajmone Marsan et al. 1984b) to GSPN (Generalized Stochastic Petri Nets) which have exponential and immediate transitions, enables to apply to the analysis of Petri nets the Markovian analysis techniques and in particular to compute the probability of each state in steady state and transient conditions.

Petri nets have two advantages over Markov models. They are a formalized language of system specification and allow, thru the computation of the reachability graph, the automatic generation of the state space.

The method used to compute the probabilities for both the transient and the steady-state case first computes the reachability graph R(PN, M(0)). The states can be vanishing, when at least one enabled transition is immediate, and tangible, when all enabled transitions are timed. The probabilities are computed only for the tangible states. This is done, for the steady-state, solving the balance equations of the Markov chain, typically by the Gauss-Seidel method and, for the transient case, computing the exponential of the rate matrix of the Markov chain.

These probabilities allow to compute the performance indices of the system thru the distribution of the number of tokens in each place of the net.

This analysis has been performed for the net of Figure 2.3. The number of markings is 128. Gauss-Seidel converged in 5 iterations to the steady-state probability for each state.

A substantial limitation of the Markovian approach is that the number of states can be exceedingly high, making computationally unfeasible the generation of the reachability graph, even for middle size models, and subsequently the computation of the probabilities.

A second limitation of the analytical approach is that it can be applied, as already remarked, basically only when the probabilistic structure of the net is Markovian.

The other main analysis technique is the direct simulation of the net. In this case, tokens are moved around according to the firing rules and the statistics are collected about the number of tokens in each place.

The simulation procedure, (Sciomachen 1986), is simple and straighforward. In a tangible marking it checks which transitions are enabled. For the transitions which were already enabled in the previous marking the time elapsed between the two markings is subtracted from their time constant. The variables required by each transition are subsequently generated and the event list is updated accordingly. At this point one can decide which transition fires first and which marking is generated. Before moving in the new marking the statistical counters of the simulation are updated.

The lenght of the simulation run is controlled by the usual criteria of statistical simulation. The search for stopping criteria explicity dependent on the structure of the Petri net is still very much an open problem (Haas and Shedler 1986a, 1986b, 1987a, 1987b). In the same papers the simulation of SPN is analyzed in the framework of Generalized Semi Markov Processes.

Over traditional simulation languages Petri nets offer the advantages that the system specifications, as they are represented in the net, and the initial marking are the only input required by the simulation procedure which generates automatically the simulation program.

The methods recalled above are common in most of the software packages for analyzing timed Petri nets. Specific features may be different depending on the environment in which the package has been developed and its main application area. For instance MEGAS (Multiple Event Graph Analysis &

Simulation), (Sciomachen 1986), developed in a manufacturing environment, has several options for dealing with conflicts, allowing the use of state dependent rules, which can represent adaptive control policies.

Moreover the timing of the transitions can vary dynamically during the execution of the net. This feature, as illustrated in Archetti et al. (1986), is important when dealing, in realistic conditions, with machines like that considered in this paper.

For the machine of Figure 2.3 the token distribution has been computed by the analytical technique and by simulation.

Table 4.1 gives the probability of having 1 token for each place of the net.

Table 4.1

Place	Probability of 1 token Analytic	Simulation
1	0.198	0.196
2	0.312	0.314
3	0.118	0.118
4	0.118	0.119
5	0.082	0.082
6	0.082	0.082
7	0.289	0.295
8	0.358	0.353
9	0.076	0.079
10	0.109	0.106
11	0.112	0.115
12	0.145	0.143
13	0.418	0.413
14	0.357	0.361
15	0.116	0.116
16	0.116	0.116
17	0.152	0.150
18	0.152	0.151
19	0.021	0.020
20	0.021	0.021
21	0.152	0.154
22	0.152	0.152
23	0.141	0.141
24	0.133	0.135
25	0.080	0.081
26	0.105	0.102

The agreement between the analytical and simulation results is more than satisfactory.

It is important to note the relation between the place invariants and the token distribution. Let P be the steady-state probability vector of having 1 token in each place of the net and X any place invariant vector. Then the sum of the probabilities of having 1 token in the places belonging to each place invariant is related to the corresponding invariant by the relation XP = 1.

The verification of this property is given in Table 4.2 where for each invariant it is computed the scalar product XP. In the Table the values P(i), representing the probability of having 1 token in place i, are the results obtained by the analytical procedure; the same results hold for the simulation results.

Table 4.2

Invariant	XP
X1	P(1)+P(2)+P(21)+P(22)+P(25)+P(26) = 0.999
X2	P(3)+P(4)+P(9)+P(10)+P(17)+P(18)+P(23)+P(25) = 1.
X3	P(7)+P(13)+P(17)+P(23) = 1.
X4	P(8)+P(14)+P(18)+P(24) = 1.
X5	P(5)+P(6)+P(11)+P(12)+P(17)+P(18)+P(23)+P(24) = 0.999
X6	P(3)+P(4)+P(15)+P(16)+P(19)+P(20)+P(21)+P(22)+P(25) +P(26) = 1

CONCLUDING REMARKS

The example discussed in this paper and the growing research interest in the use of Petri nets to model manufacturing systems bear witness to their great potential in the design and management of the automated factory.

The main advantages of Petri nets over traditional simulation languages can be summarized in the following points. First, the availability of a design and simulation language which can be used at different levels of abstraction and detail. Different analysis technique using the same data structure can be employed from the initial macrosimulation for capacity planning to the system control at workcell level. Second, the automatic generation from the net of the simulation program, with a drastic reduction in the time and cost of program development.

The main obstacle to a broader use of Petri net based tools is the lack of a generally agreed upon methodology for constructing models out of the system specifications. For this reason, the model in this paper as well as the more complex ones of Sciomachen et al. (1987) and Abraham and Sciomachen (1986), have been presented in such a way as to outline a general methodology for building Petri net models. This methodology holds, in the authors' opinion, beyond the issue, specifically addressed in this paper, of modelling CNC machines.

REFERENCES

Abraham, C., Sciomachen, A. (1986). Planning for Automatic Guided Vehicle Systems by Petri Nets. RC 12288, IBM T.J. Watson Research Center, Yorktown Heights, New York.

Agerwala, T. (1979). Putting Petri Nets to Work. IEEE Computer, vol.12, 1979.

Ahmadi, J., Grotzinger, S., Johnson, D. (1986). Emulation of Concurrency in Circuit Card Assembly Machines. RC 12161, IBM T.J. Watson Research Center, Yorktown Heights, New York.

Ajmone Marsan, M., Balbo, G., Ciardo, G., Conte, G. (1984a). A Software tool for the automatic analysis of Generalized Stochastic Petri Net Models. Proceedings of "International Conference on Modelling Techniques and Tools for Performance Analysis", Paris 1984.

Ajmone Marsan, M., Balbo, G., Conte, G. (1984b). A Class of Generalized Stochastic Petri Nets for the performance evaluation of multiprocessor systems. ACM Transactions on Computer Systems. vol.2 #2, pp.93-122.

Alla, H., Ladet, P., Martinez, J., Silva-Suarez, M. (1985). Modeling and Validation of complex systems by coloured Petri Nets; application to a flexible manufacturing system. Proceedings of International Workshop on Timed Petri Nets, IEEE Computer Press, Torino 1985.

Archetti, F., Fagiuoli, E., Sciomachen, A. (1985). Performance Evaluation of a transfer line via Stochastic Petri Nets. Conv. ANIPLA, Genova 1985.

Archetti, F., Grotzinger, S., Sciomachen, A. (1986). The design of a Petri net based tool for the performance evaluation of pick and place machines. RC 12359, IBM T.J. Watson Research Center, Yorktown Heights, New York.

Archetti, F., Fagiuoli, E., Sciomachen, A. (1987). Computation of the Makespan in a transfer line with station breakdowns using Stochastic Petri Nets. To appear on Computer and Operation Research.

Beck, C., Krogh, B. (1986). Models for Simulation and Discrete Control of Manufacturing Systems. Proceedings IECON, IEEE Computer.

Billington, J., Wilbur-Ham, M.C. (1987). PROTEAN: A high level Petri Net tool. Petri Net Newsletter, 26 Aprile 1987.

Chiola, G. (1985). A Software Package for the analysis of Generalized Stochastic Petri Net Models. Proceedings of International Workshop on Timed Petri Nets, IEEE Computer Press, Torino 1985.

Cumani, A. (1985). ESP. A Package for evaluation of Stochastic Petri Nets with phase-type distributed transition times. Proceeding of International Workshop on Timed Petri Nets, IEEE Computer Press, Torino 1985.

Dubois, D., Stecke, K. (1983). Using Petri Nets to represent production Processes. Proceedings 22 IEEE Conference on Decision and Control, pp.1062-1067.

Dugan, J.B., Bobbio, A., Ciardo, G., Trivedi, (1985). The Design of a unified package for the solution of

Stochastic Petri Net Models. IEEE Computer Press, Torino 1985.

Grotzinger, S., Sciomachen, A., (1987). A Petri net characterization of ahigh speed pacement machine. RC 12935, IBM T. J. Watson Research Center, Yorktown Heights, New York.

Haas, P., Shedler, G. (1986a). Regenerative Stochastic Petri Nets. Performance Evaluation 6, pp.189-204.

Haas, P., Shedler, G. (1986b). Simulation with simultaneous events. RJ 5158, IBM Almaden Research Center, S.Jose, CA.

Haas, P., Shedler, G. (1987a). Stochastic Petri Nets with simultaneous transition firings. RJ 5464, IBM Almaden Research Center, S.Jose, CA.

Haas, P., Shedler, G. (1987b). Stochastic Petri Net Rapresentation of Discrete Event Simulation. IBM RC 5646 (57145), 715/87.

Kamath, M., Viswanadham, N. (1986). Application of Petri Net based models in the Modelling and Analysis of Flexible Manufacturing Systems. Proceedings IECON 1986, IEEE Computer.

Komoda, N., Kera, K., Kubo, T., (1984). An Autonomous Decentralized Control System for Factory Automation. IEEE Computer, December 1984.

Marabet, A.A. (1986). Dynamic Job Shop Scheduling: an Operating System based design. Flexible Manufacturing Systems: Methods and Studies, A. Kusiak Ed., North Holland, 1986.

Magott, J. (1987). New NP Complete Problems in Performance Evaluation of Concurrent Systems Using Petri Nets. IEEE Transactions on Software Engineering, vol.SE, n.5, May 1987.

Martinez, J. Alla, H., Silva-Suarez, M. (1986). Petri Nets for the Specification of FMS. Modelling and Design of Flexible Manufacturing Systems. A. Kusiak Ed., North Holland, 1986.

Molloy, M.K. (1981). On the integration of delay and throughput measures in distributed processing models. PhD Thesis, UCLA.

Molloy, M.K. (1982). Performance Analysis using Stochastic Petri Nets. IEEE Transaction on Computers, C31, N.9, pp.913-917.

Molloy, M.K., (1986). The Stochastic Petri Net Analyzer system design tool for Bit-mapped workstations. TR-86-12, University of Texas, Dept. of Computer Science, 1986.

Murata, T. (1984a). A Petri Net based Factory Automation Controller for Flexible and Maintainable Control Specifications, Proceedings IECON 84.

Murata, T. (1984b). Petri Nets and their application: an introduction. Management and Office Information Systems, Plenum Press, 1984.

Naharari, Y., Viswanadham, N., (1984). Analysis and synthesis of Flexible Manufacturing Systems using Petri Nets. Proceedings of the First ORSA/TIMS Conference on FMS, Ann Arbor, 1984.

Peterson, J.L. (1981). Petri Net Theory and the Modeling of Systems. Prentice Hall Inc., 1981.

Ramchadami, C. (1974). Analysis of Asynchronous Concurrent Systems by Timed Petri Nets. PhD Thesis MIT 1974 MAC-TR-120.

Reisig, W., (1982). Petrinetze. Springer Verlag, 1982.

Sciomachen, A. (1986). The Design of a Petri Net based Package for Modeling and Analysis of Manufacturing Systems. RC 12218, IBM T.J. Watson Research Center, Yorktown Heights, New York.

Sciomachen, A., Grotzinger, S., Archetti, F. (1987). Petri Net based emulation for a highly concurrent Pick and Place Machine. RC 12940, IBM T.J. Watson Research Center, Yorktown Heights, New York.

Sifakis, J., (1977). Use Petri Nets for Performance Evaluation. Measuring, Modelling and Evauating Computer Systems. H. Beilner & E. Gelenbe Eds., North Holland 1977.

Sifakis, J. (1980). Performance Evaluation of Systems using Nets. Proc. of the third Int. Symposium IFIP Working Group 7.3, Ed. H. Beilner and E. Gelembe.

Thuriot, E., Courvoisier, M. (1983). Implementation of a centralized synchronization concept for production systems. Proceedings Real-Time Systems Symposium, Arlington, VA, 1983.

Torino (1985). Proceedings of the International Workshop on Timed Petri Nets. IEEE Computer Press, 1985.

Valette, R., Courvoisier, M., Mayeux, D. (1983a). Control of Flexible Production Systems and Petri Nets. European Workshop on Application and Theory of Petri Nets, Varenna 1982, Springer Verlag, vol.63.

Valette, R., Courvoisier, M., Mayeux, D. (1983b). A Petri Net based programmable logic controller. Computer Applications in Production and Engineering (CAPE 83), North Holland.

The SMARTIE Framework for Modelling Discrete Dynamic Systems

Geert-Jan Houben, Jan L.G. Dietz, Kees M. van Hee

1. INTRODUCTION

A major problem in software and systems engineering is the precise specification of the system to be analysed or designed. A formal model of the system to be build can be considered as a specification of the system, restricted to the aspects considered in the model.

A computer-model is an implementation of a formal model. It can be used to simulate the behaviour of a modelled system. In case this system is an information system a computer-model can be used as a prototype of the system. Users or potential users of an information system usually are unable to understand a formal model of the system. With a prototype of the system they can see if their requirements are translated correctly by the system designers.

The systems we are dealing with are discrete dynamic systems. Such a system is at each moment in one of a set of states. At some moments it performs a transition to another, not necessarily different state. The number of transitions in each finite time interval is finite. A transition is triggered by one or more actions. The system may produce by each transition one or more reactions. Actions are coming from the environment of the system or they may be created by the system itself and fed back to the system.

Many real systems, including information systems, are discrete dynamic systems.

In literature there are many approaches to model discrete dynamic systems. Finite state machines, Markov chains and Petri nets are well-known examples of generic models. In Dietz and van Hee (1986) a framework, called SMARTIE, is developed. It is an extension of finite state machines combined with a modelling language based on predicate logic. In Harel (1986) another generalization of finite state machines combined with a graphical modelling language is presented. A different and less formal approach is found in Jackson's System Development (Jackson 1983). In Jackson (1983) several interesting examples are presented. In Sridhar and Hoare (1985) some of these examples are modelled using the language of Hoare CSP. In that paper it is suggested that CSP could provide a formal basis for the JSD method.

We feel that our approach is a powerful alternative. In van Hee, Houben and Dietz (1987) this is already demonstrated by treating some examples. In that paper we also focused on a modelling language and a diagram technique.

In this paper we give the description of our model for a discrete dynamic system (dds) in Section 2. In Section 3 we briefly describe our modelling language and in Section 4 we give some examples. The examples include a banking system, a queueing system and a heating system. In Section 5 some conclusions are drawn.

2. MODEL OF DISCRETE DYNAMIC SYSTEMS

A dds is determined by a seven-tuple, the components of which will not all be known. Some of the components will be fully specified during the design phase of a dds, while others will become known during the operational phase of the system.

In this section we define the components of a dds and its behaviour. In this paper we do not define the aggregate of a dds. However, since it turns out that such an aggregate is a dds itself, our model allows decomposition and integration of dds'ses.

We use the following notations :

- $P(X)$ denotes the set of all finite subsets of a set X.
- Δ denotes the symmetric difference-operator, i.e. $a \Delta b = (a \cup b) \backslash (a \cap b)$.
- dom and rng are functions that assign domain and range to a function, respectively.
- $X \rightarrow Y$ denotes the set of all functions with domain contained in X and range contained in Y.
- $(\cup j \in J : A_j)$ denotes $\{x \mid \exists j \in J : x \in A_j\}$.
- $(\cap j \in J : A_j)$ denotes $\{x \mid \forall j \in J : x \in A_j\}$.

Furthermore, we use the usual notations of set theory and symbolic logic, including $\rightarrow$ for implication. Often we write f_x instead of $f(x)$ for a function application. We frequently use the Δ-operator and the fact that this operator is commutative and associative. Therefore, we may define for some set of sets X, $\Delta X = X_1 \Delta X_2 \Delta \cdots \Delta X_n$ for some enumeration $X_1, \ldots, X_n$ of X. Similarly for some set-valued function X, $(\Delta i \in \text{dom}(X) : X_i) = X_{i_1} \Delta X_{i_2} \Delta \cdots \Delta X_{i_n}$ for some enumeration $i_1, \ldots, i_n$ of dom(X). Let $\mathbb{N}$ denote the set of natural numbers. Let $\mathbb{R}^+$ denote $\{x \mid x \in \mathbb{R} \wedge x \geq \varepsilon\}$ for some fixed $\varepsilon > 0$, where $\mathbb{R}$ denotes the reals. $\mathbb{R}^+$ will be used as the time domain of dds'ses.

Definition 2.1.

A discrete dynamic system is a seven-tuple

$$<S, M, A, R, T, I, E>$$

where

- S is a set-valued function,

  for $i, j \in \text{dom}(S)$ we have $i \neq j \rightarrow S_i \cap S_j = \emptyset$,

  dom(S) is called the set of store indices,

  for $j \in \text{dom}(S)$: S_j is called the state base of store j,

  $\overline{S_j} = P(S_j)$ is called the state space of store j.

- M is a function,
  dom(M) is called the set of <u>processor indices</u>,
  for $i \in$ dom(M) : $M_i = \langle MC_i, MR_i \rangle$
  where:
  M_i is called the <u>motor of processor i</u>,
  MC_i and MR_i are functions,
  MC_i is called the <u>change function of processor i</u>,
  MR_i is called the <u>response function of processor i</u>.

- A is a set-valued function,
  dom(A) = dom(M),
  for $i, j \in$ dom(M) : $i \neq j \rightarrow A_i \cap A_j = \varnothing$,
  for $i \in$ dom(M) and $j \in$ dom(S) : $A_i \cap S_j = \varnothing$,
  $i \in$ dom(M) : A_i is called the <u>action base of processor i</u>,
  $\overline{A_i} = P(A_i)$ is called the <u>action space of processor i</u>.

- R is a set-valued function,
  dom(R) = dom(M),
  for $i, j \in$ dom(M) : $i \neq j \rightarrow R_i \cap R_j = \varnothing$,
  $i \in$ dom(M) : R_i is called the <u>reaction base of processor i</u>,
  $\overline{R_i} = P(R_i)$ is called the <u>reaction space of processor i</u>.

- T is a function, dom(T) = dom(M) $\times$ dom(M)
  for $i, j \in$ dom(M) : $T_{ij} \in \overline{R_i} \rightarrow P(A_j \times \mathbb{R}^+)$,
  T is called the <u>transfer function</u>.

- I is a set-valued function, dom(I) = dom(M)
  for $i \in$ dom(M) : $I_i \subset$ dom(S),
  I is called the <u>interaction function</u>.

- For $i \in$ dom(M):
  * $MS_i = (\cup j \in I_i : S_j)$ is called the <u>state base of processor i</u>,
  * $\overline{MS_i} = P(MS_i)$ the <u>state space of processor i</u>,
  * $MC_i \in \overline{MS_i} \times \overline{A_i} \rightarrow \overline{MS_i}$,
  * $MR_i \in \overline{MS_i} \times \overline{A_i} \rightarrow \overline{R_i}$,
  * $\forall s \in \overline{MS_i} : MC_i(s, \varnothing) = MR_i(s, \varnothing) = \varnothing$.

- $E = \langle EU, ES, EA \rangle$ where EU, ES and EA are functions:
  * dom(EU) = dom(S) and
    for $j \in$ dom(S) : $EU_j \in P(S_j \times \mathbb{R}^+)$,
    EU_j is called the <u>external update set</u> of store j
  * dom(ES) = dom(S) and
    for $j \in$ dom(S) : $ES_j \in \overline{S_j}$,
    ES_j is called the <u>initial state of store j</u>.

* $\text{dom}(EA) = \text{dom}(M)$ and
  for $i \in \text{dom}(M) : EA_i \in P(A_i \times \mathbb{R}^+)$,
  EA_i is called the external event set of processor i.

(end of definition)

A mechanical appreciation of a dds is as follows. A dds consists of a set of processors and a set of stores. Processors are mutually connected by transaction channels and processors and stores may be connected by interconnections. The motor of a processor transforms instantaneously a set of actions into updates for the connected stores (by means of the function MC) and simultaneously it produces some set of reactions. The transformations may depend on the states of the connected stores. The state of a store may change by an update from a connected processor or by some external update. Hence an environment may influence a dds by external updates on stores and by imposing actions on the processors. The occurrence of an action at a particular moment is called an event. More than one event at a time for one processor is allowed. The output of a processor is sent to the environment of the processor and a transfer function transforms some reactions into actions with a time delay. Such a pair is sent to a processor as a new event with a time stamp equal to the sum of the processing time and the delay.

The delays are elements of $\mathbb{R}^+$ and therefore the number of transitions in a finite time interval is always finite. The events produced by a dds for itself or another processor are called internal events. They are inserted into the event agenda of the receiving processor. Initially, this agenda consists of all the external events, later on it contains also internal events. The external updates are supposed to commute; in fact we assume that each update is specified by some value from the state space of a store. If this value is denoted by s_1 and the actual state of that store is s_2, then the effect of the update will be $s_1 \Delta s_2$.

Next we define the behaviour of a dds.

Definition 2.2.
Let $<S, M, A, R, T, I, E>$ be a dds. The process of the dds is a five-tuple

$$<\tau, \alpha, \rho, \sigma, \phi>$$

where:

- $\tau \in \mathbb{N} \to \mathbb{R}$ and for $n \in \mathbb{N} : \tau_n$ is the time stamp of the n-th activation,
- $\alpha, \rho, \sigma, \phi$ are functions with

$$\text{dom}(\alpha) = \text{dom}(\rho) = \text{dom}(\phi) = \text{dom}(M) \text{ and } \text{dom}(\sigma) = \text{dom}(S) .$$

For $i \in \text{dom}(M)$:

- $\alpha_i \in \mathbb{R} \to \overline{A}_i$, $\alpha_i(t)$ is the action set of processor i at time t,
- $\rho_i \in \mathbb{R} \to \overline{R}_i$, $\rho_i(t)$ is the reaction set of processor i at time t,
- $\phi_i \in \mathbb{R} \to P(A_i \times \mathbb{R}^+)$, $\phi_i(t)$ is the event agenda of processor i at time t.

For $j \in \text{dom}(S)$:

- $\sigma_j \in \mathbb{R} \to \overline{S}_j$, $\sigma_j(t)$ is the state of store j at time t.

These functions are defined recursively:

- $\tau_0 = 0$, for $i \in \text{dom}(M) : \alpha_i(0) = \varnothing$, $\rho_i(0) = \varnothing$, $\phi_i(0) = EA_i$
  for $j \in \text{dom}(S) : \sigma_j(0) = ES_j$.

Let τ_n be defined and let the functions α, ρ, σ and ϕ be defined on the interval $[0,\tau_n]$,

- $\tau_{n+1} = \min\{t \mid \exists i \in \text{dom}(M) : \exists a \in A_i : \langle a, t\rangle \in \phi_i(\tau_n)\}$

and for $j \in \text{dom}(S)$ let δ be defined by

$$\delta_j(\tau_{n+1}) = \sigma_j(\tau_n) \,\Delta\, (\Delta \{x \mid \langle x,y\rangle \in EU_j \wedge \tau_n < y < \tau_{n+1}\})$$

and for $i \in \text{dom}(M)$ let γ be defined by:

$$\gamma_i(\tau_{n+1}) = (\cup j \in I_i : \delta_j(\tau_{n+1}))$$

and let:

- $\alpha_i(\tau_{n+1}) = \{a \mid \langle a,\tau_{n+1}\rangle \in \phi_i(\tau_n)\}$,
  for $\tau_n < t < \tau_{n+1} : \alpha_i(t) = \varnothing$;
- $\rho_i(\tau_{n+1}) = MR_i(\gamma_i(\tau_{n+1}),\alpha_i(\tau_{n+1}))$,
  for $\tau_n < t < \tau_{n+1} : \rho_i(t) = \varnothing$;
- $\phi_i(\tau_{n+1}) = \{\langle a,t\rangle \mid (\langle a,t\rangle \in \phi_i(\tau_n) \wedge t > \tau_{n+1}) \vee$
  $\vee (\exists d \in \mathbb{R}^+ : \exists k \in \text{dom}(M) : t = \tau_{n+1} + d \wedge \langle a,d\rangle \in T_{ki}(\rho_k(\tau_{n+1}))))\}$
  for $\tau_n < t < \tau_{n+1} : \phi_i(t) = \phi_i(\tau_n)$.

For $j \in \text{dom}(S)$:

- $\sigma_j(\tau_{n+1}) = \delta_j(\tau_{n+1}) \,\Delta\, (\Delta i \in \text{dom}(M) : S_j \cap MC_i(\gamma_i(\tau_{n+1}),\alpha_i(\tau_{n+1}))) \,\Delta$
  $\Delta\, (\Delta\{x \mid \langle x,\tau_{n+1}\rangle \in EU_j\})$
  for $\tau_n < t < \tau_{n+1} : \sigma_j(t) = \sigma_j(\tau_n) \,\Delta\, \Delta\{x \mid \langle x,y\rangle \in EU_j \wedge \tau_n < y \leq t\}$.

(end of definition)

Note that $\delta_j(\tau_{n+1})$ is the last state of store j before τ_{n+1} and that $\gamma_i(\tau_{n+1})$ is the last state of processor i before τ_{n+1}. The state of store j at τ_{n+1} includes also the external updates at time τ_{n+1}.

Here we do not define the aggregate of a dds, but it turns out that the aggregate is a dds itself. However, it has only one processor and one store. So it may be called a simple dds. In a top-down design-process we start in fact with a simple dds and we decompose it into a dds with more stores and processors. At the top level we do not specify much components of the dds. However, the further we decompose the system, the more details we specify. If we finally have specified at the bottom level all details of the dds, then that is also the specification of the simple dds at the top level. In this design-process the diagram technique, proposed in van Hee, Houben and Dietz (1987), can be helpful. When we observe the processes of a dds and its aggregate, then we learn that they have the same outputs and therefore the same external behaviour. So we could consider them equivalent.

In practice we only specify the first six components of a dds, since we cannot look into the future to determine EU and EA. However often we know or require some properties from these sets, for example that the time lag between two events or external updates is bounded from below by some known quantity. Such information may be used to prove properties of the behaviour of the system, i.e. of the process of the system. On the other hand we sometimes require properties of the process of a system, and then these requirements may be translated into requirements for E and therefore for the environment of the system.

In our model we assume that state transitions are executed instantaneously. This assumption is made to facilitate modeling. In practice it is often impossible to implement instantaneous transitions. There are several ways to guarantee that the time lag between two transitions is longer than the time needed to realise the transition in the real system. One of these methods is demonstrated in the second example of section 4. We think that this kind of modifications of a model is a next phase in the design process : first we model an ideal system, afterwards we take care of the limitations of implementations such as bounds on store sizes and execution times for state transitions.

Finally we note that our framework assumes the existence of absolute time. However the processors we model do not have the possibility to inspect some absolute clock. The absolute time we assume is just for the definition of the process of a system and may be used to express and prove properties of the dynamics of systems.

3. MODELLING LANGUAGE

The modelling language that we introduce in this section is one of the possible ways to describe the components of a dds, defined in the framework of Section 2. Although we feel that a large class of systems can be described in this way, we do not claim that this is true for every dds.

Our modelling language consists of two parts. The first part is a first order language L that is used to describe the state, action and reaction bases and spaces. The second part is a language PRL for production rules, that is used to describe the motor functions.

The first order language L is constructed in the usual way (cf. Chang and Lee 1973). It is extended by introducing two additional quantifiers for summation and cardinality.
The alphabet consists of:

- set of variables
- set of constants, called F_0
- set of n-ary function symbols, called F_n, for $n \in \mathbb{N}$ and $n > 0$
- set of n-ary predicate symbols, called P_n, for $n \in \mathbb{N}$,
  $P = (\cup n \in \mathbb{N} : P_n)$
- quantifiers: $\exists$, $\forall$, Σ, #
- logical operators: $\vee$, $\wedge$, $\neg$, $\rightarrow$, $\leftrightarrow$
- relational operators: $<$, $>$, $\leq$, $\geq$, $=$, $\neq$
- arithmetic operators: $+$, $-$, $*$, mod, div
- punctuation symbols: (,), :, ,, { , }, |

Terms are defined by:

- constants and variables are terms;
- if $t_1, \ldots, t_n$ are terms and $f \in F_n$, then $f(t_1, \ldots, t_n)$ is a term;
- if t_1 and t_2 are terms, then $(t_1 + t_2)$, $(t_1 * t_2)$, $(t_1 - t_2)$, $(t_1 \underline{\text{div}}\, t_2)$ and $(t_1 \underline{\text{mod}}\, t_2)$ are terms;
- if t is a term, x is a variable and Q a formula, then $(\Sigma x : Q : t)$ and $(\#x : Q)$ are terms.

Atoms are defined by:

- if $t_1, \ldots, t_n$ are terms and $p \in P_n$, then $p(t_1, \ldots, t_n)$ is an atom;
- if t_1 and t_2 are terms, then $(t_1 \leq t_2)$, $(t_1 \geq t_2)$, $(t_1 < t_2)$, $(t_1 > t_2)$, $(t_1 = t_2)$ and $(t_1 \neq t_2)$ are atoms.

Formulas are defined by:

- an atom is a formula;
- if Q and R are formulas, then $(Q \vee R)$, $(Q \wedge R)$, $(\neg Q)$, $(Q \rightarrow R)$ and $(Q \leftrightarrow R)$ are formulas;
- if Q is a formula and x is a variable, then $(\forall x : Q)$ and $(\exists x : Q)$ are formulas.

Note that, when no problems arise, parentheses are often omitted. In formulas free and bounded variables are distinguished, in the usual way. To give formulas a (formal) interpretation (cf. Lloyd 1984), we choose the set of integers as the domain of interpretation. This means that every constant is mapped to an integer and every variable is given an integer value, but this restriction to integers is only made for convenience and is not essential.

For describing the state base of a store or an action- and reaction base of a processor we choose a subset of the predicate symbols P. The bases are defined as the sets of all ground atoms with corresponding predicate symbols. Note that when we specify such a set of predicate symbols, we also specify for each predicate symbol the number of arguments that the corresponding ground atoms will have; $p(\cdot,\cdot)$ denotes that the ground atoms with predicate symbol p have two arguments. All sets of predicate symbols for base-definitions should be mutually disjoint. Remember that the state space of a processor is the union of the state bases of stores with which it is connected.

We assume that relational and arithmetic operators have their usual interpretation, as have the logical operators and quantifiers. For each processor i with action- and reaction bases defined with predicate symbol sets PA_i and PR_i respectively and a state base defined with predicate symbols PS_i, a subset PD_i of P is defined that is disjoint with PA_i, PR_i and PS_i. The predicate symbols in PD_i are used for shorthands in the description of the motor of i.

For each processor i a set of closed formulas D_i is defined. Formulas in D_i may contain predicate symbols of $PD_i \cup PA_i \cup PR_i \cup PS_i$. These formulas are considered to be axioms; they have the truth value true. These formulas serve as definitions for shorthands or as constraints on states and actions. We can for instance specify constraints for specifying that some argument values are not allowed for some predicate symbols, thus specifying as domain of an argument only a subset of the integers. The set D_i is called the <u>axiom base</u> of processor i.

We follow the closed-world assumption (cf. Reiter 1984), which implies that, given some state s and some action a all ground atoms in s and a have the truth value true, whereas all other ground atoms that can be formed by predicate symbols from the corresponding bases have the truth value false. We require that a processor i never has to deal with a state or an action that is in contradiction with D_i. It is the responsibility of the designer of a system to prove that a dds fulfills this requirement. Usually, this is done by showing that, given a state and an action, that do not contradict D_i, the new state does not contradict D_i either.

The definitions in D_i are closed formulas of the kind:

$$\forall x_1 : \cdots : \forall x_n : p(x_1, \ldots, x_n) \longleftrightarrow Q \ ,$$

where $p \in PD_i$ and Q is some formula involving at most $x_1, \ldots, x_n$ as free variables and predicate symbols from $PD_i \cup PA_i \cup PR_i \cup PS_i$. Each predicate symbol of PD_i occurs exactly once in such a formula on the left-hand side. It is again the designer's responsibility to guarantee that for each ground atom, with its predicate symbol in PD_i, a truth value can be determined w.r.t. some state s of processor i and some action a for processor i.

Given D_i, some state s of processor i and some action a for processor i, the motor M_i may change the state and therefore the formal interpretation of formulas. Such a change of state consists of additions and deletions of ground atoms with predicate symbols from PS_i. A deletion means that the negation of that atom gets truth value true. This can never cause a contradiction with a definition in D_i, but it may create contradictions with constraints. If an axiom base, a state and an action are considered to be axioms of a theory, then a transition may change this theory into a new one.

Now we can define the language of production rules PRL, for describing the motor of a processor. First we define formally the language's syntax; its semantics will be defined informally afterwards.

In PRL, formulas of L occur. The non-terminals <formula> and <atom> refer to formulas and atoms of L resp. Using EBNF-notation we define:

<rule> ::= <condition><D-part><I-part><R-part>
<condition> ::= $\models$ <formula>
<D-part> ::= $\overset{D}{\Rightarrow}$ <atom set>
<I-part> ::= $\overset{I}{\Rightarrow}$ <atom set>
<R-part> ::= $\overset{R}{\Rightarrow}$ <atom set>
<atom set> ::= [<enumerated set> | <conditional set> | <atom set> $\underline{\cup}$<atom set >]
<enumerated set> ::= $\underline{\{}$ <atom list>$\underline{\}}$
<atom list> ::= <atom> { $\underline{,}$ <atom>}
<conditional set> ::= $\underline{\{}$ <atom> $\underline{\mid}$ <formula>$\underline{\}}$

Note that '$\models$', '$\overset{D}{\Rightarrow}$', '$\overset{I}{\Rightarrow}$', '$\overset{R}{\Rightarrow}$' and the underscored symbols are terminals.

An example:

$$\models ((p(1,3) \wedge q(0)) \vee r(x,y,z))$$

$$\overset{D}{\Rightarrow} \{q(8), r(7)\} \cup \{p(2*x,y) \mid t(y) \wedge y \geq x\}$$

$$\overset{I}{\Rightarrow}$$

$$\overset{R}{\Rightarrow} \{m(x,z) \mid t(x)\}$$

There may occur free variables in a rule, i.e. in the formula of the condition or in an atom set. However, the free variables in an atom set have to occur also in the condition. Note that a variable that occurs as a free variable before and after the bar in a conditional set is bounded. With each state and action of a processor we associate an active domain. This is the set of all constants that occur in the axiom base, the state or the action. The active domain and the set of all variables occurring in the description of a motor or store are finite. A binding of a set of variables is a function with this set of variables as domain and the active domain as range.

The semantics of a rule are as follows. For each binding of the free variables of the formula in the condition of a rule, it is checked whether this formula is true, with respect to the formulas in the axiom base, the (current) state and the action. If it is true, then the atom sets of the D-, I- and R-parts are computed, where for free variables in these atom sets the values defined by the binding are substituted. The quantifications over bounded variables in closed formulas and conditional sets are computed also with respect to the active domain, so these quantifications are computable. The computation of a conditional set is as usual. When in a D-, I- or R-part no atom set occurs, then the set of ground atoms computed will be empty. For reasons of convenience we allow that, instead of writing a part without an atom set, such a part is omitted.

Denote for rule n and binding b of the free variables of the condition of the rule, the sets of ground atoms computed in the D-, I- and R-part by $D_{n,b}$, $I_{n,b}$ and $R_{n,b}$, respectively. Then we define:

$$C = (\Delta n : (\cup b : D_{n,b})) \,\Delta\, (\Delta n : (\cup b : I_{n,b})) ,$$

$$R = (\cup n : (\cup b : R_{n,b})) .$$

For a transition of a motor is now defined for state s and action a:

$$MC(s,a) = s \,\Delta\, C$$

$$MR(s,a) = R$$

so the state is changed by taking the symmetric difference of the old state and for all rules, the union, over all bindings of the free variables in the conditions in the condition of the rule, of the computed sets of the D- and I-parts, whereas the reaction is just the union over all computed sets of R-parts for all rules and bindings. Note that the distinction between D- and I-parts is only made for convenience.

4. EXAMPLES

In this section we present three examples. The first one treats a banking system similar to the example treated in Sridhar and Hoare (1985). The second one shows how to deal with queues. The last one treats a heating system.

4.1. Banking System

In this section we will give two versions of a banking system. The first version describes a very simple system in which the balances of accounts are managed. This example comes from Sridhar and Hoare (1985).

Processor 1 receives banking transactions, originating from the account holders, and makes appropriate changes to the balances in Store 1. Processor 2 periodically reads the contents of Store 1 and produces a balance report, which is sent to the bank management. Store 1 holds the balances of the accounts. For Store 1 we will use ground atoms balance(i,x), where balance(i,x) means that the balance of account i is equal to x.

Processor 1, which has interaction with Store 1, can be specified as follows:

$A = \{\text{invest}(\cdot,\cdot), \text{payin}(\cdot,\cdot), \text{withdraw}(\cdot,\cdot), \text{terminate}(\cdot)\}$

$R = \varnothing$

M:

$\models$ invest(i,x) $\wedge$ no–account(i)
I
$\Rightarrow$ {balance(i,x)}

$\models$ terminate(i) $\wedge$ balance(i, 0)
D
$\Rightarrow$ {balance(i, 0)}

$\models$ balance(i,x) $\wedge$ $\neg$invest(i,y) $\wedge$ $\neg$terminate(i)
D
$\Rightarrow$ {balance(i,x)}

I
$\Rightarrow$ {balance(i,y) | $y = x + (\Sigma w$: payin(i,w) : $w) - (\Sigma w$: withdraw(i,w) : $w)$}

$D = \{\forall i$: no–account(i) $\longleftrightarrow \neg(\exists x$: balance(i,x))$\}$

Processor 2, which has interaction with Store 1, can be specified as follows:

$A = \{\text{makereport}\}$

$R = \{\text{report}(\cdot,\cdot), \text{doreport}(\cdot)\}$

M:

$\models$ makereport $\wedge$ balance(i,x)
R
$\Rightarrow$ {report(i,x)}

% note that this rule is executed for all accounts, if a makereport action is received %

$\models$ makereport
R
$\Rightarrow$ {doreport(t)}

% note that this rule is executed only once when a makereport action is received %

Furthermore :

$T_{2,2}$: $\forall t$: doreport(t) ∈ A → $T_{2,2}$(A) = {<makereport,t>}

% via the feedback transaction channel a next makereport action is transferred, that will be received t time units later %

Store 1 can be specified as follows :

S = {balance(· ,·)}

When we consider this specification, we can notice that we assume that Processor 1 can only get some sets of actions at a time, e.g. not an invest action and a terminate action together. This implies that we have some knowledge concerning the environment, i.e. the possible sets of events.

In the second version of the banking system we will specify Processor 1 in such a way that we do not have to make assumptions about the event sets coming from the environment. Furthermore we will specify exactly the domains of the predicate symbols. Again both processors have interaction with Store 1.

Specification of Processor 2 :

A = {makereport}

R = {report(· ,·), doreport(·)}

M:

$\models$ makereport ∧ balance(i,x)

$\overset{R}{\Rightarrow}$ {report(i,x)}

$\models$ makereport

$\overset{R}{\Rightarrow}$ {doreport(t)}

Specification of Store 1 :

S = {balance(· ,·),no–account(·)}

Specification of Processor 1 :

A = {invest(· ,·), payin(· ,·), withdraw(· ,·), terminate(·)}

R = {error(·)}

M:

$\models$ inputerror(i)

$\overset{R}{\Rightarrow}$ {error(i)}

$\models$ ¬inputerror(i) ∧ invest(i,x)

$\overset{D}{\Rightarrow}$ {no–account(i)}

$\overset{I}{\Rightarrow}$ {balance(i,x)}

$$\models \neg\text{inputerror}(i) \wedge \text{terminate(i)} \wedge \text{balance}(i,x)$$
$$\overset{D}{\Rightarrow} \{\text{balance}(i,x)\}$$
$$\overset{I}{\Rightarrow} \{\text{no–account}(i)\}$$

$$\models \neg\text{inputerror}(i) \wedge \text{balance}(i,x) \wedge \neg\text{invest}(i,y) \wedge \neg\text{terminate}(i)$$
$$\overset{D}{\Rightarrow} \{\text{balance}(i,x)\}$$
$$\overset{I}{\Rightarrow} \{\text{balance}(i,y) \mid y = x + (\Sigma w : \text{payin}(i,w) : w) - (\Sigma w : \text{withdraw}(i,w) : w)\}$$

If D is the union of the axiom bases of both processors, then D can be specified as follows :

$$\begin{aligned}
D = \{ & \forall i : \text{inputerror}(i) \leftrightarrow \\
& \quad (\exists\, x,y : \text{invest}(i,x) \wedge \text{balance}(i,y)) \vee \\
& \quad (\exists\, x : (\text{payin}(i,x) \vee \text{withdraw}(i,x) \vee \text{terminate}(i)) \wedge \text{no–account}(i)) \vee \\
& \quad (\exists\, x,y : \text{invest}(i,x) \wedge \text{invest}(i,y) \wedge x \neq y) \vee \\
& \quad (\exists\, x,y : \text{invest}(i,x) \wedge (\text{payin}(i,y) \vee \text{withdraw}(i,y) \vee \text{terminate}(i))) \vee \\
& \quad (\exists\, x : \text{terminate}(i) \wedge (\text{payin}(i,x) \wedge \text{withdraw}(i,x))) \vee \\
& \quad (\exists\, x : \text{terminate}(i) \wedge \text{balance}(i,x) \wedge x \neq 0), \\
& \forall i,x : \text{invest}(i,x) \vee \text{payin}(i,x) \vee \text{withdraw}(i,x) \vee \text{terminate}(i) \vee \text{no–account}(i) \rightarrow \\
& \qquad i \in \mathbb{N} \wedge x \in \mathbb{N}, \\
& \forall i,x : \text{balance}(i,x) \vee \text{report}(i,x) \rightarrow i \in \mathbb{N} \wedge x \in \mathbb{Z}, \\
& \forall t : \text{doreport}(t) \rightarrow t \in \mathbb{R}^+, \\
& \forall i : \text{no–account}(i) \leftrightarrow \neg\exists x \in \mathbb{Z} : \text{balance}(i,x), \\
& \forall i,x,y : \text{balance}(i,x) \wedge \text{balance}(i,y) \rightarrow x = y, \\
& \forall i : \text{error}(i) \rightarrow i \in \mathbb{N} \}
\end{aligned}$$

One of the differences between the two specifications is that the second one imposes less constraints on the environment. The sets of events which in the first case lead to an undefined situation are accepted in the second case and imply an error message to the environment. Another difference is, that no-account(i) is a shorthand in the first specification, whereas it is a ground atom, bound to some constraints, in the second specification. We already mentioned the exact specification of the domains of the predicate symbols.

Note that in the second specification D includes

$$\forall i,x,y : \text{balance}(i,x) \wedge \text{balance}(i,y) \rightarrow x = y,$$

which is a constraint stating that for each account only one balance is valid. Of course, this constraint is met in the first specification, but this can only be easily verified due to the simple nature of the specification. In more complex cases we shall specify constraints in order to make a specification more readable, but also to make the task of proving the correctness of the specification (as mentioned in Section 3) much easier.

4.2. Queueing System

A processor will be activated instantaneously when a non-empty set of actions arrives. In practice many systems can process only one action at a time and moreover each processing takes time, so that the system is unable to handle actions arriving during the busy period. In principle it is possible to modify a given dds with instantaneously reacting processors into a system that simulates processing time and fifo-queueing order. To do this one needs to modify the processors such that they give themselves feedback events telling that the processing has finished and one has to keep an administration of the waiting actions.

Instead of modifying processors we add to the dds another dds, which is a simple dds and which behaves as a waiting room for the original dds. Since the waiting room we describe here can be defined independent of the characteristics of the original dds, the waiting room may be considered a standard dds construction.

Suppose a simple dds, called dds1, is given. Remember that every dds can be aggregated into a simple dds. The composition of dds1 and the dds, called waiting room, is also a dds. This dds is called dds2. For simplicity we assume that the action base of dds1 consists of ground atoms with only one unary predicate symbol : p. We also assume that the motor of dds1 produces upon arrival of an action a reaction that contains a ground atom ready(d), where ready is a predicate symbol not used elsewhere in dds1 and where d is an integer indicating the time needed for processing the just arrived action. This quantity d may depend on the state of dds1 and on the received action. The transfer function T interprets d as a delay. We assume that T transforms such a reaction into a pair $<ok, d>$, where ok is a ground atom of the action base of the waiting room and where d is the delay.

So we have for dds1 :

$A = \{p(\cdot)\}$

$\text{ready}(\cdot) \in R$

M:

. . .

$\models p(x) \wedge$ "d is the time of processing for $p(x)$"

$\overset{R}{\Rightarrow} \{\text{ready}(d)\}$

. . .

We can specify the waiting room, consisting of one processor and one store which have interaction, as follows :

$S = \{ps(\cdot,\cdot)\}$

$A = \{pa(\cdot), ok\}$

$R' = \{pr(\cdot)\}$

M:

$\models pa(x) \wedge \max(k) \wedge \text{rank}(x,m)$

% rank(x,m) means that in the action set there are m ground atoms with a constant smaller than or equal to x; max(k) means that in the state the maximal constant occurring in a ground atom is k, i.e. the highest scheduling number is k; if there is no ground atom in the state, then max(0) holds, i.e. the queue is empty%

$\overset{I}{\Rightarrow} \{ps(x, k+m)\}$

% this rule specifies that upon receiving an action, this action is stored with a scheduling number equal to the maximal number in the store plus the rank of the action atom in the set of atoms in the action%

$\models pa(x) \wedge \max(0) \wedge \mathrm{rank}(x, 1)$

$\overset{R}{\Rightarrow} \{pr(x)\}$

% if the queue is empty, then the action can be sent to dds1 directly%

$\models ok \wedge ps(x,k) \wedge \neg\exists y : ps(y,k-1)$

$\overset{D}{\Rightarrow} \{ps(x,k)\}$

% upon receiving an *ok* signal of dds1, the action to which the *ok* reflects should be deleted; note that $ps(x,k)$ is deleted after $p(x)$ is processed by dds1%

$\models ok \wedge ps(x,k) \wedge \exists y : ps(y,k-1) \wedge \neg\exists y : ps(y,k-2)$

$\overset{R}{\Rightarrow} \{pr(x)\}$

% if, upon receiving an *ok* signal of dds1, there is a next action, then that action is sent to dds1; note that at the same time the third rule makes that $ps(y,k-1)$ is deleted%

If D is the union of the axiom bases of dds1 and the waiting room, then D can be specified as follows :

D :

$\{\forall x : pa(x) \vee pr(x) \vee ps(x) \vee p(x) \rightarrow x \in \mathbb{N},$

$\forall d : \mathrm{ready}(d) \rightarrow d \in \mathbb{R}^+,$

$\forall y, m : \mathrm{rank}(y, m) \leftrightarrow (pa(y) \wedge (\# z : pa(z) \wedge z \le y) = m),$

$\forall k : \max(k) \leftrightarrow (\exists y : ps(y,k)) \wedge \neg(\exists y : ps(y,k+1)),$

$\max(0) \leftrightarrow \forall y, k : \neg ps(y,k)\}$

The transfer function will transform $pr(x)$ into $<p(x), \varepsilon>$. So, if 1 is the abbreviation for dds1 and w that for the waiting room, then we can state :

$\forall d : \mathrm{ready}(d) \in A \rightarrow T_{1,w}(A) = \{<ok, d>\},$

$\forall x : T_{w,1}(\{pr(x)\}) = \{<p(x), \varepsilon>\}.$

Now we have a standard dds construction for modelling a processor with non-instantaneous processing and fifo-queueing order.

4.3 Heating System

First we will describe the heating system in natural language. This example originates from the problem set for the 4th International Workshop on Software Specification and Design.

The controller of an oil hot water home heating system regulates in-flow of heat, by turning the furnace on and off. The controller activates the furnace whenever the home temperature falls below t_r - 2 degrees, where t_r is the desired temperature set by the user. The activation procedure is as follows :

(1) the controller signals the motor to be activated;

(2) the controller monitors the speed and once the speed is adequate it signals the ignition and oil valve to be activated;

(3) the controller monitors the water temperature and once the water temperature has reached a predefined value it signals the circulation valve to be opened; the heated water then starts to circulate through the house;

(4) once the home temperature reaches $t_r + 2$ degrees, the controller deactivates the furnace by first closing the oil valve and then, after 5 seconds, stopping the motor.

In addition the system is subject to the constraint that the minimum time for a furnace restart after a prior operation equals 5 minutes.

This heating system can be described with one processor and one store. The specification of the store is :

S = { furnaceoff, startingmotor, heatingwater, furnaceon, stopping, coolingdown, $t_r(\cdot)$ }

The processor, that has of course interaction with that store, can be specified as follows :

A = { hometemp($\cdot$), motorspeed($\cdot$), watertemp($\cdot$), stepstopmotor, cooldown, setdestemp($\cdot$) }

R = { checkhometemp, activatemotor, checkmotorspeed, activateignition, checkwatertemp, opencircvalve, closeoilvalve, stepstopmotor, stopmotor, cooldown, error }

M:

$\models$ in?
$\stackrel{R}{\Rightarrow}$ {error}

%in? means that the input, i.e. the action set, is not meaningful (in this state)%

$\models \neg$in? $\wedge$ furnaceoff $\wedge$ hometemp(t) $\wedge$ $t_r(t') \wedge t \geq t'-2$
$\stackrel{R}{\Rightarrow}$ {checkhometemp}

$\models \neg$in? $\wedge$ furnaceoff $\wedge$ hometemp(t) $\wedge$ $t_r(t') \wedge t < t'-2$
$\stackrel{D}{\Rightarrow}$ {furnaceoff}
$\stackrel{I}{\Rightarrow}$ {startingmotor}
$\stackrel{R}{\Rightarrow}$ {activatemotor, checkmotorspeed}

$\models \neg$in? $\wedge$ startingmotor $\wedge$ motorspeed(s) $\wedge$ $s <$ *adequatespeed*
$\stackrel{R}{\Rightarrow}$ {checkmotorspeed}

$\models \neg$in? $\wedge$ startingmotor $\wedge$ motorspeed(s) $\wedge$ $s \geq$ *adequatespeed*
$\stackrel{D}{\Rightarrow}$ {startingmotor}
$\stackrel{I}{\Rightarrow}$ {heatingwater}
$\stackrel{R}{\Rightarrow}$ {activateignition, checkwatertemp}

$\models \neg \text{in?} \wedge \text{heatingwater} \wedge \text{watertemp}(t) \wedge t < predefinedvalue$

$\overset{R}{\Rightarrow} \{\text{checkwatertemp}\}$

$\models \neg \text{in?} \wedge \text{heatingwater} \wedge \text{watertemp}(t) \wedge t \geq predefinedvalue$

$\overset{D}{\Rightarrow} \{\text{heatingwater}\}$

$\overset{I}{\Rightarrow} \{\text{furnaceon}\}$

$\overset{R}{\Rightarrow} \{\text{opencircvalve, checkhometemp}\}$

$\models \neg \text{in?} \wedge \text{furnaceon} \wedge \text{hometemp}(t) \wedge t_r(t') \wedge t \leq t' + 2$

$\overset{R}{\Rightarrow} \{\text{checkhometemp}\}$

$\models \neg \text{in?} \wedge \text{furnaceon} \wedge \text{hometemp}(t) \wedge t_r(t') \wedge t > t' + 2$

$\overset{D}{\Rightarrow} \{\text{furnaceon}\}$

$\overset{I}{\Rightarrow} \{\text{stopping}\}$

$\overset{R}{\Rightarrow} \{\text{closeoilvalve, stepstopmotor}\}$

$\models \neg \text{in?} \wedge \text{stopping} \wedge \text{stepstopmotor}$

$\overset{D}{\Rightarrow} \{\text{stopping}\}$

$\overset{I}{\Rightarrow} \{\text{coolingdown}\}$

$\overset{R}{\Rightarrow} \{\text{stopmotor, cooldown}\}$

$\models \neg \text{in?} \wedge \text{coolingdown} \wedge \text{cooldown}$

$\overset{D}{\Rightarrow} \{\text{coolingdown}\}$

$\overset{I}{\Rightarrow} \{\text{furnaceoff}\}$

$\overset{R}{\Rightarrow} \{\text{checkhometemp}\}$

$\models \text{setdestemp}(t) \wedge t_r(t')$

$\overset{D}{\Rightarrow} \{t_r(t')\}$

$\overset{I}{\Rightarrow} \{t_r(t)\}$

The transition function T can be specified as follows :

$$T(\{\text{stepstopmotor}\}) = \{<\text{stepstopmotor}, 5\ sec.>\}$$

$$T(\{\text{cooldown}\}) = \{<\text{cooldown}, 5\ min.>\}$$

The axiom base D of the processor can be specified as follows :

$$
\begin{aligned}
D = \{ & \exists\, t : t_r(t)\, , \\
& in? \leftrightarrow (\exists\, t,t' : \text{hometemp}(t) \wedge \text{hometemp}(t') \wedge t \neq t') \vee \\
& \quad (\exists\, s,s' : \text{motorspeed}(s) \wedge \text{motorspeed}(s') \wedge s \neq s') \vee \\
& \quad (\exists\, t,t' : \text{watertemp}(t) \wedge \text{watertemp}(t') \wedge t \neq t') \vee \\
& \quad (\exists\, t,t' : \text{setdestemp}(t) \wedge \text{setdestemp}(t') \wedge t \neq t') \vee \\
& \quad (\exists\, t : \text{hometemp}(t) \wedge \neg(\text{furnaceoff} \vee \text{furnaceon})) \vee \\
& \quad (\exists\, s : \text{motorspeed}(s) \wedge \neg\text{startingmotor}) \vee \\
& \quad (\exists\, t : \text{watertemp}(t) \wedge \neg\text{heatingwater}) \}
\end{aligned}
$$

When we consider this specification we can notice that we assume that the system itself is always on, i.e. there is no possibility for the user to turn off the system.

Neither do we consider errors in the operation of the system. As far as the receiving of actions is concerned we have stated in the constraints that the action sets are such that an action like hometemp(t) is only received in a state in which such an action is expected. When such an action is received in another state, then we signal this as an error. We also signal an error, whenever e.g. two different hometemperatures are sent. We assume that the system starts in a state with two ground atoms : furnaceoff and $t_r(t)$ for some temperature t. This means that at the start the furnace is off and a desired temperature is known.

Of course, we also assume that whenever a message is sent in order to learn for instance the motorspeed, then there is definitely coming a message with this information. As always with dds'ses the decision, on what to describe explicitly in the dds and what to assume for the environment of the dds, is based on the definition of the system and thus of its environment.

5. CONCLUSIONS AND FUTURE RESEARCH

A framework is developed for the formal description of systems of a large class, including information systems. In this framework data modelling and process specification are completely integrated. Hierarchical decomposition is possible.

The language for system description that we proposed is powerful, but may be replaced by others. For instance, the data modelling can be replaced by the relational model, the entity relationship model, a binary model or by a functional data model. The process modelling may be replaced by any third generation programming language or by a functional language.

The framework is also useful for the development of simulation models of physical systems. The framework may be used for formulating and proving temporal properties of a dds. This issue is a current research topic.

Another research topic is the extension of the framework to allow for the creation and starvation of copies of a dds. This extension will allow for the application of object-oriented programming techniques.

We are also studying on a software design environment based on our framework in order to be able to derive prototypes directly from high level system specifications.

REFERENCES

Chang, C.L., Lee, R.C.T. (1973). Symbolic Logic and Mechanical Theorem Proving. Academic Press, 1973.

Dietz, J.G., van Hee, K.M. (1987). A Framework for the Conceptual Modeling of Discrete Dynamic Systems. Proc. of Temporal Aspects in Information Systems, 1987.

Harel, D. (1986). Statecharts: A Visual Approach to Complex Systems. CS 86-02, The Weizmann Institute of Science, 1986.

van Hee, K.M., Houben, G.J., Dietz, J.L.G. (1987). Modelling of Discrete Dynamic Systems Framework and Examples. Eindhoven University of Technology, 1987.

Jackson, M. (1983). System Development. Prentice Hall, 1983.

Lloyd, J.W. (1984). Foundations of Logic Programming. Springer Verlag, 1984.

Reiter, R. (1984). Towards a Logical Reconstruction of Relational Databases. In M.L. Brodie, J. Mylopoulos, J.W. Schmidt (eds.), On Conceptual Modeling. Springer Verlag, 1984.

Sridhar, K.T., Hoare, C.A.R. (1985). Oxford University Computing Laboratory, 1985.

Ward, P.T., Mellor, S.J. (1985). Structured Development for Real-Time Systems. Yourdon Press, 1985.

A HIERARCHICAL FRAMEWORK FOR DISCRETE EVENT SCHEDULING IN MANUFACTURING SYSTEMS

Stanley B. Gershwin
Massachusetts Institute of Technology, Cambridge, Massachusetts, USA
Boston University, Boston, Massachusetts, USA

1. INTRODUCTION

Operating policies for manufacturing systems must respond to important discrete events that occur during production such as machine failures, setups, demand changes, expedited batches, etc. These feedback policies must be based on realistic models, and they must be computationally tractable. In this paper, we develop a class of hierarchical scheduling and planning algorithms whose structure is systematically based on the characteristics of the specific kind of production that is being controlled. The levels of the hierarchy correspond to classes of events that have distinct frequencies of occurrence.

Computational tractability is an important concern because of the complexity of the system. Even for a very small, deterministic, idealization of a production system, the computational effort for combinatorial optimization renders it impractical for real-time control. Any control scheme must be based on a simplified representation of the system and a heuristic solution of the scheduling problem.

There have been many hierarchical scheduling and planning algorithms, some quite practical and successful. However, there has been no systematic justification of this structure. The main contribution of this paper is a framework for studying and synthesizing such a structure.

This work extends a formulation by Kimemia and Gershwin (1983) in which only two kinds of events were considered: production operations on parts and failures and repairs of machines. Operations occurred much more often than failures, and this allowed the use of a continuous representation of material flow. A dynamic programming formulation led naturally to a feedback control policy. The state of the system had two parts: a vector of real numbers represented the surplus, the cumulative difference between production and requirements. The discrete part of the state represented the set of machines that are operational. The object was to choose the production rate vector as a function of the state to keep the surplus near 0.

The production rate (the continuous control variable) was restricted by linear inequality constraints that depended on the repair state. They represented the instantaneous capacity of the system, and they expressed the idea that no machine, while it is operational, may be busy more than 100% of the time; and no machine, while it is not operational, may be used at all. The present paper describes the extension of this work to the widest possible variety of phenomena and decisions in a manufacturing environment.

Figure 1.1 illustrates some of the issues that are considered here. It is a graph of the cumulative production and demand for one part type (j) among many that share one machine. A long term production rate (u_j^1) is specified for this part type, and its integral is represented by the solid straight line. It is not possible to follow this line exactly because the machine is set up for Type j parts only during a set of time intervals. During such intervals, the medium term production rate u_j^2 must be greater than u_j^1, because during the other intervals -- while it is set up for other parts -- u_j^2 is 0. The integral of u_j^2 (the dashed line) is staircase-like, close to the integral of u_j^1.

The dashed line cannot be realized either. The machine is unreliable, and while it is down, its production rate u_j^3 is 0. Consequently, while it is up and set up for Type j, it must be operated at a short term rate u_j^3 greater than that of the dashed line. The dotted line, which represents this phenomenon, is again staircase-like, and is close to the dashed line. Finally, the actual cumulative production graph (which requires too much resolution to be plotted) is a true staircase. It has vertical steps at the instants when parts are loaded, and it is flat otherwise. It is very close to the dotted line.

This paper formalizes this hierarchy, and extends it to an arbitrary number of levels and several machines.

Literature Survey

There is a large literature in scheduling (Graves, 1981). Many papers are based on combinatorial optimization/integer programming methods (Lageweg, Lenstra, and Rinnooy Kan, 1977 and 1978; Papadimitriou and Kannelakis, 1980) or mixed integer methods (Afentakis, Gavish, and Karmarkar, 1984; Newhouse, 1975a and 1975b; Wagner and Whitin, 1958). Because of the difficulty of the problem, authors are limited to analyzing computational complexity, or proposing and analyzing heuristics.

An important class of problem formulations is that of hierarchical structure (Bitran, Haas, and Hax, 1981; Dempster et al., 1981; Graves, 1982; Hax and Meal, 1975; and others). The goal is to replace one large problem by a set of many small ones because the latter is invariably easier to solve. These methods are often used but there is no general, systematic way of synthesizing hierarchies for large classes of stochastic scheduling problems.

Multiple time scale problems have recently been studied in the control theory (Saksena, O'Reilly, and Kokotovic, 1984) and Markov chain literature (Delebecque, Quadrat, and Kokotovic, 1984). We use insights from these methods to develop a systematic justification for hierarchical analysis. This paper also makes use of, and extends the work of Kimemia and Gershwin (1983). A recent survey (Maimon and Gershwin, 1987) describes this and several related papers.

Outline

Section 2 describes the manufacturing systems that we are considering. It establishes terminology and discusses the basic concepts for the present approach: capacity and frequency separation. Section 3 builds on the frequency separation to derive a small set of results that form the foundation of the hierarchy. Control in the hierarchy is described in detail in Section 4. Sections 5 and 6 present the two building blocks: the staircase strategy and the hedging point strategy. A simple example appears in Section 7, and conclusions are drawn in Section 8.

2. PRODUCTION EVENTS AND CAPACITY

In this section, we discuss the discrete events that occur during the production process. We define terminology to help describe these events. We categorize events in two ways: the frequency with which they occur; and the degree of control that decision-makers can exert over them. We define capacity, and show how capacity is affected by production events.

2.1 Definitions

A *resource* is any part of the production system that is not consumed or transformed during the production process. Machines -- both material transformation and inspection machines, workers, pallets, and sometimes tools -- if we ignore wear or breakage -- can be modeled as resources. Workpieces and processing chemicals cannot.

An *activity* is a pair of events associated with a resource. The first event corresponds to the start of the activity, and the second is the end of the activity. Only one activity can appear at a resource at any time. For example, the *operation* of drilling the 3/8" hole in part type 12, serial number 947882927 that started at 10:03 this morning and ended at 10:07 is an activity. Other examples include machine failures, setups (i.e., changes of tooling, etc.), preventative maintenance, routine calibration, inspection, and training sessions. We use the same term to refer to a set of such pairs of events. For example, drilling 3/8" holes in type 12 parts is an activity; specifically, an operation.

Let i be a resource and j an activity. Define $\alpha_{ij}(t)$ to be the state of resource i. This is a binary variable which is 1 if resource i is occupied by activity j at time t, and 0 otherwise. Since at most only one activity may be present at a resource at a given time,

$$\sum_j \alpha_{ij}(t) \leq 1. \text{ for all } i \tag{1}$$

Every activity has a *frequency* and a *duration*. To define frequency, let $N_{ij}(T)$ be the total number of times that resource i is occupied by activity j in (0,T). Then define

$$u_{ij} = \frac{1}{T} N_{ij}(T). \tag{2}$$

This is the frequency with which type j activities occur at resource i.

In the following, we do not indicate a resource (i) explicitly in the subscript of u. This allows the flexibility of either considering the index j to include a specific resource (in which case j might mean "operation 30 on a Type 12 part at Machine 5") or any resource (in which case j might mean "the required operation on a Type 12 part at the current machine"). When u has only the j subscript, (2) holds only when activity j actually takes place at resource i. If it does not take place at resource i, (2) is meaningless, and if it takes place at more than one resource i, it must hold for each i. (This implies a "conservation of flow" condition, since $u_j = u_{1j} = u_{2j}$ if j goes to both resource 1 and resource 2.)

The vector u is the *activity rate vector*. It satisfies $u_j \geq 0$. Let τ_{ij} be the average duration of activity j at resource i. Then τ is the *activity duration matrix*. It satisfies $\tau_{ij} \geq 0$. (We can now say that (2) holds only when $\tau_{ij} > 0$.) Durations may be random or deterministic, but we assume that they are not under the control of the decision-maker.

Observation: If the system is in steady state,

$$\tau_{ij} u_j = E\alpha_{ij} \tag{3}$$

Proof:

Consider a sample history of the system. The total time that resource i is occupied by activity j in (0,T) is

$$\int_0^T \alpha_{ij}(t)dt. \tag{4}$$

The average duration satisfies

$$\tau_{ij} = \frac{\int_0^T \alpha_{ij}(t)dt}{N_{ij}(T)} = \frac{\frac{1}{T}\int_0^T \alpha_{ij}(t)dt}{u_j} . \tag{5}$$

If the system is in steady state, then the time average of a quantity is the same as its expected value, so the numerator is $E\alpha_{ij}$ and (3) is proved. (This can also be viewed as an instance of Little's law.) The assumption that the system is in steady state is an important one. In later sections, the dynamics of the system is divided into subsets, each considered over different time scales. Each subset has a different time period for steady state

Since only one activity may occur at a resource at one time, the fraction of resource i's time that is spent on activity j is $\tau_{ij}u_j$. This is called the *occupation* of resource i by activity j.

Example: Type 1 parts arrive at Machine 1 at a rate of 1 per hour (u_1). They undergo operations that take 20 minutes (τ_{11}). Therefore Machine 1 is occupied by making Type 1 parts for 1/3 of its time.

2.2 Capacity

From (1),

$$1 \geq E\sum_j \alpha_{ij}(t) = \sum_j \tau_{ij}u_j \text{ for all resources i.} \tag{6}$$

This is the fundamental capacity limitation: no resource can be occupied more than 100% of the time.

Example: In addition to the Type 1 parts, we wish to send Type 2 parts to Machine 1 for an operation that takes 25 minutes (τ_{12}). There is a demand of one Type 2 part every 35 minutes (u_2). This is not possible because it violates (6).

The set of all activity rate vectors u that satisfies (6) is the *capacity set* Ω. It is important to observe that *capacity is a set* -- a polyhedron -- and not a scalar. Here we have defined capacity as a constant set. In later sections, capacity is described as a function of the state of the system. This means that *capacity is a* time-varying, *stochastic set.*

2.3 Frequency Separation

Dynamic models always have two parts: a constant part and a time-varying part. In all dynamic models, there is something that is treated as unchanging over time: some parameters, and, most often, the structure of the model. For example, the model described in Sections 2.1 and 2.2 is a conventional one in which there are static quantities (u_j, τ_{ij}), a static structure, and dynamic quantities ($\alpha_{ij}(t)$, $N_{ij}(t)$).

Recently, the dichotomy between static and dynamic has been extended to systems with multiple time scales, modeled as differential equations or Markov chains. At one end of the scale, there are quantities that are treated as static. The other variables are divided into groups according to the speed of their dynamics. Because of this grouping, it is possible to simplify the computation of the behavior of these systems. Approximate but accurate techniques have been developed to calculate the effects of the slower and faster dynamics of adjacent groups on each group of variables.

The essential idea is: when treating any dynamic quantity, treat quantities that vary much more slowly as static; and model quantities that vary much faster in a way that ignores the details of their variations (such as replacing fast-moving quantities by their averages; or by Brownian noises with appropriate parameters.) This is the central assumption of the hierarchical decomposition presented here.

Assumption 1: The activities can be grouped into sets J_1, J_2, ... such that for each set J_k, there is a characteristic frequency f_k satisfying

$$f_1 \ll f_2 \ll \dots \ll f_k \ll f_{k+1} \ll \dots \tag{7}$$

The activity rates satisfy

$$j \in J_k \Rightarrow f_{k-1} \ll u_j \ll f_{k+1}. \tag{8}$$

Figure 2.1 represents two kinds of production that satisfy this assumption. The horizontal axis represents frequency and the vertical axis represents occupation of some critical resource. Because of Assumption 1, all the event frequencies occur at distinct clusters.

The time period over which a component of the system reaches steady state depends on the frequency classes of the activities that affect that component. It is on the order of $1/f_{k-1}$ if the lowest frequency activity is a member of J_k.

A capacity set can be associated with each time scale k. Consequently, *capacity is a set of time-varying, stochastic sets.*

2.4 Slow variation

In 2.1 and 2.2, u_j is treated as constant. However, it is convenient to allow u_j to be slowly varying. That is, u_j is not constant, but it changes slowly compared to the changes in α_{ij}. An important special case is where u_j is piecewise constant, and its changes occur much less often than those of α_{ij}. Equation (3) is now

$$\tau_{ij}\, u_j(t) = E\alpha_{ij}(t). \tag{9}$$

This is established in the same manner as (3), but the bounds of the integral (4) are t_1 and t, where t_1 is the time of the most recent change in u_j, and t is the current time. The quantity $u_j(t)$ satisfies

$$N_{ij}(t) = \int_0^t u_j(s)ds \text{ for } \tau_{ij} > 0, \text{ or}$$

$$N_{ij}(t) - N_{ij}(t_1) = \int_{t_1}^t u_j(s)ds = (t - t_1)u_j(t_1). \tag{10}$$

The assumption here is that many occupations of resource i by activity j occur in the interval (t_1, t): enough so that

$$E\alpha_{ij}(t) = \frac{1}{t-t_1}\int_{t_1}^t \alpha_{ij}(s)ds. \tag{11}$$

2.5 Degree of control

Events may or may not be under the control of the decision-maker. Figures 2.2-2.4 represent a variety of activities with different degrees of control. Figure 2.2 shows the two repair states of a machine: operational and down. In this case, the times at which the transitions occur are beyond the control of the production personnel.

Figure 2.3 represents the operation states of a flexible machine. It can work on a family of four parts, and setup is not required. That is, after doing an operation on one part, the time required to do an operation on another part depends only on the new part, and not the identity of the part that preceded it. While the machine is in the idle state, it may be used to do an operation on any of the parts. When to make the transition, and what state to visit next, are entirely at the discretion of the manager. Once that decision has been made, however, the manager loses control. The time required to perform the operation may or may not be known, but it cannot be chosen, and the next state must be the idle state.

Figure 2.4 displays the configuration states of a machine which can do operations on three families of parts. There is a substantial setup time to switch the machine from operations on one family to another. While the system is set up for any one family, it can remain that way indefinitely. The manager can choose when to switch out of the current family and which family to switch into next. However, the system then goes to the appropriate setup state. (While it is there, tools are changed, calibration is performed, test parts are made, etc.) It stays in that state for a length of time which is not under the control of the manager. (Again, it may or may not be known, but it cannot be chosen.) After that, the system goes to the new family state, and the series of events repeats.

2.6 Effects of events

The goal of the factory is to produce in a way that satisfies demand at least cost. The only events that directly further this goal are the production events, and only if they are chosen correctly. The direct effects of all the other events work against this goal.

When any activity occurs, it prevents all other activities from occurring at the same resource. Thus a low frequency, high occupation activity is a major disruption to the system. During such an activity, the resource it occupies is unavailable for a very long time (as seen by the high frequency events). This may not simply shut down all production; instead, it may temporarily restrict only some kinds of production. Such disruptions greatly complicate the scheduling problem.

2.7 Purpose of the decomposition

It is possible to represent the scheduling problem as an integer programming problem, particularly if time is discretized. However, this almost always leads to a problem which cannot practically be solved even in the absence of random events. The goal of the approach described below is to formulate the problem in a way that will provide an approximate feedback solution for the stochastic scheduling problem.

The solution approach is based on a reformulation of the problem in which the large set of binary variables that indicate the precise times when events occur is replaced by a small set of real variables representing the rates that events occur. This is a good approximation because of the large difference in frequencies of these events. Eventually, the binary variables are calculated, but by a much simplified procedure.

3. THE SPECTRUM AND THE HIERARCHY

In this section, we define the variables of the hierarchy and what calculations take place at each of the levels. In the following sections, we propose problem formulations for those calculations.

3.1 Definitions

The structure of the hierarchy is based on Assumption 1: that events tend to occur on a discrete spectrum. Classes of events have frequencies that cluster near discrete points on the spectrum. The control hierarchy is tied to the spectrum. Each level k in the hierarchy corresponds to a discrete point on the spectrum and thus to a set of activities. This point is the characteristic frequency f_k (and $1/f_k$ is the characteristic time scale) of those activities.

At each level of the hierarchy, events that correspond to higher levels (i.e., lower frequencies, and lower values of k) can be treated as discrete and constant or slowly varying. Events that correspond to lower levels can described by continuous (real) variables. These variables can be treated as though they are deterministic.

The approach is to define a set of rate or frequency variables for every activity. These quantities represent the behavior of the system in an aggregated way. At each level, we calculate optimal values for those aggregate variables. Optimal, here, means that they must be close, on the average, to the corresponding values chosen at the higher levels. However, they must respond to events that occur at their level.

Define the *level* $L(j)$ of activity j to be the value of k in Assumption 1 associated with this activity. That is,

$$L(j) = k \text{ if } j \in J_k \tag{12}$$

in (8). We choose the convention that less frequent activities are higher level activities and have lower values of k; lower levels have higher values of k.

In the following we define Level k quantities. These are values of system states as perceived by an observer who is not able to distinguish individual events -- that is, changes in α -- that happen much more frequently than f_k. The frequencies of high frequency events, as perceived by this observer, depend on the current states of low frequency activities, and expectations must similarly be conditioned on the current states of low frequency activities.

Let $\alpha_{ij}^k(t)$ be the *level k state of resource i.* This is defined only for activities j whose level is k or higher. If $L(j) \le k$,

$$\alpha_{ij}^k(t) = \alpha_{ij}(t). \tag{13}$$

Define α^k as the vector whose components are α_{ij}^k. Its dimensionality depends on k.

Let E_k be the *level k expectation operator.* It is the conditional expectation, given that all level m quantities ($m \le k$) remain constant at their values at time t. That is, we treat $\alpha_{ij}^m(t)$ as constant.

Let u_j^k be the *level k rate of activity j.* This is defined only if the level of activity j is lower than k, that is, $L(j) > k$. The level k rate of activity j is the frequency that a level k observer would measure that activity j occurs while all level m events ($m \le k$) are held constant at their current values. This rate satisfies

$$u_j^k(t) = \frac{E_k \alpha_{ij}(t)}{\tau_{ij}} \tag{14}$$

and

$$u_j^k(t) \ge 0. \tag{15}$$

The conditioning event of E_k is a subset of that of E_{k-1}. This is because the set of quantities held constant for E_{k-1} is a subset of that for E_k. Consequently,

$$E_{k-1} E_k \alpha_{ij} = E_{k-1} \alpha_{ij}. \tag{16}$$

Taking the level k-1 expectation of (14):

$$E_{k-1} u_j^k = E_{k-1} \frac{E_k \alpha_{ij}(t)}{\tau_{ij}}. \tag{17}$$

But this is equal to $\frac{E_{k-1}\alpha_{ij}(t)}{\tau_{ij}}$ according to (16). This implies that

$$E_{k-1}u_j^k = u_j^{k-1}. \tag{18}$$

That is, the level k-1 rate of an activity is the level k-1 expectation of the level k rate of the activity. This is a very important observation, because it relates quantities at different levels of the hierarchy.

If $L(j) > k$, level k of the hierarchy calculates u_j^k. How that calculation is performed depends on the degree of control of activity j. If activity j can be initiated by the decision-maker rather than by nature, then u_j^k is chosen to satisfy (18).

All activities j appear in three different guises in the hierarchy. At their own level ($k=L(j)$), they appear as pairs of discrete events (the start and the end of the activity). This is, of course, exactly what they are. No approximate representation is possible. At higher levels in the hierarchy ($k<L(j)$), however, their details are ignored, and they are represented by rates (u_j^k). At lower levels ($k>L(j)$), they are treated as constant at their current values.

Controllable activities are chosen from top down. That is, a rate u_j^1 is chosen initially. Then ($k>1$) is chosen to satisfy (18) and other conditions (according to the *hedging point strategy* of Section 6) for increasing values of k until $k = L(j)$. At that point, α_{ij} is chosen to satisfy (14) according to the *staircase strategy* described in Section 5.

On the other hand, (14) and (18) have different interpretations when activity j is not controllable (for example, machine failures). In that case, the expectations are statistical operations, in which data are collected and sample means are found. The rate $u_j^{L(j)-1}$ is calculated from (14) by observing the values of α_{ij}. If $L(j)<k-1$, (18) is repeated for decreasing values of k.

3.2 Capacity in the hierarchy

For each k, the sum in (1) can be broken into two parts:

$$\sum_{L(j)>k} \alpha_{ij} \le 1 - \sum_{L(j)\le k} \alpha_{ij}^k \tag{19}$$

in which (13) is applied to the high-level sum on the right side. If we take a level k expectation of (19), the right side is not affected. From (14),

$$\sum_{L(j)>k} \tau_{ij}u_j^k \le 1 - \sum_{L(j)\le k} \alpha_{ij}^k \tag{20}$$

This equation is the basic statement of capacity in the hierarchy. It limits the rates at which low level events can occur as a function of the current states of high level events. If any high level activity is currently at resource i, that resource is not available for any low level events. In that case, the right side of (20) is 0 and all u_j^k that have a positive coefficient must be zero. If none of the high level activities in (20) are currently taking place, this inequality becomes

$$\sum_{L(j)>k} \tau_{ij}u_j^k \le 1. \tag{21}$$

Capacity is thus a function of hierarchy level and, since it depends on the state of the system, a stochastic function of time. We define the *level k capacity set* as

$$\Omega^k(\alpha^k) = \left\{ u^k \,\middle|\, \sum_{L(j)>k} \tau_{ij}u_j^k \le 1 - \sum_{L(j)\le k} \alpha_{ij}^k \quad \forall\ i; \qquad u_j^k \ge 0 \quad \forall\ j,\ L(j)>k \right\}. \tag{22}$$

This set is the constraint on the hedging point strategy. It limits the choice of rates u^k as a function of the current state of the system. Note that the condition

$$u^k \in \Omega^k(\alpha^k) \tag{23}$$

is a necessary but not sufficient condition. That is, $\Omega^k(\alpha^k)$ was constructed so that every sequence of events must satisfy (23). However, we have not demonstrated that for every u^k that satisfies (23) there corresponds a feasible sequence of events. We assume sufficiency in the following, however.

4. CONTROL IN THE HIERARCHY

The goal of the hierarchical scheduler is to select a time to initiate each controllable event. This is performed by solving one or two problems at each level k. We emphasize control -- i.e., scheduling and planning -- here. Data-gathering and processing is also an important function of the hierarchy, but is not discussed in this paper. The hierarchy is illustrated in Figure 4.1.

Problem 1: (The hedging point strategy)

Find u_j^k (for all j, $L(j)>k$) satisfying (18) and (20) (and possibly other conditions).

Problem 2: (The staircase strategy)

Find α_{ij}^k (for all j, $L(j)=k$) satisfying

$$E_{k-1}\alpha_{ij}^k = \tau_{ij}u_j^{k-1} \tag{24}$$

(and possibly other conditions).

At the top level of the hierarchy ($k=1$), required rates of some of the controllable activities are specified, for example, production rates and maintenance frequencies. Other rates may not be specified, such as setup frequencies. We assume that rates of uncontrollable events are known. The frequency associated with the top level is 0. Consequently, there is no Problem 2 at that level, and Problem 1 reduces to a static optimization. The function of Problem 1 here is to choose all the rates that were not specified. The vector u^1 is the target rate vector for level 2.

If there are any controllable events at level $k > 1$, we solve Problem 2. (An example is the change in setup of a machine.) Controllable events are thereby initiated in such a way that their rates of occurrence are close to the target rates that are determined at level k-1.

Then we solve Problem 1 to determine the level k rates of occurrence u_j^k of all activities j whose frequencies are much higher than f_k. These rates are refinements of the target rates determined at level k-1: u_j^{k-1}. They differ from the higher level rates in that they are affected by the level k discrete events. These events, if they are controllable, were chosen by Problem 2 at this level. However, even if the level k events are not controllable, the level k rates differ from the higher level rates. These rates are then the targets for level k+1.

For example, if at level k we choose setup times, the production rates must be calculated so that they are appropriate for the current setup. If we are making Type 1 parts at the rate of 4 per day, but the necessary machine is only set up for that part on Tuesdays, then we must work at a rate of 20 per day on Tuesday and 0 Type 1 parts per day during the rest of the week.

Similarly, the activities associated with level k may not be controllable, such as machine failures. It is still necessary to refine the production rates. If the overall requirements for Type 1 parts are 20 per day, and the machine is down 10% of the time, *and failures occur several times per day*, then the appropriate strategy is to operate the machine at a rate of 22.2 parts per day while it is up. Note that this only makes sense if failures are much more frequent than setups and much less frequent than operations. If not, related but different calculations must be performed in a different order. That is, a different hierarchy is appropriate.

An important feature of this hierarchy is that rates u_j^k are always chosen to be within the current capacity of the system. When a level m event occurs ($m \leq k$), the capacity set (22) changes. Problem 2 is then recalculated so that the new rates remain feasible. As mentioned earlier, this is necessary for feasibility. In all the simulation experiments that we have performed, it appeared to be sufficient as well.

5. THE STAIRCASE STRATEGY

The staircase strategy was introduced by Gershwin, Akella, and Choong (1985) and Akella, Choong, and Gershwin (1984), although stated somewhat differently from here. It was used to load parts in a simulation of a flexible manufacturing system.

Instead of treating the statement of Problem 2 in Section 4 directly, we choose starting times for events α_{ij}^k to satisfy (2), or rather

$$N_{ij}^k(t) \sim \int_0^t u_j^k(s)ds \tag{25}$$

where $N_{ij}^k(t)$ is the number of times activity j occurs at resource i during [0,t]. This expression is only approximate because the left side is an integer and the right side is a real number. The objective is to develop an algorithm which keeps the error in (25) less than 1. This is because, approximately,

$$E_{k-1}\alpha_{ij}^k(t) = \frac{1}{t-t_1}\int_{t_1}^t \alpha_{ij}(s)ds = \frac{1}{t-t_1}\Big(N_{ij}^k(t)-N_{ij}^k(t_1) \Big)\tau_{ij} \tag{26}$$

in steady state. If the times to start activities are chosen to satisfy (25), then

$$E_{k-1}\alpha_{ij}^k(t) = \tau_{ij}u_j^k(t_1) \tag{27}$$

The difference between (25) and (24) is that a simple algorithm can be devised to implement (25). It is called the *staircase strategy* because of the graph of $N_{ij}^k(T)$.

Staircase strategy: For all activities j such that L(j) = k, perform activity j at resource i as early as possible after the *eligibility rule* is satisfied.

$$\textit{Eligibility rule: } N_{ij}^k(T) \leq \int_0^T u_j^k dt \tag{28}$$

If there were only one activity in the system, it would be initiated as soon as (28) were satisfied with equality. Immediately afterward, the left side of (25) would exceed the right side by exactly 1. The difference would then start to diminish until, again, (28) is satisfied with equality. Thus, the error in (25) would never grow larger than 1. Figure 5.1 represents this strategy, and illustrates the term "staircase." The solid line represents the right side of (28), and the dashed line represents the left side. Note that the change in slope of the solid line poses no difficulties for this strategy.

Example: If activity j is an operation on Type A parts at Machine 6, attempt to load a Type A part into the machine whenever (28) is satisfied.

In reality, there are two complications. First, because there are other activities, activity j may not be the only one to satisfy (28) at any instant. Therefore, there must be a mechanism or an additional eligibility rule for selecting one. Consequently, we can no longer assert that (25) is satisfied with an error no larger than 1.

Second, there are relationships among activities other than non-simultaneity. For example, some manufacturing operations may not be performed unless the system is set up in a certain way. That is, in order to perform an operation on Type 1 parts, the system must be set up for them. The most recent setup activity must have been one that is appropriate for Type 1 parts. This leads to additional eligibility rules.

Example: If activity j is an operation on Type A parts at Machine 6, attempt to load a Type A part into the machine whenever (28) is satisfied and Machine 6 is set up for Type A and the part that has been waiting longest for Machine 6 that can be produced in its current configuration is Type A.

Methods for implementing this strategy can be developed based on the methods of Ramadge and Wonham (1985), Maimon and Tadmor (1986).

6. THE HEDGING POINT STRATEGY

The hedging point strategy was introduced by Kimemia and Gershwin (1983) and refined by Gershwin et al. (1985) for a restricted version of the scheduling problem discussed here. In that problem, there were only two activities: operations and failures. The hedging point strategy was used to calculate the production rates of parts in response to repairs and failures of machines.

In the present context, the purpose of the problem is to find u_j^k (for all j such that $L(j) > k$) to satisfy (18) and (23) (and possibly other conditions). That is, we find the optimum frequencies of controllable events whose frequencies are much higher than f_k. These frequencies are chosen in response to changes in low frequency activities: those whose values change at a frequency roughly f_k or slower.

6.1 Surplus

We introduce x_j^k, the *activity j surplus*. This quantity represents the excess of occurrences of activity j as determined by u_j^k over the number of occurrences required by u_j^{k-1}. The surplus is illustrated in Figure 6.1. It satisfies

$$x_j^k(t) = \int_0^t u_j^k(s)ds - \int_0^t u_j^{k-1}(s)ds \tag{29}$$

or

$$\frac{dx_j^k}{dt} = u_j^k - u_j^{k-1}. \tag{30}$$

To keep u_j^k near u_j^{k-1}, we must keep x_j^k near 0. We therefore define a strictly convex function g such that $g(0) = 0$; $g(x) \geq 0 \; \forall \; x$; and $\lim_{||x||\to\infty} g(x) = \infty$; and we seek u_j^k to minimize

$$E_{k-1}\int_0^T g(x_j^k(t))dt \tag{31}$$

in which T is long enough so that the dynamic programming problem has a time-invariant solution $u_j^k(x^k,\alpha^k)$. Thus T is much greater than $1/f_k$. If (31) is small, then $x_j^k(t)$ must be small for all t. Equation (29) then implies that $u_j^k(t)$ is near u_j^{k-1}.

6.2 Capacity constraints

The activity rate vector $u^k(t)$ must satisfy the stochastic capacity constraints

$$u^k(t) \in \Omega^k(\alpha^k(t)) \tag{32}$$

where $\Omega^k(\alpha^k(t))$ is given by (22). This means that the activity rates of all high frequency activities are restricted in a way that depends on the current states of activities whose frequencies are roughly f_k or less. Those whose frequencies are much less than f_k can be treated as constant at their present values, but the variations of those that change at a frequency comparable to f_k must be considered.

Because Kimemia and Gershwin were dealing with machine failures and repairs, they could treat $\alpha^k(t)$ as the state of a Markov process. Here, however, some components of $\alpha^k(t)$ are chosen by the scheduler according to the staircase strategy of Section 5. For the purpose of determining the frequencies of high-frequency activities, we treat $\alpha^k(t)$ as though it is generated by some exogenous stochastic process with transition rate matrix λ^k:

$$\lambda^k_{\alpha\beta}\delta t = \text{prob}\ \{\alpha^k(t+\delta t)=\beta \mid \alpha^k(t)=\alpha\},\ \alpha\neq\beta\ ; \qquad \lambda^k_{\alpha\alpha} = -\sum_{\beta\neq\alpha}\lambda^k_{\alpha\beta}. \tag{33}$$

By treating all level k events this way, we are ignoring information that could be used, in principal, to improve the performance function (31). Since the time for the next event is known and not random, the optimal trajectory should be different. This requires further study.

6.3 Other constraints

Some activities are non-controllable, such as machine failures. Their frequencies cannot be chosen; they are given quantities. Thus, if $\mathcal{N}$ is the set of uncontrollable activities,

$$u^k_j \text{ specified, } j \in \mathcal{N}. \tag{34}$$

Other activities require special constraints because of their special nature. For example, when a resource may have more than one configuration, and setups require significant time, setup frequencies are constrained to satisfy a set of equality constraints. Assume resource i has configurations 1, ..., C(i). Denote $j=(iab)$ as the activity of changing the configuration of resource i from a to b. Then u^k_{iab} is the level k frequency of changing the configuration of resource i from a to b. These frequencies must satisfy

$$\sum_a u^k_{iab} = \sum_a u^k_{iba}. \tag{35}$$

since the frequency of changing into setup b must be the same as the frequency of switching out of setup b. Related formulations appear in Gershwin (1986) and Choong (1987).

We summarize all such miscellaneous constraints as

$$m(u^k(t)) = 0. \tag{36}$$

6.4 Problem statement

Here we present a compact statement of the problem. It is a dynamic programming problem whose states are $x^k(t)$ and $\alpha^k(t)$ and whose control is $u^k(t)$. (The rates u^{k-1} are treated as exogenous constants.)

Find the feedback control law $u^k(x^k(t),\alpha^k(t),t)$ to minimize (31) subject to (30), (32), and (36) in which α^k is the state of an exogenous Markov process, with parameters λ^k. The initial conditions at $t=0$ are $x^k(0)$, $\alpha^k(0)$. T is very large.

6.5 Solution

Kimemia and Gershwin (1983) derived a Bellman's equation for this problem:

$$0 = \min \left\{ g(x^k) + \frac{\partial J}{\partial x}(u^k - u^{k-1}) + \frac{\partial J}{\partial t} + \sum_{\beta} \lambda_{\alpha\beta} J[x^k,\beta,t] \right\} \tag{37}$$

in which $J[x^k(t),\alpha^k(t),t]$ is the cost-to-go function, the cost incurred during (t,T) if the initial conditions are $x^k(t)$ and $\alpha^k(t)$ at time t. The minimization in (37) is performed at every t subject to (32) and (36). If such a J function could be found to satisfy this nonlinear partial differential equation, the optimal control u^k could be determined from the indicated minimization.

If J were known, determining u^k would reduce to solving

$$\left. \begin{array}{l} \min \frac{\partial J}{\partial x} u^k \\ \text{subject to (32) and (36).} \end{array} \right\} \tag{38}$$

If $m(u^k)$ is a linear function, this is a linear programming problem.

Akella and Kumar (1986), Bielecki and Kumar (1987), and Sharifnia (1987) have obtained analytic solutions for versions of this problem in which x^k and u^k are scalars. In no other cases are exact solutions to this problem known. Numerical solutions are equally unavailable because of the "curse of dimensionality." To overcome this difficulty, Akella et al. (1984) show that a quadratic approximation of J can produce excellent performance.

Kimemia and Gershwin ran several simulations to test a simple hierarchical policy: solve (38) at every time instant to determine u^k, and then load parts (in a manner somewhat more complex and less effective than the staircase strategy of Section 5) so that the rate of loading parts was close to u^k. This worked well until the solution of (38) changed abruptly. (This is an important possibility since (38) is a linear program.) Very often, it changed abruptly again at the next time instant, and this led to reduced performance.

Gershwin et al. (1985) avoided this chattering by observing a behavior similar to that of a closely related problem of Rishel (1975). The continuous part of the state, x^k, is restricted to reduced dimensional surfaces whenever u^k would otherwise chatter. In the present problem, chattering is avoided by adding linear equality constraints to (38) whenever x^k reaches certain planes.

This step has the additional benefit of reducing computational effort. It is no longer necessary to solve (38) at every time instant. Instead, a series of computations is performed at every time t_c when there is a change in α^k. At those instants (38) is solved, and then solved repeatedly with additional constraints, as described above. The outcome of these calculations is a piecewise constant function of t, $u_j^k(t;\alpha^k(t_c^+))$, defined for $t>t_c$. This function is the set of target rates for level k+1. When α^k changes, the function is recalculated.

There are two kinds of states α^k: feasible and infeasible. *Feasible* states are those for which $u^{k-1} \in \Omega^k(\alpha^k(t))$. All other states are infeasible. If α^k is feasible and constant for a long enough period, the strategy drives x^k to the value that minimizes $J(x^k,\alpha^k,t)$. In steady state, this is a constant which we call the *hedging point*. We have assumed that T is large enough so that the system can be assumed to be in steady state.

The hedging point represents a safety level of the surplus. Infeasible states are certain to occur eventually. While α^k is infeasible, x^k must decrease, and possibly become negative. The hedging point represents a compromise between a cost for positive x^k and a cost for negative x^k. When the activities considered are production operations on parts, for example, the tradeoff is between production that is ahead of demand (and therefore can lead to inventory) and production that is behind demand (and therefore leads to starved downstream resources or unhappy customers). The hedging point need not be positive. Bielecki and Kumar show that it can sometimes be 0.

7. SIMPLE EXAMPLE

In this section, we illustrate the ideas developed in this paper with a two-part, two-machine system. There are only two phenomena in this system: failures and operations. The former are much less frequent, but of much greater duration, than the latter. This is an example of the methods of Kimemia and Gershwin (1983) and Gershwin, Akella, and Choong (1985). An extension of this system, in which setup plays a role, is described in Gershwin (1987).

7.1 Description of System

Figure 7.1 illustrates the two-machine system. In this system, Machine 1 is perfectly flexible. That is, it can do operations on either part type, without time lost for changeover. It is unreliable, however: it fails at random times and stays down for random lengths of time Machine 2 is perfectly reliable, but totally inflexible. It can only make Type 1 parts. Thus Machine 1 is shared among the two part types and Machine 2 is devoted entirely to Type 1.

The data that are specified are the demand rates for the parts, the failure (p) and repair (r) rates, and the durations of the operations (τ_{11}, τ_{12}, and τ_{21}, where τ_{ij} is the duration of an operation on a Type j part at Machine i). To simplify the problem, we assume that the demand rate for Type 1 parts is broken down by the machine at which the operation is performed, so that the specified demand rates are d_{11}, $d_{12}=d_2$, and d_{21}.

For this problem, Assumption 1 becomes:

> τ_{11}, τ_{12}, τ_{21}, $1/d_{11}$, $1/d_2$, $1/d_{21}$ are the same order of magnitude.
>
> These quantities are all smaller than $1/r$, $1/p$, which are the same order of magnitude.

7.2 Level 1: Hedging point strategy

The states of the system are α, the repair state of Machine 1, an exogenous random variable; and x_{11}, x_{12}, and x_{21}, the surpluses. The control variables are u_{ij}, the level 1 flow rate of Type j parts to Machine i (ij = 11, 12, 21).

Here, (30) becomes

$$\dot{x}_{ij} = u_{ij} - d_{ij} \text{ for } ij = 11, 12, \text{ and } 21. \quad (39)$$

The linear programming problem of Section 6.5, which determines u_{ij}, becomes

$$\min_{u_{ij}} \sum c_{ij}(x,\alpha)\, u_{ij} \quad (40)$$

subject to:

$$\tau_{11}u_{11} + \tau_{12}u_{12} \le \alpha \quad (41)$$

$$\tau_{21}u_{21} \le 1 \quad (42)$$

$$u_{ij} \ge 0 \quad (43)$$

where for ij = 11, 12, and 21 and for mn = 11, 12, and 21,

$$c_{ij}(x,\alpha) = \sum_{mn} A_{ijmn}(\alpha)\, x_{mn} + b_{ij}(\alpha). \quad (44)$$

Here, $c(x,\alpha)$ is the approximation of $\partial J/\partial x$. Satisfactory results have been obtained with diagonal A matrices, so we choose $A_{ijmn} = 0$ if $(mn) \neq (ij)$. The hedging point is then

$$z_{ij}(\alpha) = -\frac{b_{ij}(\alpha)}{A_{ijij}(\alpha)}.$$

The outcome of this calculation is a piecewise constant function of time $u_{ij}(t)$, as described by Gershwin et al. (1985). This function is used in the staircase strategy, below, until the repair state α changes. When that happens, a new function is calculated at this level.

7.3 Level 2: Staircase strategy

Loading a Type j part into Machine 1 is eligible if:

1. The number of Type j parts made on Machine 1 is less than

$$\int_0^t u_{1j}(s)ds, \text{ and} \tag{45}$$

2. Machine 1 is now idle.

Loading a Type 1 part into Machine 2 is eligible if:

1. The number of Type 1 parts made on Machine 2 is less than

$$\int_0^t u_{21}(s)ds, \text{ and} \tag{46}$$

2. Machine 2 is now idle.

7.4 Simulation results

Figure 7.2 demonstrates how the cumulative output follows the cumulative requirements when the system is run with this strategy.

8. CONCLUSIONS

A hierarchical scheduling and planning strategy has been described for manufacturing system. It is based on two major propositions:

1. Capacity. No resource can function more than 100% of the time.

2. Frequency separation. We assume that the spectrum of events is discrete. The frequencies of important events are grouped into distinct clusters.

This work is in its early stages. Among the important outstanding research problems are proving the conjecture that hierarchical decomposition is asymptotically optimal as times scales separate; determining how to deal with systems in which time scales are not widely separated; formulating and solving the hedging point problem with non-Markov events (such as those generated by a staircase strategy); developing sufficiency conditions for capacity. To improve on the staircase policy, new formulations of combinatorial optimization problems are required in which the objective is to load material as close as possible to a given rate.

We have not discussed at all the collection and processing of data in the hierarchy. This will require the solution of statistics problems. Some extensions include the reduction of the problem size at higher levels. This requires aggregation of activities (so that one considers, for example, large classes of part types, rather than individual types) and of resources (so that the smallest unit can be a cell or workshop or even factory, rather than a machine).

The last issue is related to the long time that parts spend in some kinds of manufacturing, particularly semiconductor fabrication. Preliminary work in extending the Kimemia-Gershwin formulation to systems with both operation and queuing delay and is described in Lou et al. (1987) and Van Ryzin (1987).

ACKNOWLEDGMENTS

This work was supported by the Defense Advanced Research Projects Agency and monitored by ONR under contract N00014-85-K-0213.

REFERENCES

P. Afentakis, B. Gavish, U. Karmarkar (1984), "Computationally Efficient Optimal Solutions to the Lot-Sizing Problem in Multi-stage Assembly Systems," *Management Science*, Vol. 30, No. 2, February 1984, pp. 222-239.

R. Akella, Y. F. Choong, and S. B. Gershwin (1984), "Performance of Hierarchical Production Scheduling Policy," *IEEE Transactions on Components, Hybrids, and Manufacturing Technology*, Vol. CHMT-7, No. 3, September, 1984.

R. Akella and Kumar (1986), "Optimal Control of Production Rate in a Failure Prone Manufacturing System, *IEEE Transactions on Automatic Control*, Vol. AC-31, No. 2, pp. 116-126, February, 1986.

T. Bielecki, and P. R. Kumar (1986), "Optimality of Zero-Inventory Policies for Unreliable Manufacturing Systems", Coordinated Science Laboratory, University of Illinois Working Paper.

G. R. Bitran, E. A. Haas, and A. C. Hax (1981), "Hierarchical Production Planning: A Single-Stage System," *Operations Research*, Vol. 29, No. 4, July-August, 1981, pp. 717-743.

Y. F. Choong (1987), MIT Ph.D. Thesis in preparation.

F. Delebecque, J. P. Quadrat, and P. V. Kokotovic (1984), "A Unified View of Aggregation and Coherency in Networks and Markov Chains," *International Journal of Control*, Vol. 40, No. 5, November, 1984.

M. A. H. Dempster, M. L. Fisher, L. Jansen, B. J. Lageweg, J. K. Lenstra, and A. H. G. and Rinnooy Kan, (1981), "Analytical Evaluation of Hierarchical Planning Systems," *Operations Research*, Vol. 29, No. 4, July-August, 1981, pp. 707-716.

S. B. Gershwin (1986), "Stochastic Scheduling and Setups in a Flexible Manufacturing System," in *Proceedings of the Second ORSA/TIMS Conference on Flexible Manufacturing Systems*, Ann Arbor, Michigan, August, 1986, pp. 431-442.

S. B. Gershwin (1987), "A Hierarchical Framework for Manufacturing System Scheduling," *Proceedings of the 26th IEEE Conference on Decision and Control*, Los Angeles, California, December, 1987

S. B. Gershwin, R. Akella, and Y. F. Choong (1985), "Short-Term Production Scheduling of an Automated Manufacturing Facility," *IBM Journal of Research and Development*, Vol. 29, No. 4, pp 392-400, July, 1985.

S. C. Graves (1981), "A Review of Production Scheduling," *Operations Research*, Vol. 29, No. 4, July-August, 1981, pp. 646-675.

S. C. Graves (1982), "Using Lagrangean Relaxation Techniques to Solve Hierarchical Production Planning Problems," *Management Science*, Vol. 28, No. 3, March 1982, pp. 260-275.

A. C. Hax and H. C. Meal (1975), "Hierarchical Integration of Production Planning and Scheduling," North Holland/TIMS, Studies in Management Sciences, Vol. 1, *Logistics*.

J. Kimemia and S. B. Gershwin (1983), "An Algorithm for the Computer Control of a Flexible Manufacturing System," *IIE Transactions* Vol. 15, No. 4, pp 353-362, December, 1983.

B. J. Lageweg, J. K. Lenstra, and A. H. G. and Rinnooy Kan (1977), "Job-Shop Scheduling by Implicit Enumeration," *Management Science*, Vol. 24, No. 4, December 1977, pp. 441-450.

B. J. Lageweg, J. K. Lenstra, and A. H. G. and Rinnooy Kan (19787), "A General Bounding Scheme for the Permutation Flow-Shop Problem," *Operations Research*, Vol. 26, No. 1, January-February 1978, pp. 53-67.

X.-C. Lou, J. G. Van Ryzin and S. B. Gershwin (1987), "Scheduling Job Shops with Delays," in *Proceedings of the 1987 IEEE International Conference on Robotics and Automation*, Raleigh, North Carolina, March-April 1987.

O. Z. Maimon and S. B. Gershwin (1987), "Dynamic Scheduling and Routing For Flexible Manufacturing Systems that have Unreliable Machines," in *Proceedings of the 1987 IEEE International Conference on Robotics and Automation*, Raleigh, North Carolina, March-April 1987. See revised version: MIT LIDS Report No. LIDS-P-1610, revised July, 1987.

O. Maimon and G. Tadmor (1986), "Efficient Low-Level Control of Flexible Manufacturing Systems," MIT LIDS Report No. LIDS-P-1571.

E. F. P. Newhouse (1975a), "Multi-Item Lot Size Scheduling by Heuristic, Part I: With Fixed Resources," *Management Science*, Vol. 21, No. 10, June 1975, pp. 1186-1193.

E. F. P. Newhouse (1975b), "Multi-Item Lot Size Scheduling by Heuristic, Part I: With Variable Resources," *Management Science*, Vol. 21, No. 10, June 1975, pp. 1194-1203.

C. H. Papadimitriou and P. C. Kannelakis (1980), "Flowshop Scheduling with Limited Temporary Storage," *Journal of the ACM*, Vol. 27, No. 3, July, 1980.

P. J. Ramadge and W. M. Wonham (1985), "Supervisory Control of a Class of Discrete Event Processes," Systems Control Group Report No. 8515, University of Toronto.

Rishel, R. "Dynamic Programming and Minimum Principles for Systems with Jump Markov Disturbances", *SIAM Journal on Control*, Vol.13, No.2 (February 1975).

V. R. Saksena, J. O'Reilly, and P. V. Kokotovic (1984), "Singular Perturbations and Time-Scale Methods in Control Theory: Survey 1976-1983", *Automatica*, Vol. 20, No. 3, May, 1984.

A. Sharifnia (1987), "Optimal Production Control of a Manufacturing System with Machine Failures," Department of Manufacturing Engineering, Boston University.

H. M. Wagner and T. M. Whitin (1958), "Dynamic Version of the Economic Lot Size Model,"

Management Science, Vol. 5, No. 1, October, 1958, pp. 89-96.

J. G. Van Ryzin (1987), "Control of Manufacturing Systems with Delay," MIT EECS Master of Science Thesis, MIT LIDS Report LIDS-TH-1676.

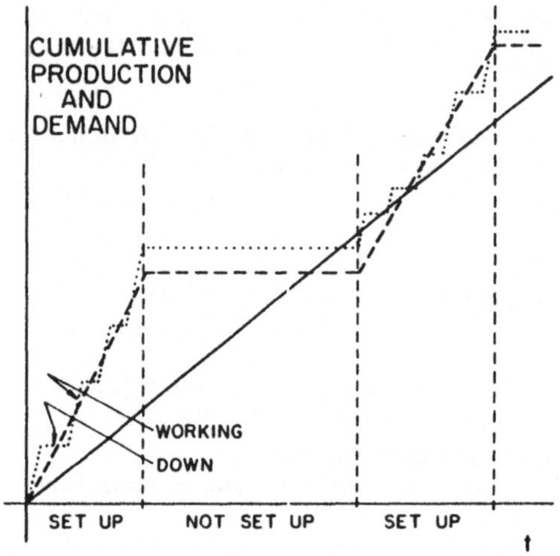

Figure 1.1 Production and Other Events

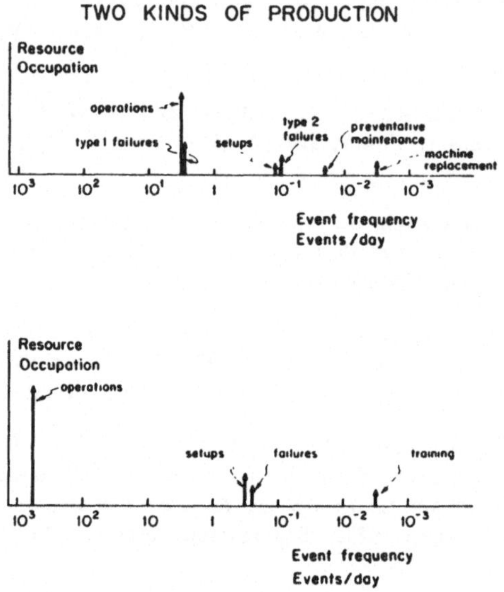

Figure 2.1 Two Spectra

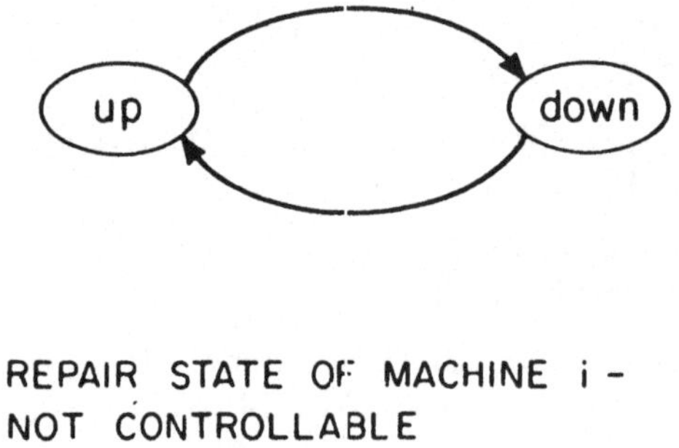

Figure 2.2 Repair States

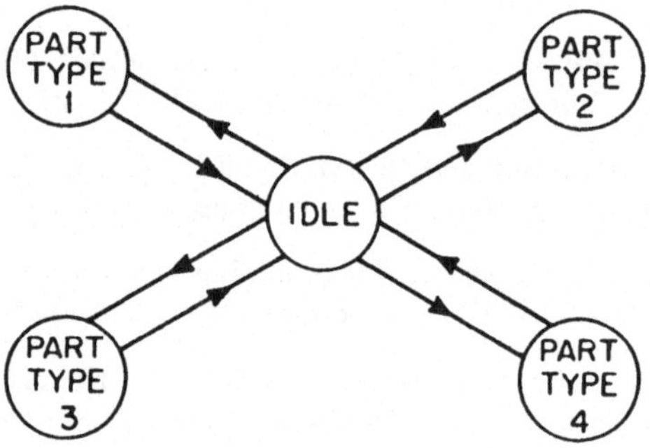

Figure 2.3 Operation States

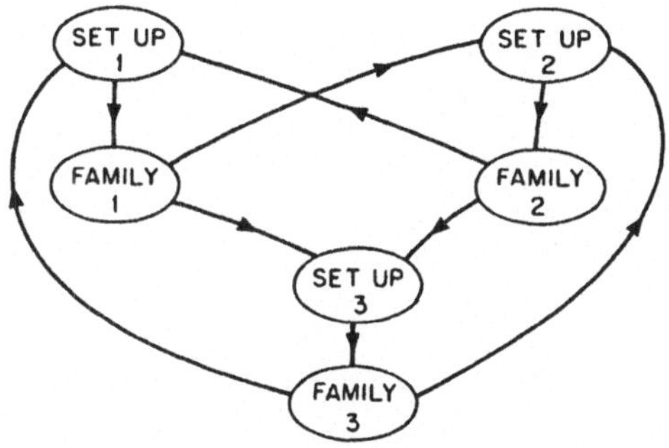

Figure 2.4 Configuration States

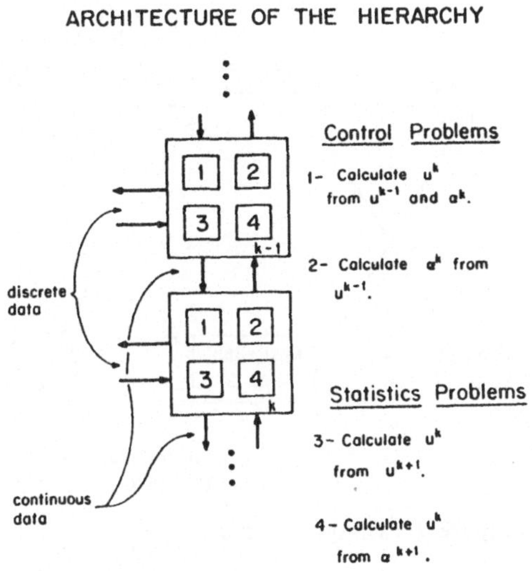

Figure 4.1 Hierarchy

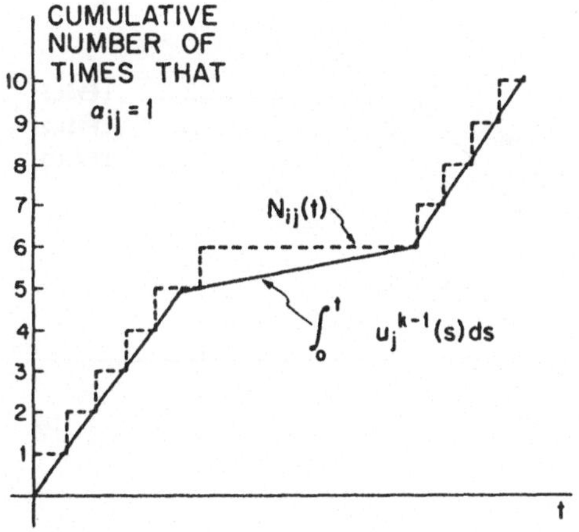

Figure 5.1 Staircase Strategy

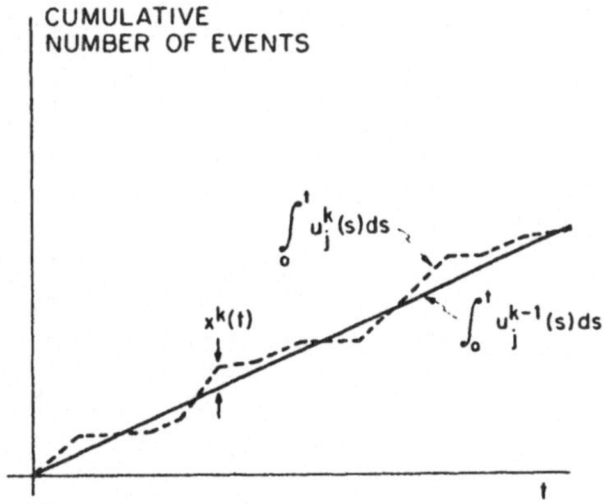

Figure 6.1 Surplus

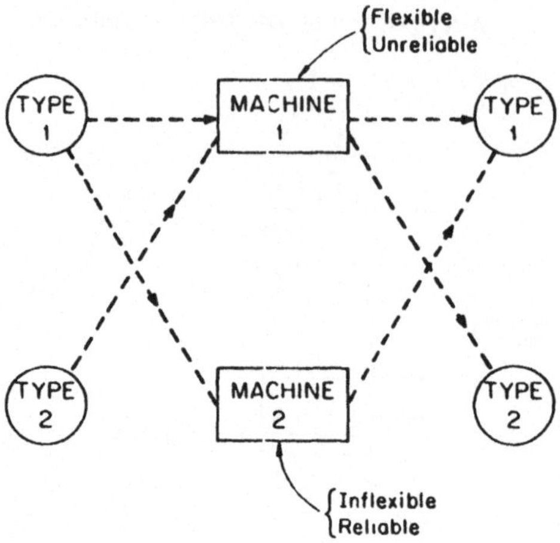

Figure 7.1 Simple System

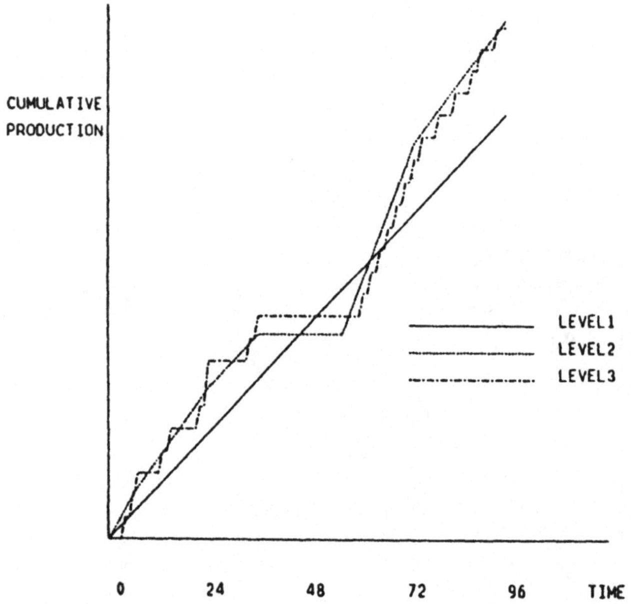

Figure 7.2 Behavior of Strategy

A Selected and Annotated Bibliography on Perturbation Analysis

by
The Perturbation Analysis Group
Harvard University
(Y. C. Ho, Editor)

Since its accidental beginning in 1977 as a by product of research on the FIAT 131 Engine Production Line Monitoring System (see reference 1 below), Perturbation Analysis (PA) has grown into a full blown research area. A total of ten present and past faculties and Ph.D. students of Harvard University have worked on the subject matter producing up to now 40 papers. The participants are **X. Cao, C. Cassandras, J. Dille, A. Eyler, W.B. Gong, P. Glasserman, Y. C. Ho, S. Li, R. Suri, and M. Zazanis**. They can be thought of as the joint authors for this annotated bibliography which is assembled to facilitate readers who are interested in this subject matter.

The bibliography is arranged in more or less historical order. We have not included here related works by other authors. However, pertinent references to these other works can be found in [reference 28] below.

1. Ho, Y.C., A. Eyler, and T.T.Chien (1979) " A Gradient Technique for General Buffer Storage Design in a Serial Production Line", *Int'l J. on Production Research* , 17, 6, pp. 557-580, 1979

2. Ho, Y.C. (1979), "Parameter Sensitivity of a Statistical Experiment", *IEEE Trans. on Auto. Control* , AC-24, 6, p. 982.

3. Ho, Y.C., A. Eyler, and T.T.Chien (1983) "A New Approach to Determine Parameter Sensitivities on Transfer Lines", *Management Science* , 29,6, pp. 700-714, 1983

These three papers started the study of Perturbation Analysis of discrete event dynamic systems. Arguments used in these papers are somewhat naive by current standards. They are included here for purposes of illustrating the development of the idea.

4. Ho, Y.C., and C.G. Cassandras (1983), "A New Approach to the Analysis of Discrete Event Dynamic Systems", *Automatica* , 19, 2, pp. 149-167.

5. Ho, Y. C., and X. R. Cao (1983), "Perturbation Analysis and Optimization of Queueing Networks", Journal of Optimization Theory and Applications, 40, 4, pp. 559-582.

These papers represent the first work of perturbation analysis on general queueing networks. Much research was stimulated by them.

6. Ho, Y.C., X. R. Cao, and C.G. Cassandras (1983), "Infinitesimal and Finite Perturbation Analysis for Queueing Networks", *Automatica* , 19, 4, pp. 439-445.

This is the first paper on the finite perturbation analysis rules and experimental results. The importance of event order change was recognized then.

7. Cao, X. R., and Y. C. Ho, (1984), "Estimating Sojourn Time Sensitivity in Queueing Networks Using Perturbation Analysis", Technical Report, Division of Applied Science, Harvard University; also *Journal of Optimization Theory and Applications* , Vol. 53, 3, 353-375, 1987.

The paper is the first work which discovers the discontinuity of the sample performance function and proposes the interchangibility problems for discrete event systems. Perturbation analysis algorithms are developed for sojourn time sensitivity estimation. Experimental results are presented. An earlier version of this paper entitled "Perturbation Analysis o Sojourn Times in Queueing Networks" was submitted to and reviewed by *Operations Research* in 1984. Eventually, a more rigorous version appeared as [ref.26]

8. Cao, X.R. (1985), "Convergence of Parameter Sensitivity Estimates in a Stochastic Experiment", *IEEE Trans. on Automatic Control* , Vol. AC-30, 9, pp. 834-843.

The paper is the first work which formalizes mathematically the problem of interchangibility of the expectation and the differentiation for discrete event systems. Conditions are found under which this interchangibility holds. It is proved that under these conditions the sample derivative is the best estimate of the derivative of the expected value among three kinds of estimates discussed in the paper.

9. Cao, X.R. (1987), "First-Order Perturbation Analysis of a Single Multi-Class Finite Source Queue", *Performance Evaluation.* Vol.7, 31-41, 1987

The contribution of this paper is twofold: First, it gives an example which shows that the interchangibility does not hold for the throughput of multiclass systems. Second, it provides an algorithm which yields the exact estimate for the throughput sensitivity of a multiclass system using first order perturbation analysis.

10. Cassandras, C.G., and Y.C. Ho (1985), "An Event Domain Formalism for Sample Path Perturbation Analysis of Discrete Event Dynamic Systems", *IEEE Trans. on Automatic Control*, Vol. AC-30, 12, pp. 1217-1221.

Consistent formalism is provided for the earlier results on Infinitesimal PA.

11. Suri,R. and M. A. Zazanis, (1985) "Perturbation Analysis Gives Strongly Consistent Sensitivity Estimates for the M/G/1 Queue", to appear in *Management Science*

This paper is among the earliest studies of consistency of IPA. It considers, for an M/G/1 queueing system, the sensitivity of mean system time of a customer to a parameter of the arrival or service distribution. It shows analytically that (i) the steady state value of the perturbation analysis estimate of this sensitivity is unbiased, and (ii) a perturbation analysis algorithm implemented on a single sample path of the system gives asymptotically unbiased and strongly consistent estimates of this sensitivity.

12. Ho, Y.C. and Cao, X.R. (1985), "Performance Sensitivity to Routing Changes in Queueing Networks and Flexible Manufacturing Systems Using Perturbation Analysis", *IEEE J. on Robotics and Automation* , Vol. 1, pp. 165-172.

This paper, among other things, shows that despite claim to the contrary, regular IPA rules can be applied to yield correct estimates of performance sensitivity to routing probabilities.

13. Cao, X. R., (1985), "On Sample Performance Functions of Jackson Queueing Networks", to appear in *Operations Research* .

The paper proposes the concepts of Sample Performance Functions and Sample Derivatives and proves that the interchangibility holds for the average time required to service one customer in any finite period as a function of the mean service time in a Jackson queueing network; and the perturbation analysis estimate of the sensitivity of throughput is a strongly consistent estimate.

14. Cassandras, C. G. (1985), "Error Properties of Perturbation Analysis for Queueing Systems", to appear in *Operations Research* .

IPA is placed in the context of a family of PA estimation procedures, showing the tradeoff between increased accuracy and state memory costs. The GI/G/1 model is analyzed to characterize the error properties of the simplest PA procedures which, under certain conditions, provide unbiased performance sensitivities. Extensions to tandem queueing networks and blocking effects are included.

15. Cao, X. R. (1987), "The Convergence Property of Sample Derivatives in Closed Jackson Queueing Networks", submitted to *Journal of Applied Probability* (The result was also presented in a technical report of Harvard University, 1986).

The paper proves that the sample elasticity of throughput with respect to the mean service time obtained by perturbation analysis converges in mean to that of the steady state mean throughput as the number of customers served goes to infinity.

16. Cao, X. R. (1987), "Realization Probability in Closed Jackson Queueing Networks and Its Application", to appear in *Advances in Applied Probability* . Sept. 1987

This paper introduces the concept of realization probability for closed Jackson networks. This new concept provides an analytical solution to the sample elasticity of the system throughput and some other sensitivities. Using realization probability and the ergodicity of the system, it is proved that the sample elasticity of throughput with respect to the mean service time obtained by perturbation analysis also converges with probability one to that of the steady state mean throughput as the number of customers serves goes to infinity.

17. Ho, Y.C. and Yang, P.Q. (1986) "Equivalent Networks, Load Dependent Servers, and Perturbation Analysis - An Experimental Study" *Proceedings of the Conference on Teletraffic Analysis and Computer Performance Evaluation* , O.J. Boxma, J.W.Cohen, H.C. Tijms (Eds), North Holland 1986.

This paper derives the PA algorithm for load dependent queueing networks and shows that the idea of PA can be applied to aggregated systems.

18. Cassandras, C. G. (1987), "On-Line Optimization for a Flow Control Strategy", to appear in *IEEE Trans. on Automatic Control.*

It is shown that a direct extension of PA, tracking queue lengths in addition to event times, can be used to estimate performance sensitivities in a simple state-dependent routing environment. This is done at the expense of state memory along the observed sample path. When a state memory constraint is imposed, the estimates become biased, but may still be sufficiently accurate.

19. Zazanis, M. A. and R. Suri (1985), "Comparison of Perturbation Analysis with Conventional Sensitivity Estimate for Stochastic Systems", submitted to *Operations Research*.

This paper examines the Mean Squared Error (MSE) of PA estimates and compares it to that of estimates obtained by conventional methods. We consider two different experimental methods that are commonly used: (i) independent replications and (ii) regenerative techniques. The analytic results obtained establish the asymptotic superiority of PA over conventional methods for both of these experimental approaches. Furthermore, it shows that PA estimates have a mean square error which is of order $O(1/t)$ where t is the duration of the experiment in a regenerative system, whereas classical finite difference estimates have a mean square error which is at best $O(1/t^{1/2})$

20. Zazanis, M. A. and R. Suri (1985), "Estimating First and Second Derivatives of Response Time for GI/G/1 Queues from a Single Sample Path", submitted to *Queueing Systems: Theory and Applications* ..

A PA algorithm is developed for estimating second derivatives of the mean system time for a class of G/G/1 queueing systems, with respect to parameters of the interarrival and service distribution, from observations on a single sample path. The statistical properties of the algorithm are investigated analytically and it is proved that the estimates obtained are strongly consistent.

21. Cao, X. R. (1986), "Sensitivity Estimates Based on One Realization of a Stochastic System", *Journal of Statistical Computation and Simulation* Vol.27, 211-232, 1987.

The paper shows that the perturbation analysis estimate corresponds to the estimate of the difference of two random functions using the same random variable: thus, its variance is smaller than other one sample path based sensitivity estimates such as the likelihood ratio estimate.

22. Cao, X. R. (1987), "Calculation of Sensitivities of Throughputs and Realization Probabilities in Closed Queueing Networks with Finite Buffers", manuscript to be submitted

The paper derives equations for realization probability for systems with finite buffers and shows that the infinitesimal perturbation analysis estimate is generally biased for these systems. However, examples indicate the bias is usually very small.

23. Cao, X. R. (1987), "Realization Probability in Multi-Class Closed Queueing Networks", submitted to *European Journal of Operations Research*

The paper discusses the concept of realization probability for multiclass closed networks.

24. Cao, X. R., and Y. Dallery (1986), "An Operational Approach to Perturbation Analysis of Closed Queueing Networks", *Mathematics and Computers in Simulation* , Vol. 28, pp. 433-451.

The paper develops and operational definition of realization probability and proves the sensitivity equations using operational assumptions.

25. Cao, X.R., and Y.C. Ho (1987), "Sensitivity Estimate and Optimization of Throughput in a Production Line with Blocking", to appear in *IEEE Trans. on Automatic Control* Vol. AC-32, # 11, 1987

The paper proves that the perturbation analysis estimate is strongly consistent for systems with finite buffer capacities but no simultaneous blocking. The perturbation analysis estimate is used in optimization of a production line. It is shown that perturbation analysis enables us to use the Robbins-Monro procedure instead of the conventional Kiefer-Wolfowitz procedure.

26. Cao, X.R., and Y.C. Ho (1986), "Perturbation Analysis of Sojourn Times in Closed Jackson Queueing Networks", submitted to *Operations Research* ,

The paper proves the convergence theorems for the perturbation analysis estimate of sojourn times in closed Jackson networks.

27. Cassandras, C.G. and Strickland, S.G. (1987), "Perturbation Analytic Methodologies for Design and Optimization of Communication Networks", submitted to *IEEE J. of Selected Areas in Communications* .

Simple PA algorithms are used to estimate performance sensitivities for communication network models. Of particular interest is the application of IPA in estimating marginal delays in links modeled as G1/G/1 queues. These estimates are used in conjunction with a distributed minimum delay algorithm to optimize routing in a quasi-static environment.

28. Ho, Y.C. (1987), "Performance Evaluation and Perturbation Analysis of Discrete Event Dynamic Systems ", *IEEE Trans. on Automatic Control*, AC-32, 6, July 1987, 563-572

This paper contains probably the most complete references on PA and related matters as of 12/86.

29. Ho, Y.C. (1987). "PA Explained" to appear *IEEE Trans. on Automatic Control.*

This note explains in simplest term via an example the essence of PA and answers intuitively the question "How can one infer the performance of a discrete event system operating under one parameter value from that of another with a different parameter value? Don't the two sample paths behave totally differently?"

30. Zazanis, M.A. (1987) "Unbiasedness of Infinitesimal Perturbation Analysis Estimates for Higher Moments of the Response Time of an M/M/1 Queue" Technical report 87-06 Northwestern University, 1987. submitted to *Operation Research*

This paper uses classical markovian analysis to establish the unbiasedness of IPA estimates for the M/M/1 system and refutes another public claim of the limitations of IPA. The restrictive markovian assumption is the price paid for the simplicity of the arguments used.

31. Suri, R., and J. Dille (1985), "A Technique for On-line Sensitivity Analysis of Flexible Manufacturing Systems", *Annals of Operations Research*, 3, pp. 381-391.

The PA approach is applied to flexible manufacturing systems (FMS). We give a simulation example illustrating how our perturbation analysis could be used on-line on an FMS to improve its performance, including reducing its operating cost. Experimental results are also presented validating the estimates obtained from this technique.

32. Suri, R., and Y.T. Leung (1987), "Single Run Optimization of Discrete Event Simulations - An Empirical Study using the M/M/1 Queue". Technical Report #87-3, Department of Industrial Engineering, University of Wisconsin, Madison.

This study proposes a stochastic optimization method to optimize a simulation model in a single simulation run. Two algorithms are developed and evaluated empirically using an M/M/1 queue problem. Experimental results show that an algorithm based on IPA provides extremely fast convergence as compared with a traditional Kiefer-Wolfowitz based method.

33. Cassandras, C. G., and Strickland, S. G. (1987), "An 'Augmented Chain' Approach for On-Line Sensitivity Analysis of Markov Processes", submitted to 26th IEEE Conference Decision and Control (also to Trans. on Automatic Control).

This paper presents a new way of estimating performance sensitivities of Markov processes by direct observation. The parameters considered are discrete (integer-valued), e.g. queue capacities, thresholds in routing policies and number of customers of a specific class in a closed network model. The main idea is to construct an "augmented chain" whose state transitions are observable when the process itself is observed.

34. Zazanis, M.A. (1987), "An Expression for the Derivative of the Mean Response Time of a GI/G/1 Queue", Technical report 87-08 Northwestern University (see also "Extension to GI/G/1 systems with a scale parameter", Technical report 87-07).

In this paper an expression is given for the derivative of the mean virtual waiting time in a GI/G/1 queue with respect to the service rate.

35. Suri, R. and M.A. Zazanis (1987), "Infinitesimal Perturbation Analysis and the Regenerative Structure of the GI/G/1 Queue", *Proc. 1987 IEEE Decision and Control Conference* , LA, Calif. to appear

The strong consistency of IPA estimates for the mean response time is shown using the regenerative structure of the GI/G/1 queue. The analysis throws some light on the conditions which are required for the consistency of IPA estimates in general systems with regenerative structure.

36. R. Suri (1987), "Infinitesimal Perturbation Analysis For General Discrete Event Dynamic Systems" *J. of ACM* , July 1987

This is the final version of the paper first presented in 1983 at IEEE Decision and control conference which sets forth IPA in a general setting under deterministic similarity assumptions.

37. Gong, Weibo, and Y.C. Ho (1987), "Smoothed (conditional) Perturbation Analysis of Discrete Event Dynamic Systems", to appear in *IEEE Trans. on Automatic Control.*

We show that by using the smoothing properties of conditional expectation, the problem of interchange between expectation and differentiation can be resolved to give consistent PA estimates for problems heretofore proclaimed to be unsolvable by PA, e.g. derivatives of throughput w.r.t. mean service time in multiclass queueing networks, mean number of customers served in a busy period w.r.t. mean service time, etc.

38. Ho, Y.C. and Shu Li (1987), "Extensions of the Infinitesimal Perturbation Analysis Technique of Discrete Event Systems", submitted to *IEEE Trans. on Automatic Control.*

In this paper, we show another general approach to circumvent the difficulty of discontinuities in PM(θ,ω) w.r.t. q for Markov systems. This technique also puts in perspective earlier work on finite PA showing it to be one member among a range of possible approximations from the crudest to the exact for handling the discontinuity problem. Robustness of these approximations is discussed and experimental supports are illustrated.

39. Glasserman, P. (1987), "IPA Analysis of a Birth-Death Process " submitted to *Operations Research Letters* .

This note shows that the regular IPA rules can be applied to a birth-death process to yield correct sensitivity estimates despite written claim to the contrary.

40. P. Heidelberger, X. Cao, M. Zazanis, R. Suri "Convergence Results for Infinitesimal Perturbation Analysis Estimates" to appear in *Management Science* , 1988

ANALOG EVENTS AND A DUAL COMPUTING STRUCTURE USING ANALOG AND DIGITAL CIRCUITS AND OPERATORS

Tamás Roska
Computer and Automation Institute, Hungarian Academy of Sciences
Uri-u 49 Budapest H-1014

1. INTRODUCTION

Despite the impressive and further increasing power of digital electronic circuits and systems proving their success in applications like computing, communication and control there are signs and facts which clearly show the inherent limits of the exclusiveness of the digital way of operation. Some of these facts are as follows.

(i) Complex algorithms (even NP complete) are solved with appropriate success and accuracy in a few time constants using analog arrays (Tank et al. 1986), (Chua et al. 1985) and there are some "smart analog" components showing practical advantages of the dual operations (Lineback 1986).

(ii) The physical and informational view of computation provides a broader understanding of electronic information processing or decision circuits and systems (Csurgay 1983).

(iii) There are well defined theoretical and practical limitations of the capability of digital simulation of large-scale analog physical circuits and systems (Roska 1983) as well as of biological systems (Conrad 1974).

(iv) A simple discrete-time approximation with delay elements and nonlinear memoryless readout maps of many practical continuous nonlinear operators is an appealing realization possibility (Boyd et.al. 1985).

(v) The summarized experiences concerning the behavioral manifestations of the human cerebral asymmetry show and suggest a combined analytic-holistic or perhaps digital-analog way of operation in solving difficult tasks (Bradshaw et al. 1983).

Paradoxically, the widely publicized so called "non von Neumann" architectures are very specialized forms proposed by von Neumann where even the analog basic operations were considered too (von Neumann 1958).

Based on the above results and facts, as well as keeping in mind the standard theory of modelling (Zeigler 1976), a novel dual computing structure is proposed using analog, digital and joint digital-analog modules and operators.

Section 2 contains the motivating facts, models and statements. In section 3 one crucial concept of the model, namely, the analog event is introduced and the principles of the physical realization of some important nonlinear operators as computational elements are discussed. The general framework of the novel computing structure is shown in Section 4 while some important parts of the model including the joint processor are presented in more details in Section 5. In Section 6 the conclusions are summarized along with some proposals.

2. MOTIVATING FACTS AND RESULTS

Analog and hybrid computations are old topics. Recently, however, based on the advantages of VLSI possibilities it turned out that regular structures with arrays of operational anplifiers shows surprising characteristics. They are able to realize complex algorithms like the nonlinear programming (Chua et al. 1985) or even the traveling salesman problem and complex signal decision algorithms (Tank et al. 1986). Hence, even a digitally NP complete problem can be solved in a few time constants with a quite useful accuracy using "neural" electronic circuits (Hopfield et al.). Furthermore, there are new "smart analog" devices on the market which combine the digital and analog functions for communication, data conversion and signal processing applications (Lineback 1986). The new BIMOS (combined bipolar - MOS) technolgies are especially suited for digital-analog dual structures.

Based on the physics of computation (Mead et al. 1980) considering the physical realization and the information content of an I/O operator constructed using artificial electronic circuits and systems it turned out (Csurgay 1983) that the hardware and the software has a unified and well-defined meaning independent of the way of operation (analog or digital).

Considering the complexity of the digital simulation of electronic circuits and systems it has been shown (Roska 1983) that increasing the complexity of the analog circuit or system to be simulated, if the digital simulator's complexity is increasing in the same rate then the complexity of the simulation is increasing too. Hence, there are well-defined limitations of the digital realization of analog functions.

A recent result (Boyd et al. 1985) shows a possibility, at least theoretically, for approximating many practical (with a fading memory) continuous nonlinear operators (even described by partial differential equations) by a simple discrete-time structure. It consists of unit delay elements and one multi-input polinomial nonlinear readout map. An extended class of the latter (any continuous nonlinear map) can be approximated by a novel memory structure (Roska 1987).

Finally, the rich research experiences concerning the behavioral manifestations of the human cerebral asymmetry provide a genuine source for understanding a dual way of processing. In what follows, based mainly on (Bradshaw et al. 1983), an attempt is made to summarize some crucial characters of cerebral asymmetry from our special point of view.

Namely, the left hemisphere (LH) and the right hemisphere (RH), which were considered as performing the analytic and holistic processing, here are considered as if they were the digital and analog type ones, respectively. Later we modify this strict view.

LH	RH
dual facts	
- analytic (breaking into parts and elements, etc.)	- holistic (global)
- differential	- integral
- sequential processing and temporal resolution of information (sequencing, discrimination of duration, temporal order and rhythm)	- immediate, perception of the relations of the parts and the whole
- verbal abilities especially at motor level	- performing abilities
- matching of conceptually similar objects	- matching of structurally (pictures, curves etc) similar objects
- events of high rate of change (50 msec)	- events of small rate of change
- information ordered in time	- information ordered in space

-dual encoding hypothesis in memory theory

verbal, symbols	pictorial, images

(tipical dual representation: metaphores)

<u>unique fact</u>

preeminence in motor control

<u>contradictional facts</u>

LH	RH
-superiority at visuospatial tasks requiring fine spatial acuity (performing by analytic extraction of significant features or elements)	
	-possesses considerable linguistic power, especially receptive
-linguistic specialization only on limited fields (e.g computational and combinatorial algorithms that characterize abstract syntax and phonology)	

2. ANALOG EVENTS AND THE PHYSICAL REPRESENTATION OF NONLINEAR COMPUTING OPERATORS.

Digitally coded logical events play a dominant role in present day information processing and decision systems. The actual processing machines operate on the elementary events specifically on the digital bits. This is one of the reason why this machines are so accurate and reliable on one hand and why they are so ineffective on the other hand in some specific other tasks (e.g. selecting some features and optimizing some cost functions).

Next, we introduce and define the analog event, its detection and storing process and the elementary operators on it. We are doing it to show, in parallel, the corresponding facts for the digital-logical events.

analog event	digital-logical event
signals	
$x(t) \in \mathcal{F}_T$	$\underset{\sim}{x}(\hat{t}) \in R^n$
$\mathcal{F}_T : C^0$, nonzero on $[0,T]$	$\hat{x}^{1,2}$: threshold values for binary codes (0,1)
	$\hat{t}$: sampling time instant
$x(t) \rightarrow \Sigma_{\mathcal{F}_T}$: $\hat{x}_j$ (t) ; a finite set of event functions, $j = 1,2..m$	$\underset{\sim}{x}(\hat{t}) \rightarrow \Sigma_{ABC}$: finite coded set of symbols
unique detection	
$\| x(t) - \hat{x}_j(t) \| \leq \varepsilon$ on $[0,T]$	$\vert x_i(\hat{t}) - \hat{x}^{1,2} \vert \leq \varepsilon$ $i = 1,2,...n$
unique detection devices	
nonlinear memoryless two-ports	comparator
storing	
events within a class: - elementary I/O operators of a unified input - elementary inputs of a unified I/O operator - reed-out map (look-up table) the class is represented by a digital code	1,0 codes

<u>elementary operators</u>

$\int dt$, d/dt, $\| \|$, sgn, etc. nonlinear progr., A/D, etc.	AND, OR, NOT +, -, x, /
real-time	sequential

<u>algorithmic elements</u>

?	elements of recursive functions (on integers as coded events)

The last question mark denotes those elementary continuous nonlinear operators which plays the role of the elementary recursive functions in the digital-logical computational paradigm (see e.g. the interpretation of the Church thesis in (Lewis et al. 1981)).

Concerning the physical representation of nonlinear operators as computational elements four concepts are summarized as follows.

(i) The electronic circuit or system as a physical object realizes an operator which worth to consider either as a solution of a differential equation or as a minimum of a well defined energy function. The first two cited examples of Section 2 are characteristic special cases. Due to the inherent constraints in the circuit the minimum of an appropriate energy function is not necessarily an unconstrained minimum.

(ii) A convenient realization of the detection of an analog event within a class of events could be a parallel connection of nonlinear (memoryless) two-ports having the inverse characterictics of the analog events. Hence, the minimum signal output selects the appropiate event.

(iii) Realizing a multi input memoryless discrete time operator with single-input single-output memories a systematic decomposition procedure (Roska 1987) based on Kolmogorov's approximation theorem reduces drastically the surface, the time or the power needed to perfom the prescribed operator.

(iv) Biological considerations (Conrad 1974) show that within well-defined problem classes non-programmable analog physical-chemical processes are extremly economic, hence, before decomposing a complex operator (or a decision system) into programmed computational structures it is worth to analyse whether the natural elementary parts of the complex operator are realizable (or not) by non-programmable electronic structures.

4. THE GENERAL FRAMEWORK

It is quite easy to prove that both the exclusively analog or digital computational decision structures have serious inherent disadvantages (a few of them have been cited above). Next, we introduce a novel computing structure (see the next page).

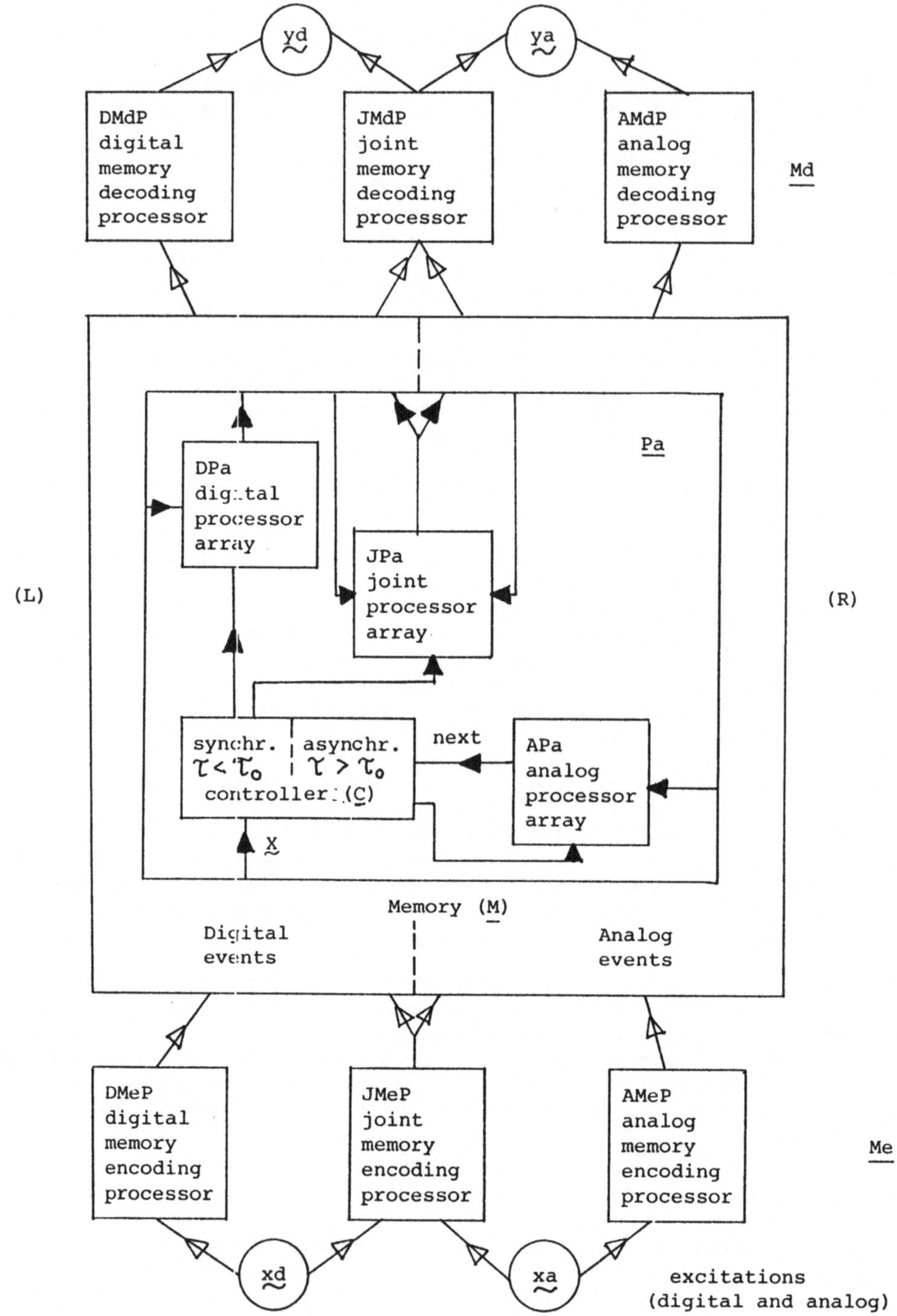

FIGURE 1 The general framework

This structure

(i) prefers a given dual way of operation (digital and analog) and

(ii) tries to capitalize the special division of labour shown by the human cerebral asymmetry (as summarized in Section 2).

We call the framework in Figure 1 "the asymmetrical, controlled, dual (digital-analog) computing structure" (shortly ACD structure).

It conists of five main parts:

(i) the memory encoding part (Me)
(ii) the dual (digital and analog) memory (M),
(iii) the controller (C)
(iv) the three types of processing arrays (Pa)
(v) the memory decoding part (Md)

The controller (C) is a standard digital finite state machine with an inherent finite memory making unique digital-logical decisions (next state functions and output functions).

The memory (M) has a dual structure, a digital and an analog part.

All the other three parts have basically three building blocks. These are the digital, the analog and the joint digital-analog processors. All the directed arcs in Figure 1 represent the flow of either digital-logical or analog events.

Considering Figure 1 some relations with the summarized behaviours of cerebral asymmetry (Left-Right) can be realized. These are the digital (motor) control, the threshold time to select between the digital and analog events as well as the sychronous and asynchronous control (based on high vs. small rate of change), the dual memory encoding and the three types of processors (digital, analog and joint).

The holistic processing of the right hemisphere and the particularly effective non-programmable (or soft programmable) analog biological primary structures (Conrad 1974) suggest the use of the analog processors operating on analog events (a special case is the analog value) under analog programs.

Hence, some of the left hemisphere characteristics are incorporated in the conventional digital processors and the right hemisphere ones in the analog processors. Both processors and arrays of them can work in series or parallel mode. The sequential mode is dedicated to the digital processor.

The contradictional facts concerning the hemispheric division and the joint actions of the two hemispheres motivate the introduction of the joint (digital-analog) processor. The main characteristics of the joint processor are as follows.

(i) Operators working on analog events are groupped in classes.

(ii) Within these classes the interconnections between the operators and the setting of some parameters of the operators are digitally programmable.

(iii) Decisions between the classes are done either by digital programs or by analog operators (analog programs)

The memory (M) has two parts. Besides the conventional digital part the analog part contains the analog events within a class of events (represented as defined in Section 3).

In the memory encoding part (Me) the joint memory encoding processor, based on the dual memory encoding hypothesis, determines the digital code of the class and the analog event within the class.

5. Details concerning the realization of some parts of the model

In what follows the realizability , the programmability and/or the uniqueness of some proposed structures are shown.

5.1. Analog event detection

The stucture of Figure 2 contains nonlinear memoryless 2-port elements (NEi), detectors detecting signals below a small threshold and a conventional digital decoder. NEi have the inverse characteristics of the analog events within a given class of events.

The structure of Figure 2 is unique, because, unlike linear dynamical systems, nonlinear memoryless operators are unique modulo scaling and delay (Theorem 1 in (Boyd et al. 1983)). However, delay is set by the controlled starting instant of the analog events and scaling is given by the fixed peak value of them. Hence, the cited Theorem assures the uniqueness of the detecting structure. The practical approximation errors of the inverse characteristics can be taken into account by the threshold values of the detectors.

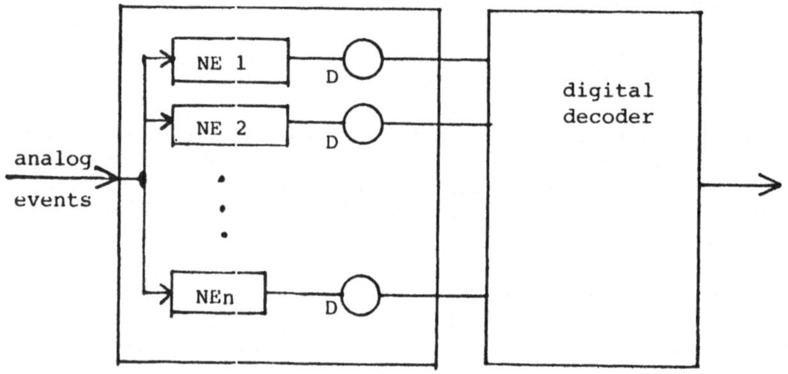

FIGURE 2 Detecting analog events

5.2 A programmable nonlinear memoryless multi-input element (PNE)

Recently, it has been shown (Roska 1987) that any multi-input nonlinear memoryless operator can be approximated (with any given precision) by a nested structure of single-input single-output digital memories and a few adders (Figure 3). The practical finite physical parameters of the realization (area, energy, power, time) are quite appealing. In a fully analog realization adders are replaced by operational amplifiers with feedback resistors approximating ψ and χ.

$$f(x_1,\ldots,x_n) = \sum_{q=1}^{2n+1} \chi_q \left(\sum_{p=1}^{n} \psi^{pq}(x_p) \right) .$$

$$M_{pq}: \psi^{pq}(x_q) \; ; \quad M_q : \chi_q() \; ;$$

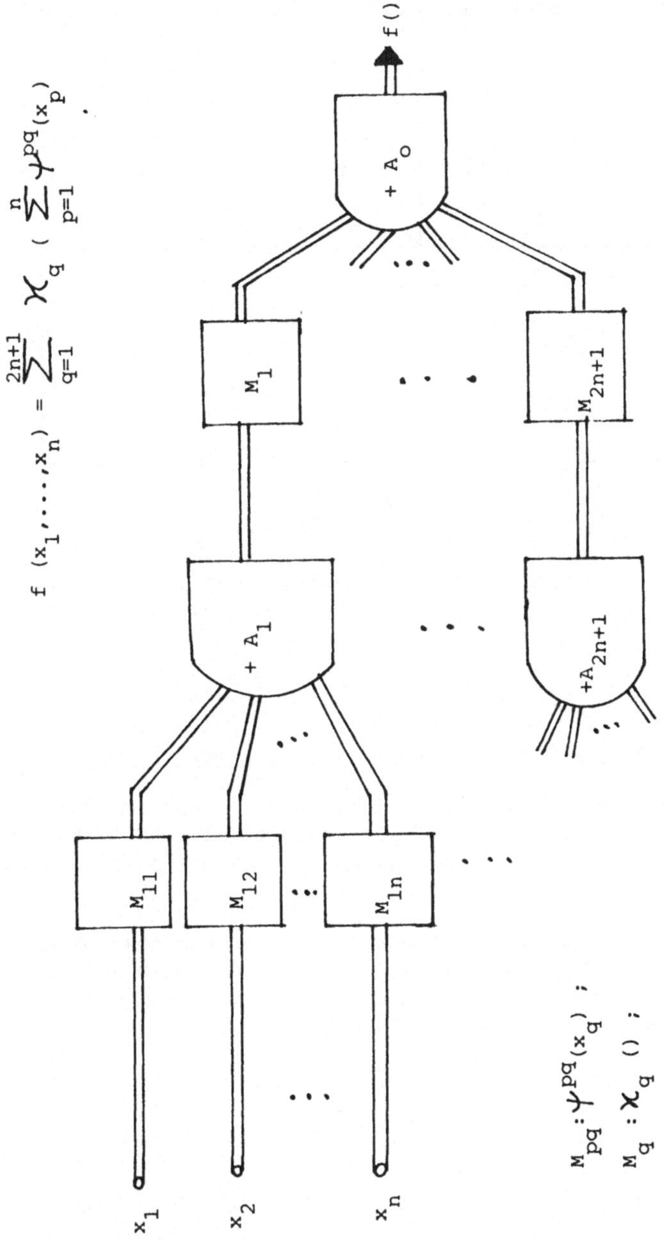

FIGURE 3 A programmable multi-input nonlinear element (PNE)

5.3 A prototype programmable dynamic analog structure (PDA)

According to Theorem 4 in (Boyd et al. 1985) , if N is any time invariant discrete-time operator (l l) with fading memory (N is continuous and the effects of the input are decreasing with time i.e. unique steady state) then the system of Figure 4 with f() being a polynomial can approximate N with any prescribed error. We generalize this structure allowing f() to be an approximation of any multi-input continuos funcion, moreover, it is programmable using PNE-s. On the other hand without using A/D and D/A converters the signals are of finite precision.

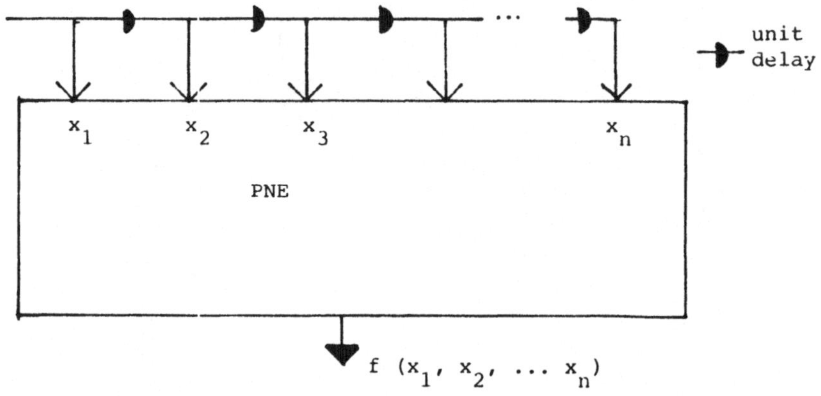

FIGURE 4 A prototype programmable dynamic analog structure (PDA)

5.4 A programmable regular feedbach structure (PRF)

Providing a programmability with PNE-s for the regular feedback structure similar that of (Hopfield et al. 1985) the system of Figure 5 seems a quite natural solution. APi is an analog processor with A/D-D/A elements.

This structure which can be seen on the next page not only provides programmability but, because the use of PNE-s allows to realize any memoryless multivariable non-symmetric feedback function, extends substantially the class of operators.

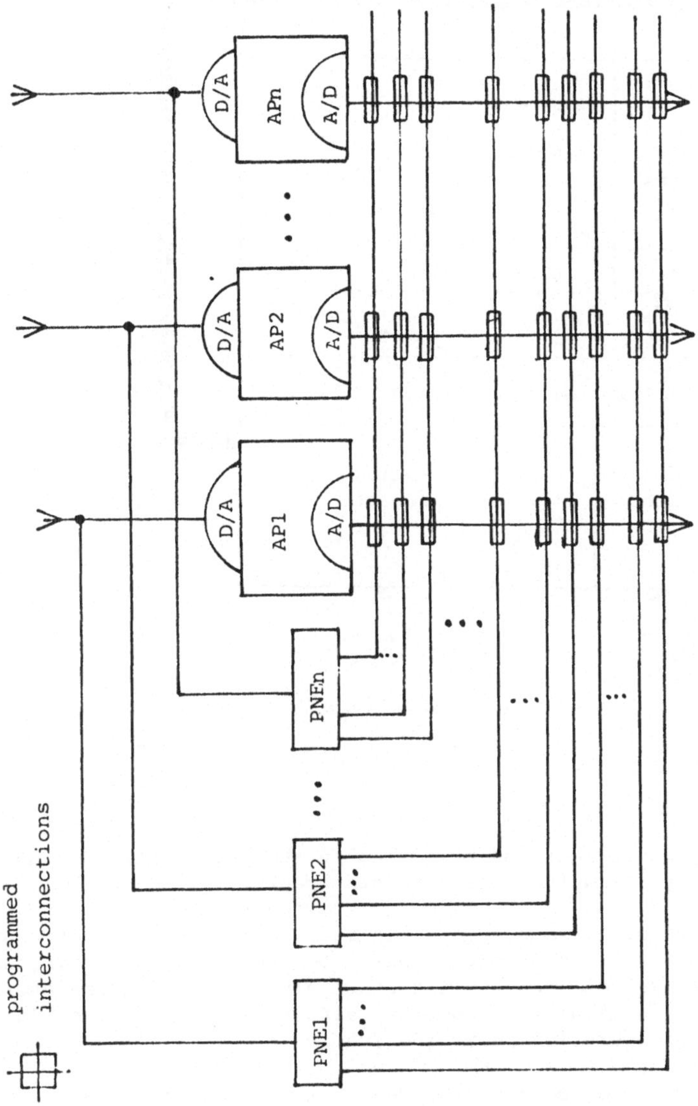

FIGURE 5 A programmable regular feedback structure (PRF)

5.5 The joint processor

Figure 6 shows the structure of a joint processor. The general control is carried out by the digital processor. The classes of the analog operators are organized according to the classes of the analog events.

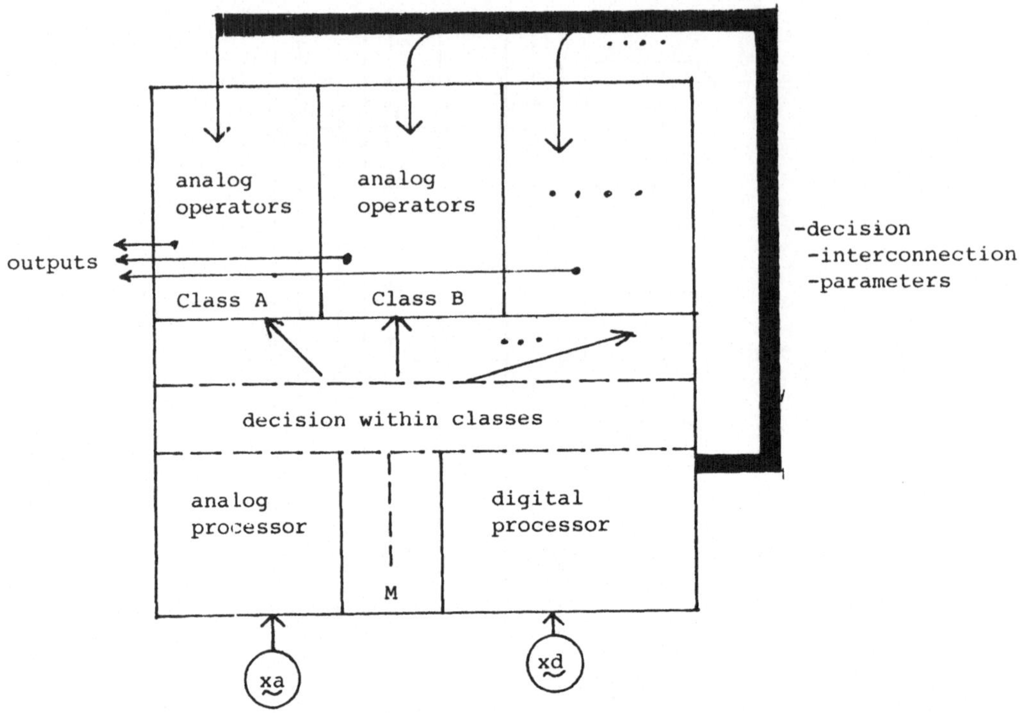

FIGURE 6 A joint (digital-analog) processor (JP)

6. IN CONCLUSION

Based on the investigations presented in the paper there are quite strong evidences supporting the following statements.

(i) Human cerebral asymmetry, as it is understood now, strongly advices to finish with the exclusiveness of the digital computing structures as well as not only to introduce also the analog operators, however, in some particularly complex situations a genuine joint digital-analog way of operation seems inevitable.

(ii) The asymmetric controlled dual (ACD) computing structure suggested in the paper is a starting attempt to build up such a complex model. Although analog events are allowed it is eventually a discrete event structure.

(iii) Some parts of the ACD structure has been elaborated and its realizability, uniqueness and programmability properties were partly shown without determining the exact class of operators solvable by the models. It is also certain that concerning the finite physical parameters (area, time, power, etc.) in some tasks this structure has definite advantages against the exclusive digital solution.

(iv) It is a quite important open question within the framework suggested hier which are the equivalent representations and the formal realizability conditions.

(v) The ACD structure is far not a model of the human cerebral asymmetry (although perhaps some investigations can be made with it). Conversely, it has been an experiment to build up a computational structure for the artificial electronic circuits and systems which reflects some genuine characteristics of the manifestations of cerebral asymmetry.

ACKNOWLEDGEMENTS

The inspiring results of and discussions with Dr. Árpád Csurgay is gratefully acknowledged. Thanks are due to Dr. Maria S. Kopp for her kind help in providing summarizing literatures on human cerebral asymmetry. The work was supported by the Academic Research Fund (AKA) of the Hungarian Academy of Sciences.

REFERENCES

Boyd, S. and Chua, L.O. (1983). Uniqueness of a basic nonlinear structure. IEEE Trans. Circuits and Systems, CAS-30:648-651.

Boyd, S. and Chua, L.O. (1985). Fading memory and the problem of approximating nonlinear operators with Volterra series. IEEE Trans. Circuits and Systems, CAS-32:1150-1161.

Bradshaw, J.L. and Nettleton, N.C. (1983). Human cerebral asymmetry (Chapter Nine). Prentice Hall, Englewood Cliffs.

Chua, L.O. and Lin, G-N. (1985). Nonlinear programming without computation. IEEE Trans. Circuits and Systems, CAS-31: 182-188.

Conrad, M. (1974). The limits of biological simulation. J. theor. Biol., 45:585-590.

Csurgay, Á. (1983). Fundamental limits in large scale circuit modeling. Proc. Eur. Conf. Circ. Theory and Design., 6th, Stuttgart, September 1983. VDE-Verlag, Berlin.

Hopfield, J.J. and Tank, D.W. (1985). "Neural" computation of decisions in optimization problems. Biol. Cybern., 52:141-152.

Lewis, H.R. and Papadimitriou, C.H. (1981). Elements of the theory of computations. Prentice Hall, Englewood Cliffs.

Lineback, J.R. (1986). "Smart analog" - IC-s due from Texas startup. Electronics, (January 20):21-22.

von Neumann, J. (1958). The computer and the brain. Yale University Press, New Haven.

Mead, C.A. and Conway, L.A. (1980). Introduction to VLSI systems. Addison Wesley, Reading.

Roska, T. (1983). Complexity of digital simulators used for the analysis of large scale circuit dynamics. Proc. Eur. Conf. Circ. Theory and Design., 6th, Stuttgart, September 1983. VDE-Verlag, Berlin.

Roska, T. (1987). A canonical memory-adder realization of nonlinear memoryless operators - approximation of look-up tables. Proc. Int. Conf. Nonlin. Osc. 11th, Budapest, August 1987. Bólyai Math. Soc. , Budapest.

Tank, D.W. and Hopfield, J.J. (1986). Simple "neural" optimization networks: an A/D converter, signal decision circuit and a linear programming circuit. IEEE Trans. Circuits and Systems, CAS-32:533-541.

Zeigler, B.P. (1976). Theory of modeling and simulation. John Wiley, New York.

ROBUST IDENTIFICATION OF DISCRETE-TIME STOCHASTIC SYSTEMS*

Han-FU CHEN and Lei GUO
Institute of Systems Science, Academia Sinica, Beijing, China

1. INTRODUCTION

Since a real system can hardly be modelled by an exact linear deterministic or stochastic system and in this case a small disturbance may cause instability of adaptive algorithms (Egardt, 1980, Riedle et al 1984, Rohrs, 1982), it is of great importance to analyse the influence of the unmodelled dynamics contained in the system upon the behavior of the adaptive control system. For recent years there has been a vast amount of research devoted to this issue (Anderson et al, 1986, Bitmead et al, 1986, Goodwin et al, 1985, Hill et al, 1985, Ioannou et al 1984a,b, Ioannou et al, 1985 and Kosut et al 1984).

The authors have analysed the robustness of identification and adaptive control algorithms in (Chen & Guo, 1986a,b) for discrete- and continuous-time stochastic systems respectively, when the extended least squares (ELS) identification is applied. Roughly speaking, there it is shown that the estimation error and the deviation of the tracking error from its minimum value is small when the unmodelled dynamics is small in a certain sense.

In this paper we are concentrated on the robustness issue of identification for the discrete-time stochastic system which consists of a modelled part being a CARMA process and of an unmodelled part η_n, i.e. the system is described by

$$A(z)y_n=B(z)u_n+C(z)w_n+\eta_n \tag{1}$$

where

$$A(z)=I+A_1z+\cdots+A_pz^p, \quad p\geq 0 \tag{2}$$

$$B(z)=B_1z+B_2z^2+\cdots+B_qz^q, q\geq 1 \tag{3}$$

$$C(z)=I+C_1z+\cdots+C_rz^r, \quad r\geq 0 \tag{4}$$

are matrix polynomials in shift-back operator z with unknown matrix coefficient

$$\theta^{\tau}=(-A_1\cdots -A_p B_1\cdots B_q C_1\cdots C_r) \tag{5}$$

*The project supported by the National Natural Science Foundation of China and by the TWAS Research Grant No.87-43.

but with known upper bounds p,q,r for orders and where the unmodelled dynamics η_n is dominated by

$$\|\eta_n\| \leq \varepsilon \sum_{i=0}^{n-1} a^{n-i} (\|y_i\| + \|u_i\| + \|w_i\| + 1) \tag{6}$$

with $a \in (0,1), \varepsilon \geq 0$.

The driven noise $\{w_n\}$ in the modelled part of the system is assumed to be a martingale difference sequence with respect to a non-decreasing family of σ-algebras $\{\mathcal{F}_n\}$ with properties

$$\sup_n E(\|w_{n+1}\|^2 / \mathcal{F}_n) < \infty \tag{7}$$

$$0 < \alpha = \liminf_{n\to\infty} \frac{1}{n} \sum_{i=0}^{n-1} \|w_i\|^2 \leq \limsup_{n\to\infty} \frac{1}{n} \sum_{i=0}^{n-1} \|w_i\|^2 = \beta < \infty \text{ a.s.} \tag{8}$$

The purpose of the paper is to show three things:

1) The estimation error $\theta - \theta_n$ has an upper bound proportional to $\sqrt{k}$ where k is the condition number of the matrix

$$\sum_{i=1}^{n} \phi_i \phi_i^{\tau}$$

where θ_n is the estimate for θ given by ELS and

$$\phi_n^{\tau} = (y_n^{\tau} \cdots y_{n-p+1}^{\tau} u_n^{\tau} \cdots u_{n-q+1}^{\tau} y_n^{\tau} - \phi_{n-1}^{\tau} \theta_n \cdots y_{n-r+1}^{\tau} - \phi_{n-r}^{\tau} \theta_{n-r+1}) \tag{9}$$

2) In stochastic adaptive control in order simultaneously to get consistent parameter estimates and optimal or suboptimal control performances a small dither is usually added to the system (Caines & Lafortune,1984,Chen,1984,Chen & Caines,1985).Later,it has been shown that better results can be achieved if the excitation is put to the system input (Chen & Guo,1986,1987)rather to the output.Here we prove that for the system with unmodelled dynamics considered in this paper the condition number of $\sum_{i=1}^{n} \phi_i \phi_i^{\tau}$ is bounded in n if the desired control is disturbed by a dither with constant variance.Thus,in this case the estimation error is of order ε.

3) For results mentioned above the positive realness of $C^{-1}(z) - \frac{1}{2}I$ is required.Further,we remove this condition for single input single output systems and get results similar to the previous ones.

2. ROBUSTNESS OF ELS ESTIMATION

Let the unknown matrix θ be estimated by the ELS algorithm

$$\theta_{n+1} = \theta_n + a_n P_n \phi_n (y_{n+1}^{\tau} - \phi_n^{\tau} \theta_n)$$

$$P_{n+1}=P_n-a_nP_n\phi_n\phi_n^\tau P_n, \quad a_n=(1+\phi_n^\tau P_n\phi_n)^{-1} \tag{10}$$

with $P_o=dI, d=mp+\ell q+mr$ and θ_o arbitrary, where ϕ_n is given by (9), m and ℓ are the dimensions of y_n and u_n respectively.

Denote by λ^n_{max} and λ^n_{min} respectively the maximum and minimum eigenvalue of

$$\sum_{i=0}^{n-1}\phi_i\phi_i^\tau+\frac{1}{d}I.$$

Theorem 1. For the system described by (1)-(8) suppose that $C^{-1}(z)-\frac{1}{2}I$ is strictly positive real. Then

$$\limsup_{n\to\infty}\|\theta_n-\theta\|^2\leq c_okd\left(1+\frac{1+2\beta}{\alpha}\right)\varepsilon^2 \tag{11}$$

where

$$k=\limsup_{n\to\infty}\lambda^n_{max}/\lambda^n_{min} \tag{12}$$

and c_o is a constant depending on C(z) and the real number a in (6).

Proof. We just point out the key steps of the proof and refer to (Chen & Guo, 1986a).

Set

$$\xi_{n+1}=y_{n+1}-w_{n+1}-\theta^\tau_{n+1}\phi_n \tag{13}$$

$$\tilde\theta_n=\theta-\theta_n, \qquad r_n=1+\sum_{i=0}^{n-1}\|\phi_i\|^2 \tag{14}$$

Then

$$C(z)\xi_{n+1}=\tilde\theta^\tau_{n+1}\phi_n+\eta_{n+1} \tag{15}$$

Using the strictly positive realness condition we can show that

$$\mathrm{tr}\,\tilde\theta^\tau_{n+1}P^{-1}_{n+1}\tilde\theta_{n+1}$$
$$\leq O(\log^{1+\delta}r_{n+1})+c_o\varepsilon^2((1+2\beta)(n+1)r_{n+1})-c_1\sum_{i=o}^{n}\|\tilde\theta^\tau_{i+1}\phi_i\|^2 \tag{16}$$

where $\delta>0, c_1>0$ are constants depending on C(z) only. Next, we can prove

$$\liminf_{n\to\infty}(r_{n+1}/n)\geq\alpha \tag{17}$$

The desired estimation follows from (16)(17). #

We now consider when k defined by (12) is bounded.

Set

$$\phi_n^o=(y_n^\tau\cdots y_{n-p+1}^\tau u_n^\tau\cdots u_{n-q+1}^\tau w_n^\tau\cdots w_{n-r+1}^\tau)^\tau \tag{18}$$

with $\phi_i^o=0$ for $i<0$ and denote by λ_{max}^{on} and λ_{min}^{on} respectively the maximum and minimum eigenvalue of

$$\sum_{i=0}^{n-1}\phi_i^o\phi_i^{o\tau}+\frac{1}{d}I.$$

<u>Lemma 1</u>. If conditions of Theorem 1 hold then

$$k\leqslant\frac{4k_o}{1-2rc_2d(1+2k_o)\varepsilon^2} \tag{19}$$

where $k_o=\limsup_{n\to\infty}\lambda_{max}^{on}/\lambda_{min}^{on}<\infty$ and c_2 is independent of ε and n but possibly depends on ω.

Proof. By (16)(17) it follows that

$$\sum_{i=0}^{n}\|\tilde{\theta}_{i+1}^\tau\phi_i\|^2\leqslant c_3(1+\frac{1+2\beta}{\alpha})\varepsilon^2 r_{n+1}+0(\log^{1+\delta}r_{n+1}) \tag{20}$$

where c_3 depends on $C(z)$ and a only.Then by (6)(17)(20) from (15) we can find $c_2>0$ such that

$$\sum_{i=0}^{n}\|\xi_{i+1}\|^2\leqslant c_2\varepsilon^2 r_{n+1}+0(\log^{1+\delta}r_{n+1})\quad \text{a.s.} \tag{21}$$

Starting from (21) we can conclude (19).For details we refer to (Chen & Guo,1986a). #

We now introduce the dither v_n to the desired control u_n^s. Let $\{v_n\}$ be a sequence of ℓ-dimensional iid random vectors independent of $\{w_n\}$ and such that

$$Ev_n=0,\ Ev_nv_n^\tau=\mu I,\ \mu>0, \|v_n\|^2<\sigma_v^2\ \forall n>0 \tag{22}$$

Without loss of generality assume

$$\mathcal{F}_n=\sigma\{v_i,w_i,0\leqslant i\leqslant n\} \tag{23}$$

and set

$$\mathcal{F}_n'=\sigma\{v_{i-1},w_i,0\leqslant i\leqslant n\} \tag{24}$$

Let the desired control u_n^s be $\mathcal{F}_n'$-measurable and let the control u_n applied to the system (1) be defined by

$$u_n=u_n^s+v_n \tag{25}$$

We need an auxiliary system

$$A(z)z_n=B(z)u_n+C(z)w_n \tag{26}$$

with the same $\{u_n\}$ and $\{w_n\}$ as those for system (1).Set

$$\psi_n^o=(z_n^\tau \cdots z_{n-p+1}^\tau u_n^\tau \cdots u_{n-q+1}^\tau w_n^\tau \cdots w_{n-r+1}^\tau)^\tau \tag{27}$$

$$\psi_n^1=(\det A(z))\psi_n^o, \quad \phi_n^1=(\det A(z))\phi_n^o$$

and denote by $\lambda_{min}(X)$ the minimum eigenvalue of a matrix X.

Lemma 2. If A(z),B(z) and C(z) have no common left factor and A_p is of full rank with $A_o=I$ by definition and if

$$\sum_{i=0}^{n}\|u_i^s\|^2=O(n)$$

then for sufficiently large n

$$\lambda_{min}(\sum_{i=0}^{n-1}\psi_i^1\psi_i^{1\tau})\geq\gamma n \tag{28}$$

where $\gamma>0$ and it may depend on ω.

Proof. The proof is essentially the same as that for (46) in (Chen & Guo,1986c) if we note that ε,δ and α of (Chen & Guo,1986c) equal 0,0 and 1 respectively in the present case. //

Theorem 2. Suppose that A(z),B(z) and C(z) have no common left factor and A_p is of full rank with $A_o=I$,control defined by (25) is applied to the system (1) and that

$$\sum_{i=0}^{n-1}(\|y_i\|^2+\|u_i\|^2)\leq Mn, \forall n \tag{29}$$

with M possibly depending on ω only.Then

$$k_o \triangleq \limsup_{n\to\infty}\lambda_{max}^{on}/\lambda_{min}^{on}\leq\frac{c_6}{c_5(\gamma-c_4\varepsilon^2)} \tag{30}$$

and

$$\lambda_{min}^{on}\geq c_5(\gamma-c_4\varepsilon^2)n \tag{31}$$

where c_i,i=4,5,6 are constants,$\varepsilon<\sqrt{(\gamma/c_4)}$ and γ is given in (28). If,in addition,$C^{-1}(z)-\frac{1}{2}I$ is strictly positive real,then

$$\limsup_{n\to\infty}\|\theta_n-\theta\|^2\leq c_7\varepsilon^2 \tag{32}$$

where

$$c_7=4c_o c_6 d(1+\frac{1+2\beta}{\alpha})/(c_5(\gamma-c_4\varepsilon^2)(1-2\gamma c_2 d\varepsilon^2)-4\gamma c_6 c_2 d).$$

Proof. Set

$$\phi_n^\eta=(\eta_n^\tau \eta_{n-1}^\tau \cdots \eta_{n-p+1}^\tau 0\cdots 0)^\tau \tag{33}$$

By (1) and (26) we have

$$\begin{pmatrix} A(z) & & 0 \\ & \ddots & \\ 0 & & A(z) \end{pmatrix} (\phi_n^o-\psi_n^o)=\phi_n^\eta \tag{34}$$

hence

$$\phi_n^1=\psi_n^1+\zeta_n \tag{35}$$

where

$$\zeta_n=\begin{pmatrix} \mathrm{Adj}A(z) & & 0 \\ & \ddots & \\ 0 & & \mathrm{Adj}A(z) \end{pmatrix} \phi_n^\eta$$

From (6)(33) and condition (29) it is easy to see

$$\sum_{i=0}^{n-1} \|\zeta_n\|^2 \leq c_4\varepsilon^2 n \text{ for some } c_4>0 \tag{36}$$

For any $x\in R^d$ from (35) it follows that

$$\|x^\tau\psi_n^1\|^2 \leq 2\|x^\tau\phi_n^1\|^2+2\|x^\tau\zeta_n\|^2$$

and hence

$$\lambda_{min}(\sum_{i=0}^{n-1}\psi_n^1\psi_n^{1\tau}) \leq 2\lambda_{min}(\sum_{i=0}^{n-1}\phi_n^1\phi_n^{1\tau})+2c_4\varepsilon^2 n.$$

From this and (28) it follows that

$$\lambda_{min}(\sum_{i=0}^{n-1}\phi_i^1\phi_i^{1\tau}) \geq \tfrac{1}{2}(\gamma-2c_4\varepsilon^2)n. \tag{37}$$

Let

$$\det A(z)=a_o+a_1 z+\cdots+a_{mp}z^{mp}.$$

By the Schwarz inequality and the fact that $\phi_i^o=0$ for $i<0$ it is easy to see that

$$\lambda_{min}(\sum_{i=0}^{n}\phi_i^1\phi_i^{1\tau})=\inf_{\|x\|=1}\sum_{i=1}^{n}(x^\tau\phi_i^1)^2 \leq (mp+1)\sum_{j=0}^{mp}a_j^2\lambda_{min}^{on} \tag{38}$$

Then (31) follows from (37)(38), while (30) from (29)(31). Finally, putting (30) into (19) leads to an estimate for k which together with (11) implies (32). #

3. REMOVAL OF SPR CONDITION

In this section, we consider the single input single output system for which $C^{-1}(z)-\frac{1}{2}$ is unnecessarily strictly positive real, but assume that we can find a polynomial

$$D(z)=1+d_1z+\cdots+d_rz^r$$

so that $(D(z)/C(z))-\frac{1}{2}$ is strictly positive real.

Instead of y_n, u_n, ϕ_n it is natural to use their prefiltered value y_n^f, u_n^f and ϕ_n^f, where

$$\phi_n^f=(y_n^f\cdots y_{n-p+1}^f u_n^f\cdots u_{n-q+1}^f y_n^f-\phi_{n-1}^{f\tau}\theta_n\cdots y_{n-r+1}^f-\phi_{n-r}^{f\tau}\theta_{n-r+1})^\tau \quad (39)$$

$$D(z)y_n^f=y_n, \quad D(z)u_n^f=u_n \quad (40)$$

In the present case the unknown parameter θ is no longer a matrix but a vector. For estimating it we apply an algorithm modified from ELS, namely, the estimate θ_n for θ is recursively given by (see Goodwin & Sin, 1984)

$$\theta_n=\theta_{n-1}+P_n\phi_{n-1}^f(e_n+(D(z)-1)\nu_n) \quad (41)$$

$$P_n=P_{n-1}-P_{n-1}\phi_{n-1}^f\phi_{n-1}^{f\tau}P_{n-1}/(1+\phi_{n-1}^{f\tau}P_{n-1}\phi_{n-1}^f) \quad (42)$$

where

$$e_n=y_n^f-\phi_{n-1}^{f\tau}\theta_{n-1} \quad (43)$$

$$\nu_n=y_n^f-\phi_{n-1}^{f\tau}\theta_n \quad (44)$$

In the sequel by $a_n=O(b_n)$ we mean that $|a_n|<cb_n$ holds for all n and for some constant $c>0$.

<u>Theorem 3</u>. Assume that for system (1)-(8) with $m=\ell=1$ there is a known polynomial $D(z)=1+d_1z+\cdots+d_rz^r$ such that $(D(z)/C(z))-\frac{1}{2}$ is strictly positive real and that the estimate θ_n for θ is given by (41)-(44). Then

1) $$\sum_{i=1}^{n}\|\tilde{\theta}_i^\tau\phi_{i-1}^f\|^2=O\left(r_n^f\frac{\varepsilon^2}{(1-a)^2}\right)+O(\log^{1+\delta}r_n^f) \quad \text{a.s.} \quad (45)$$

where $r_n^f=1+\sum_{i=1}^{n-1}\|\phi_i^f\|^2$, $\delta>0$ and $\tilde{\theta}_n=\theta-\theta_n$

2) $\limsup_{n\to\infty}\|\tilde{\theta}_n\|^2=O(k^f\varepsilon^2/(1-a)^2)$ a.s. (46)

where it is assumed that

$$k^f\triangleq\limsup_{n\to\infty}\lambda_{max}\Big(\sum_{i=0}^{n-1}\phi_i^f\phi_i^{f\tau}+\frac{1}{d}\Big)/\lambda_{min}\Big(\sum_{i=0}^{n-1}\phi_i^f\phi_i^{f\tau}+\frac{1}{d}\Big)<\infty \quad \text{a.s.} \tag{47}$$

3) $\|\theta_n\|^2=O\Big(\log^{1+\delta}r_n^f/\lambda_{min}\Big(\sum_{i=0}^{n-1}\phi_i^f\phi_i^{f\tau}+\frac{1}{d}\Big)\Big),\ \forall\delta>0$ (48)

if $\eta_n=0$, i.e., if there is no unmodelled dynamics.

Proof. By (44) we have

$$\begin{aligned}C(z)(\nu_n-D^{-1}(z)w_n)&=\nu_n+(C(z)-1)\nu_n-C(z)D^{-1}(z)w_n\\&=y_n^f-\phi_{n-1}^{f\tau}\theta_n+(C(z)-1)\nu_n-C(z)D^{-1}(z)w_n\\&=D^{-1}(z)(1-A(z))y_n+D^{-1}(z)B(z)u_n+D^{-1}(z)C(z)w_n\\&\quad+D^{-1}(z)\eta_n+(C(z)-1)\nu_n-\phi_{n-1}^{f\tau}\theta_n-C(z)D^{-1}(z)w_n\\&=\phi_{n-1}^{f\tau}\tilde{\theta}_n+\eta_n^f\end{aligned} \tag{49}$$

where

$$\eta_n^f=D^{-1}(z)\eta_n \tag{50}$$

Set

$$\zeta_n=D(z)\nu_n-w_n \tag{51}$$

Clearly, we have

$$D(z)(\nu_n-D^{-1}(z)w_n)=\zeta_n \tag{52}$$

Combining (49)(52) yields

$$\frac{D(z)}{C(z)}(\phi_{n-1}^{f\tau}\tilde{\theta}_n+\eta_n^f)=\zeta_n \tag{53}$$

By the assumption of strictly positive realness there are constants $k_1>0, k_2>0$ such that for all $n>0$

$$s_n=\sum_{i=0}^{n}(\phi_{i-1}^{f\tau}\tilde{\theta}_i+\eta_i^f)\Big(\zeta_i-\frac{1+k_1}{2}(\tilde{\theta}_i^\tau\phi_{i-1}^f+\eta_i^f)\Big)+k_2\geq 0 \tag{54}$$

By (41)-(44) and (53) it is easy to see that

$$\zeta_n+w_n=(e_n+(D(z)-1)\nu_n)/(1+\phi_{n-1}^{f\tau}P_{n-1}\phi_{n-1}^f) \tag{55}$$

and

$$\theta_n=\theta_{n-1}+P_{n-1}\phi^f_{n-1}(\zeta_n+w_n) \tag{56}$$

Similar to (19) in (Chen & Guo,1986c) from (56) we have

$$\begin{aligned}\mathrm{tr}\ \tilde\theta^\tau_n P^{-1}_n\tilde\theta_n &\leqslant \tilde\theta^\tau_{n-1}P^{-1}_{n-1}\tilde\theta_{n-1}+\|\tilde\theta^\tau_n\phi^f_{n-1}\|^2-2\zeta_n\tilde\theta^\tau_n\phi^f_{n-1}-2w_n\tilde\theta^\tau_n\phi^f_{n-1}\\ &=\tilde\theta^\tau_{n-1}P^{-1}_{n-1}\tilde\theta_{n-1}-2(\phi^\tau_{n-1}\tilde\theta_n+\eta^f_n)(\zeta_n-\frac{1+k_1}{2}(\tilde\theta^\tau_n\phi^f_{n-1}+\eta^f_n))\\ &\quad -k_1(\tilde\theta^\tau_n\phi^f_{n-1})^2-(1+k_1)(\eta^f_n)^2+2\eta^f_n(\zeta_n-(1+k_1)\tilde\theta^\tau_n\phi^f_{n-1})-2w_n\tilde\theta^\tau_n\phi^f_{n-1}\end{aligned}$$

Summing up both sides of the last inequality we obtain

$$\begin{aligned}\tilde\theta^\tau_n P^{-1}_n\tilde\theta_n &\leqslant \tilde\theta^\tau_o P^{-1}_o\tilde\theta_o-2s_n+2k_2-k_1\sum_{i=1}^n(\tilde\theta^\tau_i\phi^f_{i-1})^2-(1+k_1)\sum_{i=1}^n(\eta^f_i)^2\\ &\quad +2\sum_{i=1}^n\eta^f_i(\zeta_i-(1+k_1)\tilde\theta^\tau_i\phi^f_{i-1})-2\sum_{i=0}^n w_i\tilde\theta^\tau_i\phi^f_{i-1}\\ &\leqslant O(1)-k_1\sum_{i=1}^n(\tilde\theta^\tau_i\phi^f_{i-1})^2-(1+k_1)\sum_{i=1}^n(\eta^f_i)^2\\ &\quad +2\sum_{i=1}^n\eta^f_i(\zeta_i-(1+k_1)\tilde\theta^\tau_i\phi^f_{i-1})-2\sum_{i=0}^n w_i\tilde\theta^\tau_i\phi^f_{i-1}\end{aligned} \tag{57}$$

By stability of C(z) from (53) we see that

$$\sum_{i=0}^n\zeta^2_n\leqslant k_3\Big(\sum_{i=0}^n((\tilde\theta^\tau_i\phi^f_{i-1})^2+(\eta^f_i)^2)\Big) \tag{58}$$

for some constant $k_3>0$ and that for any $\delta>0$

$$\begin{aligned}&2\Big|\sum_{i=1}^n\eta^f_i(\zeta_i-(1+k_1)\tilde\theta^\tau_i\phi^f_{i-1})\Big|\\ &\quad\leqslant\frac{1}{\delta}\sum_{i=1}^n(\eta^f_i)^2+2\delta\sum_{i=1}^n\zeta^2_i+2\delta(1+k_1)^2\sum_{i=1}^n(\tilde\theta^\tau_i\phi^f_{i-1})^2\\ &\quad\leqslant(1/\delta+2\delta k_3)\sum_{i=0}^n(\eta^f_i)^2+2\delta(k_3+(1+k_1)^2)\sum_{i=0}^n(\tilde\theta^\tau_i\phi^f_{i-1})^2\end{aligned} \tag{59}$$

Substituting (59) into (57) leads to

$$\begin{aligned}\tilde\theta^\tau_n P^{-1}_n\tilde\theta_n &\leqslant O(1)-(k_1-2\delta(k_3+(1+k_1)^2))\sum_{i=1}^n(\tilde\theta^\tau_i\phi^f_{i-1})^2-2\sum_{i=0}^n w_i\tilde\theta^\tau_i\phi^f_{i-1}\\ &\quad -(1+k_1-\frac{1}{\delta}-2\delta k_3)\sum_{i=1}^n(\eta^f_i)^2\end{aligned} \tag{60}$$

for any $\delta>0$.

By an argument similar to that used for (22)(30) in (Chen & Guo,1986c) we have

$$\left|\sum_{i=1}^{n} w_i\tilde{\theta}_i^\tau\phi_{i-1}^f\right|=O\left(\left(\sum_{i=1}^{n}(\tilde{\theta}_i^\tau\phi_{i-1}^f)^2\right)^\lambda\right)+O(\log^{1+\delta}r_n^f),\ \lambda\in(\tfrac{1}{2},1),\ \forall\delta>0 \tag{61}$$

Putting (61) into (60) we find

$$\tilde{\theta}_n^\tau P_n^{-1}\tilde{\theta}_n\leq O(1)-(k_1-2\delta(k_3+(1+k_1)^2))\sum_{i=1}^{n}(\tilde{\theta}_i^\tau\phi_{i-1}^f)^2$$

$$-(1+k_1-\frac{1}{\delta}-2\delta k_3)\sum_{i=1}^{n}(\eta_i^f)^2+O(\log^{1+\delta}r_n^f) \tag{62}$$

By (6)(40) and (50) it is easy to see

$$\sum_{i=1}^{n}(\eta_i^f)^2\leq(\varepsilon^2k_4/(1-a)^2)\sum_{i=1}^{n-1}((y_i^f)^2+(u_i^f)^2+n) \tag{63}$$

for some constant k_4 independent of ε and a.

Multiplying $D(z)$ to the SISO system (1) we get

$$y_n^f=(1-A(z))y_n^f+B(z)u_n^f+\eta_n^f+(C(z)D^{-1}(z)-1)w_n+w_n,$$

which means that $y_n^f-w_n$ is $\mathcal{F}$-measurable. Then by Lemma 2 of (Chen& Guo 1987) we have

$$\sum_{i=1}^{n}(y_i^f)^2=\sum_{i=1}^{n}(y_i^f-w_i)^2+\sum_{i=1}^{n}w_i+O\left(\left(\sum_{i=1}^{n}(y_i^f-w_i)^2\right)^\lambda\right),\ \tfrac{1}{2}<\lambda<1$$

and hence

$$\liminf_{n\to\infty}\frac{r_n}{n}\geq\liminf_{n\to\infty}(1/n)\sum_{i=1}^{n-1}(y_i^f)^2\geq\liminf_{n\to\infty}\frac{1}{n}\sum_{i=1}^{n-1}w_i^2=\alpha>0\quad \text{a.s.} \tag{64}$$

From (63)(64) we see

$$\sum_{i=1}^{n}(\eta_i^f)^2\leq k_5\varepsilon^2r_n^f/(1-a)^2 \tag{65}$$

for some constant $k_5>0$. Then taking δ small enough so that

$$k_1-2\delta(k_3+(1+k_1)^2)^2>0$$

and combining (62)(65) we obtain

$$\tilde{\theta}_n^\tau P_n^{-1}\tilde{\theta}_n\leq O(1)-k_6\sum_{i=1}^{n}(\tilde{\theta}_i^\tau\phi_{i-1}^f)^2+k_7\varepsilon^2r_n^f/(1-a)^2+O(\log^{1+\delta}r_n^f) \tag{66}$$

where k_6 and k_7 are positive constants independent of ε and a.

Finally, all conclusions of the theorem follow from (66):(45) holds because $\tilde{\theta}_n^{\tau}P_n^{-1}\tilde{\theta}_n \geq 0$, (46) is true if we remove the negative term on the right-hand side of (66) and (48) follows if set $\varepsilon=0$ in (66). #

We now give results similar to Theorem 2.

<u>Theorem 4</u>. If conditions of Theorem 3 are satisfied, $A(z)$, $B(z)$ and $C(z)$ have no common factor and A_p is of full rank with $A_0=1$ and if control defined by (25) is applied to system (1) and

$$\sum_{i=0}^{n-1}(y_i^2+u_i^2)\leq Ln, \quad \forall n \tag{67}$$

with L possibly depending on ω only, then there exists an $\varepsilon^*>0$ such that

$$\limsup_{n\to\infty}\|\theta_n-\theta\|\leq k_8\varepsilon^2 \qquad \text{a.s.}$$

for any $\varepsilon\in[0,\varepsilon^*)$, where k_8 is a constant independent of ε.

Proof. Defining

$$D(z)w_n^f=w_n \tag{68}$$

$$\phi_i^{fo}=(y_n^f\cdots y_{n-p+1}^f u_n^f\cdots u_{n-q+1}^f w_n^f\cdots w_{n-r+1}^f)^{\tau}$$

we have

$$\phi_i^o=D(z)\phi_i^{fo}$$

and

$$\lambda_{min}\Big(\sum_{i=0}^{n-1}\phi_i^o\phi_i^{o\tau}\Big)=\inf_{\|x\|=1}\sum_{i=0}^{n-1}(x^{\tau}\phi_i^o)^2$$

$$=\inf_{\|x\|=1}\sum_{i=0}^{n-1}\Big(\sum_{j=0}^{r}d_jx^{\tau}\phi_{i-j}^{fo}\Big)^2\leq(r+1)\sum_{j=0}^{r}d_j^2\inf_{\|x\|=1}\sum_{i=0}^{n-1}(x^{\tau}\phi_i^{fo})^2$$

$$=(r+1)\sum_{j=0}^{r}d_j^2\lambda_{min}\Big(\sum_{i=0}^{n-1}\phi_i^{fo}\phi_i^{fo\tau}\Big)$$

Hence by (31) we find

$$\lambda_{min}\Big(\sum_{i=0}^{n-1}\phi_i^{fo}\phi_i^{fo\tau}\Big)\geq\big(c_5(\gamma-c_4\varepsilon^2)n/(1+r)\big)\Big(\sum_{j=0}^{r}d_j^2\Big)^{-1} \tag{69}$$

On the other hand, by (67) it is easy to see

$$\lambda_{max}\Big(\sum_{i=0}^{n-1}\phi_i^{fo}\phi_i^{fo\tau}\Big)\leq k_9n \tag{70}$$

for some constant k_9.

Combining (69) (70) we get

$$k_o^f \triangleq \limsup_{n\to\infty} \lambda_{max}\left(\sum_{i=0}^{n-1} \phi_i^{fo}\phi_i^{fo\tau}\right)/\lambda_{min}\left(\sum_{i=0}^{n-1} \phi_i^{fo}\phi_i^{fo\tau}\right)<\infty \tag{71}$$

By definitions of ϕ_n^f and ϕ_n^{fo} and (44)(52)(68) we have

$$\phi_n^f-\phi_n^{fo}=(0\cdots C\nu_n-w_n^f\cdots\nu_{n-r+1}-w_{n-r+1}^f)^\tau$$

$$=(0\cdots CD^{-1}(z)\zeta_n\cdots D^{-1}(z)\zeta_{n-r+1})^\tau \tag{72}$$

Then applying (45)(63) to (53), from (72) we see that

$$\sum_{i=1}^{n}\|\phi_i^f-\phi_i^{fo}\|^2=O(\log^{1+\delta}r_n^f)+O(r_n^f\varepsilon^2/(1-a)^2) \tag{73}$$

which is the analogue of (21). Starting from this estimate by an argument similar to that used in Lemma 1 we get an inequality between k^f and k^f similar to (19). Finally, the desired result follows from (46)(71). #

Applications of the obtained results to robustness analysis of adaptive control and similar results for continuous time systems will be published elsewhere.

REFERENCES

Anderson, B.D.O. et al (1986), Stability of Adaptive Systems: Passivity and Avaraging Analysis, MIT Press.

Bitmead, R. & C.R. Johnson Jr. (1986), Discrete averaging and robust identification, Advances in Control & Dynamic Systems, (Ed. C.T. Leondes) Vol. XXIV.

Caines, P.E. & S. Lafortune (1984), Adaptive control with recursive identification for stochastic linear systems, IEEE Trans. Autom. Control, AC-29, 312-321.

Chen, H.F. (1984), Recursive system identification and adaptive control by use of the modified least squares algorithm, SIAM J. Control & Optimization, Vol.22, No.5, 758-776.

Chen, H.F. & P.E. Caines (1985), Strong consistence of the stochastic gradient algorithm of adaptive control, IEEE Trans. Autom. Control, Ac-30, No.2, 189-192.

Chen, H.F. & L. Guo (1986a), Robustness analysis of identification and adaptive control for stochastic systems, accepted for publication by System & Control Letters.

Chen, H.F. & L. Guo (1986b), Continuous-time stochastic adaptive control: robustness and asymptotic properties, Technical Report, Institute of Systems Science, Academia Sinica.

Chen, H.F. & L. Guo (1986c), Convergence rate of least squares identification and adaptive control for stochastic systems, International J. of Control, Vol.44, No.3, 1459-1476.

Chen, H.F. & L. Guo (1987), Asymptotically optimal adaptive control with consistent parameter estimates, accepted for publication by SIAM J. Control & Optimization.

Egardt,B.(1980),Stability analysis of adaptive control systems with disturbances,Proc.JACC San Francisco,CA.

Goodwin,G.C.& K.S.Sin (1984),Adaptive Filtering Prediction and Control,Prentice Hall.

Goodwin,G.C.,D.J.Hill, D.Q.Mayne & R.H.Middleton (1985),Adaptive robust control (convergence,stability and performance)Technical Report EE 8544.

Hill,D.J.,R.H.Middleton & G.C.Goodwin (1985),A class of robust adaptive control algorithms,Proc. of the 1985 IFAC Symposium on Identification & System Parameter Estimation.

Ioannou,P.A.& P.V.Kokotovic (1984a),Instability analysis and improvement of robustness of adaptive control,Automatica,Vol. 20,No.5,

Ioannou,P.A.& P.V.Kokotovic(1984b),Robust redesign of adaptive control,IEEE Trans.Autom.Control,AC-29,202-211.

Ioannou,P.A.& K.Tsaklis (1985),A robust direct adaptive controller,Report 85-07-1,Electrical Engineering-Systems,University of Southern California.

Kosut,R.L.& B.D.O.Anderson (1984),Robust adaptive control:conditions for local stability,Proc.23 CDC,Las Vegas.

Riedle,B.D.O.& P.V.Kokotovic (1984),Disturbance instability in an adaptive system,IEEE Trans.Autom.Control,AC-29,822-824.

Rohrs,C.E.,L.Valavani,M.Athans & C.Stein (1982),Robustness of adaptive control algorithms in the presence of unmodelled dynamics,Preprints of 21st IEEE CDC,Orlando,FL.

DERIVATIVES OF PROBABILITY MEASURES- CONCEPTS AND APPLICATIONS TO THE OPTIMIZATION OF STOCHASTIC SYSTEMS

Georg Ch. Pflug
Justus-Liebig-University
Arndtstr. 2
D-6300 Giessen, F.R.G .

1. INTRODUCTION

Consider a stochastic system with state space *S*. Assume that *Z* is the random state of the system and that μ_x, the distribution of *Z* depends on a vector x of controls. We are interested in finding the *optimal control*, that is the control which maximizes the *performance* of the system. In mathematical terms we want to find the solution of the problem

$$(P) \quad \left\| \begin{array}{l} F(x) := E_x(H(Z,x)) = \int H(Z,x)\,d\mu_x = \max! \\ x \in X \subseteq \mathbb{R}^k \quad . \end{array} \right.$$

Here H(Z,x) is a *performance-generating* function, which may depend on the control x. Applications of (P) include optimal design of service systems, optimal facility location, optimal design of communication systems, optimal traffic control, etc.

If the structure of H resp. μ_x is rather complicated, we cannot solve (P) in an analytic manner. But it is nearly always possible to simulate the system (by using a random generator for μ_x). What we get then is an *unbiased stochastic estimate* for F(x). We are however primarily interested in the solution of (P) and would like to have an unbiased stochastic estimate of the gradient of F(x), since such an estimate can be used in a *recursive stochastic gradient procedure* for the minimization of (P) (see the monograph of Ermoliev (1976) for a profound discussion of such a technique and Ho et al. (1983) for an application in Queueing Networks)

The existence of unbiased estimates of the gradient of F(x) depends on differentiability properties of $x \to H(z,x)$ (classical!) and $x \to \mu_x$.

In this paper we study different notions of derivatives of probability measures with respect to a parameter x and compare their scope and applicability. The *weak derivative* will be introduced in section 2 and the *process derivative* will be discussed in section 3. Section 4 is devoted to examples for weak derivatives. In section 5 sampling procedures are presented which allow a direct sampling of derivatives. These procedures may be used to construct unbiased estimates of the gradient of F(x). These estimates are much better than the widely used numerical stochastic gradients.

2. WEAK DERIVATIVES

Let $\mathcal{M}$ be the set of all finite signed regular measures on the Borel field $\mathcal{A}$ of a separable metric space (S,d). $\mathcal{M}$ can be endowed with a Banach-norm, the variation-norm

$$||\mu|| = \text{var}(\mu)(S) = \sup_{E_1,E_2\in\mathcal{A}} \mu(E_1)-\mu(E_2)$$

If S is not countable, then $(\mathcal{M}, || \ ||)$ is not separable. Therefore we will consider the weak topology on $\mathcal{M}$. Let $C(S)$ be the set of all bounded continuous real valued functions on S. $C(S)$ is a Banach space with norm

$$||f|| = \sup_x |f(x)|.$$

If $S=\mathbb{R}^m$, the m-dimensional Euclidean space, then we consider furthermore

$C^{(\infty)}(\mathbb{R}^m)$ the set of all test-functions

$C_k(\mathbb{R}^m)$ (for $k\geq 1$), the set of all continuous functions f, for which there are constants c_1,c_2 such that $|f(x)|\leq c_1+c_2||x||^k$.

It is well known, that $\mathcal{M}$ is contained in the topological dual of $C(S)$, with equality if S is compact (cf.Dunford-Schwartz (1957), p.265). We write $\langle g,\mu\rangle$ for the bilinear form, i.e.

$$\langle g,\mu\rangle := \int g(y)\, d\mu(y).$$

The set of all probability measures on S is denoted by $\mathcal{M}_1$. By the known Jordan-Hahn decomposition (Dunford-Schwartz (1957), p.130), every $\mu \in \mathcal{M}$ may be written as

$$\mu = c_1\mu_1-c_2\mu_2 \tag{1}$$

with $\mu_1,\mu_2 \in \mathcal{M}_1$; $c_i\geq 0$; where $c_1\mu_1$ is the positive and $c_2\mu_2$ is the negative part of μ, hence $\mu_1 \perp \mu_2$. It is clear that the representation of an element of $\mathcal{M}$ as a weighted difference of probability measures is not unique, since, for an arbitrary nonnegative measure γ the decomposition

$$\mu = (c_1+1)\,\frac{c_1\mu_1+\gamma}{c_1+1} - (c_2+1)\,\frac{c_2\mu_2+\gamma}{c_2+1}$$

would also do the job. However note that (c_1+c_2) is minimized in the representation (1), if $\mu_1\perp\mu_2$. To see this, write $c_1\mu_1-c_2\mu_2=d_1\nu_1-d_2\nu_2$. If $c_1\mu_1$ is the positive part of μ, then $c_1\mu_1(E) = \sup_{F\subseteq E} \mu(F)$ for all $E\in\mathcal{A}$. Hence $c_1=\sup_F(d_1\nu_1(F)-d_2\nu_2(F)) \leq \sup_F d_1\nu_1(F) \leq d_1$. The inequality $c_2\leq d_2$ is proved in an analogue manner.

In the following, we study applications $x \longrightarrow \mu_x$, mapping $x \in \mathbb{R}^k$ into $\mathcal{M}_1$. In particular differentiability properties will be studied.

<u>Definition 1.</u> A function $x \longrightarrow \mu_x$, mapping an open subset of $\mathbb{R}^k$ into $\mathcal{M}_1$ is called weakly differentiable at the point x, if there is a k-vektor of signed finite measures $\mu_x' := (\mu_x'^{(1)}, \quad , \mu_x'^{(k)})$; $\mu'^{(i)} \in \mathcal{M}$ such that

$$|\langle g,\mu_{x+h}\rangle - \langle g,\mu_x\rangle - \sum h_i \langle g,\mu_x'^{(i)}\rangle| = o(||h||) \tag{2}$$

for all $g \in C(S)$ as $h \to 0$. Here $o(\cdot)$ may depend on g.

The derivative μ_x' may be represented as

$$\mu_x' = \left[c_1(\dot{\mu}_x^{(1)}-\ddot{\mu}_x^{(2)}),\ c_2(\dot{\mu}_x^{(1)}-\ddot{\mu}_x^{(2)}),\ldots,c_k(\dot{\mu}_x^{(1)}-\ddot{\mu}_x^{(k)})\right]$$

where $\dot{\mu}_x^{(i)}$, $\ddot{\mu}_x^{(i)} \in \mathcal{M}_1$. We do not require that $\dot{\mu}_x^{(i)}$ and $\ddot{\mu}_x^{(i)}$ are orthogonal of each other, bearing however in mind that c_i is minimized if $\dot{\mu}_x^{(i)} \perp \ddot{\mu}_x^{(i)}$. Note that $\langle g,\mu_x'^{(i)}\rangle = 0$ for the constant function $g \equiv 1$, since $\langle g,\mu_x\rangle \equiv 1$.

We write $\mu_x' = (c, \dot{\mu}_x, \ddot{\mu}_x)$ to denote the situation that $c = (c_1,\ldots,c_k)$ $\dot{\mu}_x = (\dot{\mu}_x^{(1)},\ldots,\dot{\mu}_x^{(k)})$ $\ddot{\mu}_x = (\ddot{\mu}_x^{(1)},\ldots,\ddot{\mu}_x^{(k)})$ is the derivative of $x \longrightarrow \mu_x$ at x in the sense of (2).

The derivative obeys the following rules:

1) If $x \longrightarrow \mu_x$ and $x \longrightarrow \nu_x$ is differentiable, with derivative $(c, \dot{\mu}_x, \ddot{\mu}_x)$ resp. $(d, \dot{\nu}_x, \ddot{\nu}_x)$, then $x \longrightarrow \alpha\mu_x+(1-\alpha)\nu_x$ is differentiable with derivative

$$\left[\alpha c+(1-\alpha)d,\ \frac{\alpha c\dot{\mu}_x+(1-\alpha)d\dot{\nu}_x}{\alpha c+(1-\alpha)d},\ \frac{\alpha c\ddot{\mu}_x+(1-\alpha)d\dot{\nu}_x}{\alpha c+(1-\alpha)d}\right]$$

(Note that $\alpha\dot{\mu}_x+(1-\alpha)\dot{\nu}_x$ is in general not orthogonal to $\alpha\ddot{\mu}_x+(1-\alpha)\ddot{\nu}_x$).

2) Under the same assumptions $x \longrightarrow \mu_x * \nu_x$ (convolution) is differentiable with derivative

$$(c+d,\ \frac{c}{c+d}\dot{\mu}_x*\nu_x+\frac{d}{c+d}\mu_x*\dot{\nu}_x,\ \frac{c}{c+d}\ddot{\mu}_x*\nu_x+\frac{d}{c+d}\mu_x*\ddot{\nu}_x)$$

3) If T is a measurable transformation which maps μ_x onto μ_x^T i.e.

$$\mu_x^T(A) := \mu_x(T^{-1}(A))$$

then $x \longrightarrow \mu_x^T$ is differentiable with derivative $(c,\dot{\mu}_x^T,\ddot{\mu}_x^T)$

4) If T is a continuous application $S \longrightarrow (S',d')$ and $\mathcal{A}_o$ is the σ-algebra generated by T, then $x \longrightarrow \mu_x|\mathcal{A}_o$ is differentiable with derivative

$$(c,\ \dot{\mu}_x|\mathcal{A}_o,\ \ddot{\mu}_x|\mathcal{A}_o)$$

5) If $\mu_x \ll \mu$ with density $f_x(y)$ and $x \longrightarrow f_x(y)$ is $L^1(\mu)$- differentiable, then $x \longrightarrow \mu_x$ is also weakly differentiable. A mapping $X \longrightarrow L^1(\mu)$ is called L^1-differentiable, if there is a vector $g_x = (g_x^{(1)},\dots, g_x^{(k)})$ of L^1-functions with the property that

$$||f_{x+h} - f_x - \sum h_i g_x^{(i)}||_{L^1} = o(h) \qquad \text{for } h \to 0.$$

Of course, $c_i = ||g_x^{(i)}||_{L^1}$ and $\dot{\mu}_x$ resp. $\ddot{\mu}_x$ may be taken as

$$\frac{d\dot{\mu}_x^{(i)}}{d\mu} = \frac{1}{c_i}\max(g_x^{(i)},0)$$

$$\frac{d\ddot{\mu}_x^{(i)}}{d\mu} = \frac{-1}{c_i}\min(-g_x^{(i)},0).$$

It may happen that $\mu_x \ll \mu$ and $x \longrightarrow \mu_x$ is weakly differentiable with $\mathrm{var}(\mu_x') \ll \mu$, but $x \longrightarrow \mu_x$ is not differentiable in the L^1-sense. Consider the following example: Let $S = [0,1]$,

$$d\mu_x(y) = c(x)\cdot(1+x\cdot\sin(\tfrac{y}{x}))dy \quad ; \qquad d\mu_o(y) = dy$$

It is easy to see that $c(x) = 1+0(x^2)$. Therefore

$$\frac{1}{x}\left[f_x(y)-f_o(y)\right] = \sin\ (\tfrac{y}{x})+0(x)$$

which is not convergent in the $L_1[0,1]$-sense. But $\int_0^1 g(y)\cdot\sin(\frac{y}{x})dy \to 0$ for all measurable, bounded g (Riemann-Lebesgue Theorem).

6) If $x \to \mu_x$ is differentiable with derivative $(c,\ \dot{\mu}_x,\ \ddot{\mu}_x)$ and $x \longrightarrow g(x)$ is differentiable, then so is

$$x \longrightarrow \mu_{g(x)}$$

and has derivative

$$(c, J_g\dot{\mu}_x,\ J_g\ddot{\mu}_x)$$

where J_g is the jacobian of g.

7) Decomposition of measures.

If $\mu_x = \int \nu_x(\cdot|z)d\gamma_x(z)$ is a decomposition and both ν_x and γ_x are differentiable, then

$$(c, \frac{1}{c}\left[\int \dot{\nu}_x(\cdot|z)1_x(z) + \int \nu_x(\cdot|z)d\dot{\gamma}_x(z)\right], \frac{1}{2}\left[\int \ddot{\nu}_x(\cdot|z)d\gamma_x(z) + \int \nu_x(\cdot|z)d\ddot{\gamma}_x(z)\right]$$

8) If $x \longrightarrow \mu_x$ is weakly differentiable with derivative $(c, \dot{\mu}_x, \ddot{\mu}_x)$, then - per definitionem -

$$x \longrightarrow \langle g, \mu_x \rangle$$

is differentiable for all continuous g. However, one may slightly depart from continuity. If A is the set of discontinuities of h and satisfies $\dot{\mu}_x(A) = \ddot{\mu}_x(A) = 0$ then

$$x \longrightarrow \langle h, \mu_x \rangle$$

is differentiable, with derivative

$$c(\langle h, \dot{\mu}_x \rangle - \langle h, \ddot{\mu}_x \rangle).$$

9) If $\{\mu_x\}$ are probability measures on $\mathbb{R}^m$ and $x \in \mathbb{R}$, then weak differentiability is equivalent to the following: There is a function of bounded variation F'_x, such that for each continuity point u of F'_x

$$\lim \frac{1}{h} (F_{x+h}(h) - F_x(u)) = F'_x$$

where F_x is the distribution function of μ_x.

3. PROCESS DERIVATIVES

We have just seen that on $\mathbb{R}^m$ weak differentiability is equivalent to the differentiability of the distribution functions

$$x \longrightarrow F_x(u)$$

with respect to x. There is another notion of differentiability - somewhat dual to the above - , which is connected to the differentiability of the functions

$$x \longrightarrow F_x^{-1}(u)$$

This concept is called *process differentiability* and was used by many authors previously (see Ho et al. (1983), Suri (1987), Glynn (1987), etc.) Its relation to weak differentiability will be studied below. For simplicity we assume $S = \mathbb{R}$ and $x \in \mathbb{R}$.

<u>Definition 2.</u> A family of random variables $\{Y_x\}$ on a probability space $(\Omega, \mathcal{A}, P)$ is called a *process representation* for $\{\mu_x\}$, if Y_x has law μ_x

$$\mathcal{L}(Y_x) = \mu_x.$$

<u>Example 1.</u> If F_x is the distribution function belonging to μ_x, then

$$Y_x = F_x^{-1}(U) \qquad U \sim \text{uniform on } [0,1]$$

is always a process representation of $\{\mu_x\}$.

<u>Definition 3.</u> If $\{Y_x\}$ is a version of a process representation for $\{\mu_x\}$, for which

$$\lim_{h\to 0} \frac{1}{h}(Y_{x+h}-Y_x) =: \dot{Y}_x$$

exists almost everywhere, then $\dot{Y}_x$ is called a *process derivative* of μ_x at the point x.

It is important to stress the fact, that the law of $\dot{Y}_x$ is *not uniquely determined* by the family $\{\mu_x\}$ and hence we may only speak of "a" process derivative.

<u>Example 2.</u> Let $\{\mu_x\}$ be normal $N(0,1+x^2)$-distributions. There are two possible process representations:

1. $Y_x = Y_1+xY_2$, where Y_i are independent N(0,1)-distributions
2. $Y_x = \sqrt{1+x^2}\cdot Y$, where Y is a N(0,1)-distribution.

In the first representation, the pathwise derivative is $\dot{Y}_x = Y_2 \sim N(0,1)$ and in the second representation, the derivative is $\dot{Y}_x = \frac{x}{\sqrt{1+x^2}} Y \sim N(0,\frac{x^2}{1+x^2})$.

Remark that the second representation leads to a process derivative with smaller variance. This fact will be explained below.

Suppose that $x \longrightarrow Y_x$ is differentiable not only in the a.e. -sense but also in L_1, i.e.

$$||Y_{x+h}-Y_x-h\cdot\dot{Y}_x||_{L_1} = o(h) \qquad \text{as } h \to 0$$

Then, for each bounded, continuously differentiable function φ

$$\varphi(Y_{x+h}) = \varphi(Y_x+h\dot{Y}_x+R_h) =$$

$= \varphi(Y_x) + (h\cdot\dot{Y}_x+R_h)\cdot\varphi'(\tilde{Y}_h)$, where $R_h = Y_{x+h}-A_x-h\dot{Y}_x$ and $\tilde{Y}_h$ lies between Y_x and Y_{x+h}. Since $\tilde{Y}_h \to Y_x$ and $\dot{Y}_x+h^{-1}R_h \xrightarrow{\mathcal{L}_1} \dot{Y}_x$ it follows that

$$\frac{1}{h}\left[\varphi(Y_{x+h})-\varphi(Y_x)\right] \xrightarrow{\mathcal{L}_1} \dot{Y}_x\cdot\varphi'(Y_x)$$

In particular, $\frac{\partial}{\partial x} E(\varphi(Y_x)) = \frac{\partial}{\partial x} \langle \varphi, \mu_x \rangle = E(\dot{Y}_x \varphi'(Y_x))$. Thus, for every appropriate φ, $E(\dot{Y}_x \varphi'(Y_x))$ must be independent of the particular process representation. It is easy to check that indeed, for example 2

$$E(Y_2 \varphi'(Y_1 + xY_2)) = E\left[\frac{x}{\sqrt{1+x^2}} Y \cdot \varphi'(\sqrt{1+x^2}\, Y)\right]$$

If $x \longrightarrow Y_x$ is process differentiable at x_o, then the process behaves near x_o like

$$Y_{x_o} + (x - x_o) \cdot \dot{Y}_{x_o}$$

Thus the joint distribution of $(Y_{x_o}, \dot{Y}_{x_o})$ determines its local behavior. We have already seen that $(Y_{x_o}, \dot{Y}_{x_o})$ is not determined by $\{\mu_x\}$. Since, for any square integrable random variable Y and sub-σ-algebra $\mathcal{B}$

$$\mathrm{Var}(Y) = \mathrm{Var}\,(E(Y|\mathcal{B})) + E(\mathrm{Var}(Y|\mathcal{B}))$$

($\mathrm{Var}(Y|\mathcal{B})$ is the conditional variance), we see that $\mathrm{Var}(\dot{Y}_x \varphi'(Y_x))$ is minimized if the conditional variance $\mathrm{Var}(\dot{Y}_x|Y_x)$ is zero. Notice that in the example 2, $\mathrm{Var}(\frac{x}{\sqrt{1+x^2}} Y|Y) = 0$ for the second representation.

This idea may be extended further: Since we are mainly interested in finding good estimates for

$$\frac{\partial}{\partial x} \langle \varphi, \mu_x \rangle = E(\dot{Y}_x \varphi'(Y_x))$$

the estimate $\dot{Y}_x \varphi'(Y_x)$ can be improved (i.e. its variance can be reduced) by taking $E(\dot{Y}_x|Y_x)$ instead of $\dot{Y}_x$ itself. We call

$$E(\dot{Y}_x|Y_x)$$

a *reduced process derivative*. It may happen that there is no process representation for $\{\mu_x\}$, for which $E(\dot{Y}_x|Y_x)$ a process derivative. However there may be another process $\tilde{Y}_x$ representing $\{\tilde{\mu}_x\}$ which has $E(\dot{Y}_{x_o}|Y_{x_o})$ as process derivative at x_o and $\mu_{x_o} = \tilde{\mu}_{x_o}$. We call $\{\tilde{\mu}_x\}$ a *reduced process representation*.

Example 3. (Glynn) Let $\mu_x = \frac{2}{3}\delta_x + \frac{1}{3}\delta_x$. Then $Y+x\cdot\dot{Y}$ with $Y\equiv 0$

$$\dot{Y} = \begin{cases} 1 & \text{with prob. } 2/3 \\ -1 & \text{with prob. } 1/3 \end{cases}$$

is a process representation of $\{\mu_x\}$. Since $E(\dot{Y}|Y) \equiv \frac{1}{3}$ a reduced process derivative is 1/3 (deterministic): The process $Y+x\cdot\frac{1}{3}$ belongs to the family $\{\tilde{\mu}_x\}$, where $\tilde{\mu}_x = \delta_{x/3}$. This family has process derivative identical to the reduced process derivative of $\{\mu_x\}$.

The next theorem shows that if the joint distribution of $(Y_x,\dot{Y}_x)$ has a Lebegue density, then a differentiable process representation can be found by taking the "inverse distribution function" method. The derivative is automatically reduced.

Theorem 1. Suppose that $(Y,\dot{Y})$ has a joint Lebesgue density $f(y,z)$. Let F_h be the distribution function of $Y+h\dot{Y}$. Suppose that F_o is strictly monotone. Then

$$\lim \frac{1}{h}(F_h^{-1}(u)-F_o^{-1}(u)) = E(\dot{Y}|Y=F_o^{-1}(u)).$$

Proof. Write F_h as

$$F_h(x) = F(h,x) = P\left\{Y+h\dot{Y}\leq x\right\} = \int_{-\infty}^{x}\int_{-\infty}^{\infty} f(y-hz,z)\,dz\,dy =$$

$$= \int_{-\infty}^{\infty}\int_{-\infty}^{x-hv} f(u;v)\,du\,dv = \int_{-\infty}^{\infty} G(x-hv,v)\,dv, \text{ where}$$

$G(x,v) = \int_{-\infty}^{x} f(u,v)\,du\,dv$. Of course $\frac{\partial}{\partial x}G(x,v) = f(x,v)$. We have to show that we may differentiate under the integral sign to get

$$\frac{\partial}{\partial h}F(h,x)\Big|_{h=0} = \int_{-\infty}^{\infty} -vf(x,v)\,dv.$$

Since, for $h > 0$,

$$\int_{-\infty}^{\infty}\int_{-\infty}^{\infty}\frac{1}{h}\,|G(x-hv,v) - G(x,v)|\,dx\,dv = \int_{-\infty}^{\infty}\int_{-\infty}^{\infty}\frac{1}{h}\int_{\min(x,x-hv)}^{\max(x,x-hv)} f(u,v)\,du\,dv\,dx =$$

$$\int_{-\infty}^{\infty}\int_{-\infty}^{\infty}\int_{-\infty}^{\infty}\frac{1}{h}\,1_{[\min(u,u+hv),\max(u,u+hv)]}(x)\ f(u,v)\ dx\ dv\ du =$$

$\int_{-\infty}^{\infty}\int_{-\infty}^{\infty}|v|\ f(u,v)\ du\ dv$ and $\frac{1}{h}\ [G(x-hv,v) - G(x,v)]$ converges a.e. to $-v\cdot f(x,v)$, Scheffe's Lemma implies that $\frac{1}{h}\ [G(x-hv,v) - G(x,v)]$ converges in $L^1(dx,dv)$ to $-v\cdot f(x,v)$. The same is true for $h < 0$. Thus, for almost all x,

$\int_{-\infty}^{\infty}\frac{1}{h}\ [G(x-hv) - G(x,v)]\ dv$ converges to $\int_{-\infty}^{\infty}-vf(x,v)dv$.

Now fix a $p\in(0,1)$, such that F_0^{-1} is continuous at p. Let $F(h,x_h)\equiv p$, i.e. $x_h=F_h^{-1}(p)$. The implicit functions theorem gives

$$\frac{\partial x_h}{\partial h} = -\frac{\frac{\partial}{\partial h}F(h,x_o)\,|_{h=0}}{\frac{\partial}{\partial x}F(0,x)\,|_{x=x_o}} = \frac{\int_{-\infty}^{\infty} v\cdot f(x_o,v)dv}{\int_{-\infty}^{\infty} f(x_o,v)dx} = E(\dot{Y}\,|Y=x_o)$$

$$= E(\dot{Y}\,|Y=F^{-1}(p)).$$

□

<u>Theorem 2.</u> Let Y_h, $h\in(-\varepsilon,\varepsilon)$ be a family of random variables with distribution function F_h. Suppose that $F_h^{-1}(u)$ is Lipschitz-continuous in a neighborhood of u_o, uniformly in h. Let R_h be a random variable with the property

$$P\Big\{|R_h|>a(h)\Big\} \le b(h) \tag{3}$$

with $a(h)=o(h)$, $b(h)=o(h)$.

Let $\tilde{F}_h$ be the distribution function of Y_h+R_h. Then

$$\lim \frac{1}{h}(F_h^{-1}(u_o)-\tilde{F}_h^{-1}(u_o)) = 0.$$

<u>Proof.</u> From the inclusions

$$\{Y_h\le u\} \subseteq \{Y_h+R_h\le u+a(h)\} \cup \{|R_h|>a(h)\}$$

$$\{Y_h+R_h\le u-a(h)\} \subseteq \{Y_h\le u\} \cup \{|R_h|>a(h)\}$$

we get

$$\tilde{F}_h(u-a(h)) - b(h) \le F_h(u) \le \tilde{F}_h(u+a(h)) + b(h).$$

in a neighborhood of $u_o=F_o^{-1}(p)$. Therefore

$$\tilde{F}_h^{-1}(p-b(h))-a(h) \leq F_h^{-1}(p) \leq \tilde{F}_h^{-1}(p-b(h))-a(h)$$

Since $|\tilde{F}_h^{-1}(p-b(h)) - \tilde{F}_h^{-1}(p)| \leq$ const. b(h) we get the desired result. □

If the family $\{\mu_x\}$ is not smooth enough, then the theorem 1 does not apply. There is a simple trick to bring everything to the smooth case: a regularization.

Definition 4. Let $\{\mu_x\}$ be a family of probability measures. We call the family $\{\mu_x^{(\sigma)} = \mu_x * N(0,\sigma^2(1+(x-x_o)^2))\}$ the σ-regularized family at x_o. This family has smooth densities.

Remark. The idea behind the definition is the following: If

$$Y_x = Y_{x_o} + (x-x_o)\cdot\dot{Y}_{x_o} + R_x$$

is a differentiable process representation of $\{\mu_x\}$, then

$$\tilde{Y}_x = (Y_{x_o}+\sigma Z_1) + (x-x_o)(\dot{Y}_{x_o}+\sigma Z_2) + R_x$$

is a differentiable process representation of $\{\mu_x^{(\sigma)}\}$, where Z_1 and Z_2 are N(0,1) random variables independent of each other and everything else. The regularized variables $(Y_{x_o}+\sigma Z_1)$ resp. $(\dot{Y}_{x_o}+\sigma Z_2)$ fulfill however all smoothness conditions.

We are therefore able to state the main result.

Theorem 3. If $\{\mu_x\}$ possesses a L_1-process derivative at x_o which satisfies (3) then

(i) For the regularized family $\{\mu_x^{(\sigma)}\}$ with distribution function $F_x^{(\sigma)}$

$$\frac{\partial}{\partial x} F_x^{(\sigma)-1}(u)\Big|_{x=_o} \text{ exists for all } \sigma>0$$

and

$$\frac{\partial}{\partial x} F_x^{(\sigma)-1}(U) \text{ is uniformly integrable for U~uniform } [0,1].$$

(ii) For every $\psi\in C^{(\infty)}(\mathbb{R})$ with $\sup_j \sup_x | \psi^{(2j-1)}(x) \frac{1}{2\cdot 4\cdot\ .(2j)} | < \infty$

$x \longrightarrow \langle\psi,\mu_x\rangle$ is differentiable at $x=x_o$ and

$$\frac{\partial}{\partial x} \langle\psi,\mu_x\rangle\Big|_{x=x_o} \leq K\cdot\sup_y|\psi'(y)| \qquad \text{for a constant K.}$$

Conversely, if (i), (ii) is satisfied, then there is a pair of random varibles $(Y_{x_o}, \dot{Y}_{x_o})$ such that

$$\frac{\partial}{\partial x} \langle \varphi, \mu_x \rangle \Big|_{x=x_o} = E(Y_{x_o} \varphi'(Y_{x_o}))$$

for all $\varphi \in C^{(\infty)}$. The process

$$x \longrightarrow Y_{x_o} + (x-x_o)\dot{Y}_{x_o}$$

is a reduced process representation for $\{\mu_x\}$.

<u>Proof.</u> The first part is contained in theorems 1 and 2. In order to prove the second assertion, let

$$(Y_{x_o}^{(\sigma)}, \dot{Y}_{x_o}^{(\sigma)}) = (F_{x_o}^{(\sigma)-1}(U), \frac{\partial}{\partial x} F_x^{(\sigma)-1}(U) \Big|_{x=x_o})$$

with U~uniform [0,1]. Because of (i) $(Y_{x_o}^{(\sigma)}, \dot{Y}_{x_o}^{(\sigma)})$ is uniformly tight and we may choose a limit $(Y_{x_o}, \dot{Y}_{x_o})$ as $\sigma \to 0$. Of course $\mathcal{L}(Y_{x_o}) = \mu_{x_o}$.

Let $\mu_x^{(\sigma)} = \mu_x * N(0, \sigma^2(1+(x-x_o)^2))$ the regularization and

$$\bar{\mu}_x^{(\sigma)} = \mu_x * N(0, \sigma^2)$$

another different smoothing. Let $L(Y_x^{(\sigma)}) = \mu_x^{(\sigma)}$ and $\mathcal{L}(Z) = N(0, \sigma^2)$, Z independent of $Y_x^{(\sigma)}$. Then, for $\varphi \in C^{(\infty)}(\mathbb{R})$

$$|\langle \varphi, \mu_x^{(\sigma)} \rangle - \langle \varphi, \bar{\mu}_x^{(\sigma)} \rangle| = |E(\varphi(Y_x^{(\sigma)} + (x-x_o)\cdot\sigma Z) - \varphi(Y_x^{(\sigma)}))| =$$
$$= |x - x_o| \cdot \sigma \cdot O(1) \qquad (4)$$

Let $\varphi_\sigma(y) = \int \varphi(y-z)\, dN(0,\sigma^2)(z)$ be the regularization of φ. Then, by (4)

$$\frac{\partial}{\partial x} \langle \varphi_\sigma, \mu_x \rangle \Big|_{x=x_o} = \frac{\partial}{\partial x} \langle \varphi, \bar{\mu}_x^{(\sigma)} \rangle \Big|_{x=x_o} = \frac{\partial}{\partial x} \langle \varphi, \mu_x^{(\sigma)} \rangle \Big|_{x=x_o} + O(\sigma)$$
$$= E(\dot{Y}_x^{(\sigma)} \varphi'(Y_x^{(\sigma)})) + O(\sigma) \longrightarrow E(\dot{Y}_x \varphi'(Y_x)) \quad \text{as } \sigma \longrightarrow 0 \qquad (5)$$

A Taylor expansion of φ leads to

$$\varphi_\sigma(y) = \int \varphi(y-z)\, dN(0,\sigma^2)(z) = \varphi(y) + \sum_{j=1}^{\infty} \varphi^{(2j)}(y) \cdot \frac{\sigma^{2j}}{2\cdot 4 \cdots (2j)}$$

Since $\sup_j | \sup_x \varphi^{(2j-1)}(x) \cdot \frac{1}{2\cdot 4 \cdots (2j)} | < \infty$ we get from (ii)

$$\frac{\partial}{\partial x} \langle \varphi_\sigma, \mu_x \rangle = \frac{\partial}{\partial x} \langle \varphi, \mu_x \rangle + O(\sigma^2)$$

and therefore from (5)

$$\frac{\partial}{\partial x} \langle \varphi, \mu_x \rangle = E(\dot{Y}_{x_o} \varphi'(Y_{x_o}))$$

which is the desired identity. □

If $x \longrightarrow \mu_x$ possesses a process derivative, it may not be weakly differentiable (Simply take $Y_x \equiv x$). Conversely $x \longrightarrow \mu_x$ may be weakly differentiable without having a process derivative:

Take ξ uniformly distributed in [0,1] and

$$Y_x = \begin{cases} 0 & \xi \le x \\ 1 & \xi > x. \end{cases}$$

If A is a set with boundary ∂A, satisfying $(\dot{\mu}_x + \ddot{\mu}_x)(\partial A) = 0$ then $x \longrightarrow \mu_x(A)$ is differentiable, if $x \longrightarrow \mu_x$ is weakly differentiable (cf. property 8. of section 2).A similar statement is not valid for process derivatives.

<u>A numerical example</u>

Let μ_x be an exponential distribution with expectation x. μ_x has density

$$f_x(y) = x \cdot \exp(-xy)$$

and inverse distribution function

$$F_x^{-1}(u) = -\frac{1}{x} \ln(1-u).$$

Let Y_x be distributed according to μ_x. We look for estimates for $\frac{\partial}{\partial x} E(\sqrt{Y_x}) = \frac{\partial}{\partial x} \langle \sqrt{y}, \mu_x \rangle$. Three different methods will be considered:

1. Numerical differences

Fix a h > 0. We estimate $\frac{\partial}{\partial x} \langle \sqrt{y}, \mu_x \rangle$ by

$$Z_x^{(1)} := \frac{1}{h} (\sqrt{Y_{x+h}} - \sqrt{Y_x}) \tag{6}$$

where Y_{x+h}, Y_x are independent. Of course, (6) is a biased estimate. We may reduce the bias by making h very small but then the variance will increase.

2. Process derivatives

A process representation of $\{\mu_x\}$ is

$$Y_x = -\frac{1}{x} \ln(1-U) \qquad U \sim \text{uniform } [0,1].$$

A process derivative of $\sqrt{Y_x}$ is

$$Z_x^{(2)} := -\frac{1}{2} x^{-2/3} \sqrt{-\ln(1-U)}.$$

which is an unbiased estimate.

3. Weak derivatives

$\frac{\partial}{\partial x} \langle \sqrt{y}, \mu_x \rangle$ is estimated by

$$Z_x^{(3)} := \frac{1}{e \cdot x} (\sqrt{Y_x^{(1)}} - \sqrt{Y_x^{(2)}})$$

where $Y_x^{(1)}$ resp. $Y_x^{(2)}$ are distributed according to $\dot{\mu}_x$ resp. $\ddot{\mu}_x$. Details of the generation of these random variables are contained in the next section, example 2). This estimate is also unbiased.

The following table compares the variances of the three estimates. Notice, that in the particular case we may calculate analytically that

$$H(x) := \langle \sqrt{y}, \mu_x \rangle = \int x \sqrt{y} \exp(-xy)\, dy = \frac{1}{2} \sqrt{\pi/x}.$$

and $$H'(x) = \frac{-\sqrt{\pi}}{4\, x^{3/2}} .$$

TABLE 1:

		x=1.0	x=2.0	x=3.0
H'(x)		-0.4431	-0.1566	-0.0852
Numerical difference :				
h=0.1	$E(Z_x^{(1)})$	-0.4590	-0.1569	-0.0853
	$Var(Z_x^{(1)})$	10.8	5.37	3.57
h=0.01	$E(Z_x^{(1)})$	-0.4308	-0.1567	-0.0852
	$Var(Z_x^{(1)})$	1069.9	534.5	356.2
Process derivative:				
	$Var(Z_x^{(2)})$	0.0485	0.0061	0.0008
Weak derivative:				
	$Var(Z_x^{(3)})$	0.0226	0.0027	0.0001

We see that the weak derivative has always the smallest variance.

4. EXAMPLES FOR WEAK DERIVATIVES

4.1 Poisson distribution.

Let μ_x be a Poisson distribution with mean x, i.e.

$$\mu_x = e^{-x} \sum_{j=0}^{\infty} \frac{x^j}{j!} \delta_j$$

(δ_u is the point mass at u). $x \longrightarrow \mu_x$ is differentiable with derivative

$$\mu_x' = -e^{-x} \sum_{j=0}^{\infty} \frac{x^j}{j!} \delta_j + e^{-x} \sum_{j=0}^{\infty} \frac{x^j}{j!} \delta_{j+1}$$

μ_x' may be represented in different ways. The simplest representation is

$$\mu_x' = \dot{\mu}_x - \ddot{\mu}_x$$

where $\dot{\mu}_x = e^{-x} \sum_{j=0}^{\infty} \frac{x^j}{j!} \delta_{j+1}$, a shifted Poisson distribution and $\ddot{\mu}_x = \mu_x$. This is however not the Jordan-Hahn decomposition. The latter is

$$\dot{\mu}_x = \frac{1}{c} e^{-x} \sum_{j=\lceil x \rceil}^{\infty} \left[\frac{j-x}{x}\right] \frac{x^j}{j!} \delta_j \; ; \qquad \ddot{\mu}_x = \frac{1}{c} e^{-x} \delta_o + \sum_{j=1}^{\lfloor x \rfloor} \left[\frac{x-j}{x}\right] \frac{x^j}{j!} \delta_j$$

with

$$c = e^{-x} \sum_{j=x}^{\infty} \left[\frac{j-x}{x}\right] \frac{x^j}{j!} \delta_j \; .$$

The second representation has a smaller c, namely

x	c
0.5	0.6.65
1.0	0.3679
2.0	0.2707
3.0	0.2240
4.0	0.1954
5.0	0.1754

but the first representation has clear sampling advantages. We may sample $\ddot{\mu}_x$ as a Poisson-variable Y and $\dot{\mu}_x$ as Y+1. Of course, process derivatives do not exist, since the Poisson distribution is discrete.

4.2 <u>Exponential distribution</u>

If μ_x is an exponential distribution with density $x \cdot e^{-xy}$, a derivative is

$$\left[\frac{1}{x \cdot e}, \; x \cdot e(1-xy)e^{-xy} \cdot 1_{\{y \leq 1/x\}} dy, \; x \cdot e \cdot (xy-1)e^{-xy} 1_{\{y \geq 1/x\}} dy\right]$$

A process representation for the exponential distributions is

$$-\frac{1}{x} \ln(\xi)$$

where ξ is uniformly [0,1] distributed. A process derivative is

$$\frac{1}{x^2} \ln(\xi) .$$

This derivative does not help us in differentiating e.g. $x \longrightarrow \mu_x([a,\infty])$, the probability that a μ_x-distributed random variable is larger than a. With the weak derivative we get

$$\frac{\partial}{\partial x}\mu_x([a,\infty]) = c\left[\dot{\mu}_x([a,\infty])-\ddot{\mu}_x([a,\infty])\right].$$

4.3 Normal distribution

Let μ_x have a Lebesgue density

$$f_x(y) = \frac{1}{\sqrt{2\pi}}\,\sigma(x)\,\exp\left[-\frac{1}{2}\left[\frac{y-m(x)}{\sigma(x)}\right]^2\right]$$

and $x \longrightarrow m(x)$ and $x \longrightarrow \sigma(x)$ be differentiable.
A process representation for μ_x is

$$\sigma(x)(\xi-m(x))$$

with $\xi\sim N(0,1)$ and derivative $\sigma'(x)(\xi-m(x)) + \sigma(x)\cdot\mu'(x)$.
If g is a $C^{(1)}$-function then

$$x \longrightarrow \langle g,\mu_x\rangle$$

is differentiable with derivative

$$\frac{\partial}{\partial x}\langle g,\mu_x\rangle = \langle Lg,\mu_x\rangle = \int Lg\cdot f_x dy \tag{7}$$

where L is the differential operator

$$(Lg)(y) = g'(y)\left[y\cdot\frac{\sigma'(x)}{\sigma(x)} - \sigma(x)m'(x)\right]$$

In the space $L^1(\mathbb{R}) \cap C^{(1)}(\mathbb{R})$ (7) is the inner product. L has an adjoint operator in this space

$$\int(Lg)\cdot f_x dy = \int g\cdot L^*(f_x)dy$$

where

$$L^*(f_x) = \frac{\partial}{\partial y}\left[f_x(y)(\sigma(x)m'(x) - \frac{\sigma'(x)}{\sigma(x)}\cdot y)\right] \tag{8}$$

The weak derivatives have Lebesgue densities which are identical to the positive resp. negative part of (8).

5. SAMPLING DERIVATIVES OF PROBABILITY MEASURES

Let μ_x be a probability measure on $(S,\mathcal{A})$ and let $(c,\dot{\mu}_x,\ddot{\mu}_x)$ be a weak derivative. If a pair of random variables $(\dot{Y},\ddot{Y})$ has marginal distribution $\dot{\mu}_x$ resp. $\ddot{\mu}_x$, then

$$c\ (g(\dot{Y})-g(\ddot{Y})) \tag{9}$$

is an unbiased estimate for $\frac{\partial}{\partial x}\ \langle g,\mu_x\rangle$ for $g \in C(S)$. Of course, we are interested in estimates with small variance. Let $\dot{Z}:=g(\dot{Y})$ and $\ddot{Z}:=g(\ddot{Y})$. The marginals of $(\dot{Z},\ddot{Z})$ are the image measures $\dot{\mu}_x^g$ resp. $\ddot{\mu}_x^g$ of $\dot{\mu}_x$ resp. $\ddot{\mu}_x$. and hence fixed. We are looking for the joint distribution with given marginals which minimizes $\mathrm{Var}(\dot{Z}-\ddot{Z})$. The solution of this minimization problem is given by the following theorem.

Theorem 4. (Major,1978). Let φ be a convex function. Let (Z_1,Z_2) have marginal distribution functions F_1 resp. F_2. Then

$$\inf\{\ E(\varphi(Z_1-Z_2)\ :\ Z_1 \text{ has d.f. } F_1 \text{ and } Z_2 \text{ has d.f.} F_2\ \} =$$
$$= \int_0^1 \varphi(F_1^{-1}(u)\ -\ F_2^{-1}(u))\ du$$

Corollary. The joint distribution which minimizes $\mathrm{Var}(Z_1-Z_2)$ is that of

$$(F_1^{-1}(U),F_2^{-1}(U)) \qquad U \sim \text{uniform } [0,1].$$

Thus the minimal variance estimate for (7) is

$$c(Z_1\ -\ Z_2)$$

where $Z_1= F_1^{-1}(U)$, $Z_2= F_2^{-1}(U)$, $U \sim$ uniform$[0,1]$, F_1 d.f. of $\dot{\mu}_x^g$, F_2 d.f. of $\ddot{\mu}_x^g$.

Since the decomposition of a signed measure is not unique, we may decompose μ_x' in several ways.

$$\mu_x' = c\ (\dot{\mu}_x\ -\ \ddot{\mu}_x) \qquad \text{(Jordan-Hahn decomposition)}$$

$$\mu_x' = d\ (\dot{\nu}_x\ -\ \ddot{\nu}_x) \qquad \text{(other decomposition).}$$

Although the constant c is minimized for the Jordan-Hahn decomposition, it may happen that a different decomposition leads to an estimate for $\frac{\partial}{\partial x}\ \langle g,\mu_x\rangle$ with smaller variance. It is somewhat astonishing that in some cases we can do better than to decompose μ_x' into orthogonal parts and to use then the optimal estimate given by the Corollary. Here is the example:

Example: Let $\mu_x = (0.125+x\cdot0.1)\delta_1 + (0.125-x\cdot0.4)\delta_2 + (0.125+x\cdot0.1)\delta_3 + (0.125-x\cdot0.2)\delta_4 + (0.125+x\cdot0.1)\delta_5 + (0.125+0.2)\delta_6 + (0.125-x\cdot0.4)\delta_7 + (125+x\cdot0.5)\delta_{7.1}$

The Jordan-Hahn decomposition of μ_x' is

$$c = 1$$

$$\dot{\mu}_x = 0.1\cdot\delta_1 + 0.1\cdot\delta_3 + 0.1\cdot\delta_5 + 0.2\cdot\delta_6 + 0.5\cdot\delta_{7.1}$$

$$\ddot{\mu}_x = 0.4\cdot\delta_2 + 0.2\cdot\delta_4 - 0.4\cdot\delta_7$$

Another decomposition of μ_x' is

$$d = 1.05263 = (0.95)^{-1}$$

$$\dot{\nu}_x = 0.095\cdot\delta_1 + 0.145\cdot\delta_3 + 0.095\cdot\delta_5 + 0.19\cdot\delta_6 + 0.475\cdot\delta_{7.1}$$

$$\ddot{\nu}_x = 0.38\cdot\delta_2 + 0.05\cdot\delta_3 + 0.19\cdot\delta_4 + 0.38\cdot\delta_7.$$

Let F_1 resp. F_2 be the distribution functions of $\dot{\mu}_x$ resp. $\ddot{\mu}_x$ and G_1 resp. G_2 the distribution functions of $\dot{\nu}_x$ resp. $\ddot{\nu}_x$. We may estimate

$$\frac{\partial}{\partial x}\int y \, d\mu_x(y)$$

(which has by the way the constant value 1.25, easily obtainable by direct calculation) either by

$$D_1 := 1\cdot(F_1^{-1}(U) - F_2^{-1}(U)) \qquad U \sim \text{uniform } [0,1]$$

or by

$$D_2 = 1.05263\cdot(G_1^{-1}(U) - G_2^{-1}(U))$$

A somewhat lengthy calculation shows that

$$E(D_1) = E(D_2) = 1.25 \quad \text{(as it should be!)}$$

but

$$\mathrm{Var}(D_1) = 2.4415 > 2.3198 = \mathrm{Var}(D_2) \;!$$

For the practical implementation of optimization algorithms one is often satisfied with a reasonable, though not optimal estimate of $\frac{\partial}{\partial x}\langle g,\mu_x\rangle$. The simplest way is to take the Jordan-Hahn decomposition $\mu_x' = c(\dot{\mu}_x - \ddot{\mu}_x)$ and to sample $\dot{Y}$ resp. $\ddot{Y}$ independently from $\dot{\mu}_x$ resp. $\ddot{\mu}_x$. Of course, the variance could be decreased, if $g(\dot{Y})$ and $g(\ddot{Y})$ has positive correlation, but it is not easy to sample from such a joined distribution.

In principle, any known method for random number generation can be used to sample $(\dot{Y},\ddot{Y})$ from $(\dot{\mu}_x,\ddot{\mu}_x)$. (For the generation of non-uniform variables see Devroye(1986)). The problem discussed in this section is however the following:

Is it possible to modify a generator for a μ_x-distributed random varible Y in an easy way to get a random number generator for the derivative $(\dot{Y},\ddot{Y})$?

We begin with <u>discrete</u> distributions. Let $\mu_x = \sum_{i=1}^{m} p_i(x) \cdot \delta_{u_i}$, where δ_u is the point mass at u. Alternatively, we may write

$$P\{Y=u_i\} = p_i(x).$$

Of course, $x \longrightarrow \mu_x$ is weakly differentiable iff $x \longrightarrow p_i(x)$ is differentiable for all i. Let

$\dot{p}_i^{(j)}(x) := \frac{\partial}{\partial x_j} p_i(x)$ and $a^+ := \max(a,0)$; $a^- := -\min(a,0)$.

The derivative is:

$$c_j = \sum_{i=1}^{m} (\dot{p}_i^{(j)}(x))^+ = \sum_{i=1}^{m} (\dot{p}_i^{(j)}(x))^-$$

$$\dot{\mu}_x^{(j)} = \sum_{i=1}^{m} \frac{1}{c_j} (\dot{p}_i^{(j)}(x))^+ \cdot \delta_{u_i} \ ; \qquad \ddot{\mu}_x^{(j)} = \sum_{i=1}^{m} \frac{1}{c_j} (\dot{p}_i^{(j)}(x))^- \ \delta_{u_i}$$

Consider the following graph for illustration. Let x be univariate.

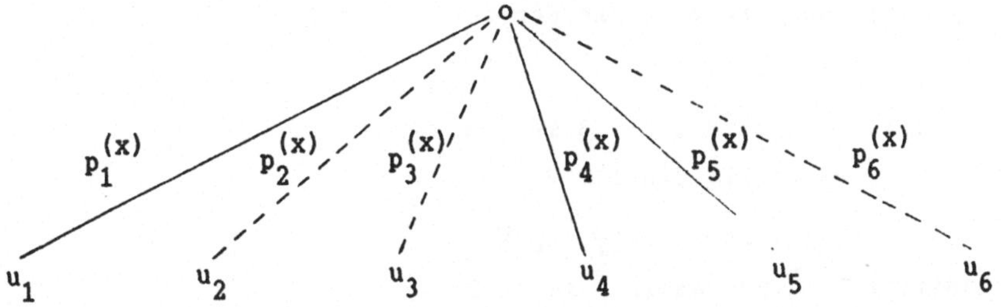

To sample μ_x, we have to choose one of the i-th arc with probability $p_i(x)$. Suppose that the solid arcs indicate that $\frac{d}{dx} p_i(x) \geq 0$ and the dashed arcs correspond to $\frac{d}{dx} p_i(x) < 0$. We sample $\dot{\mu}_x$ resp. $\ddot{\mu}_x$ by

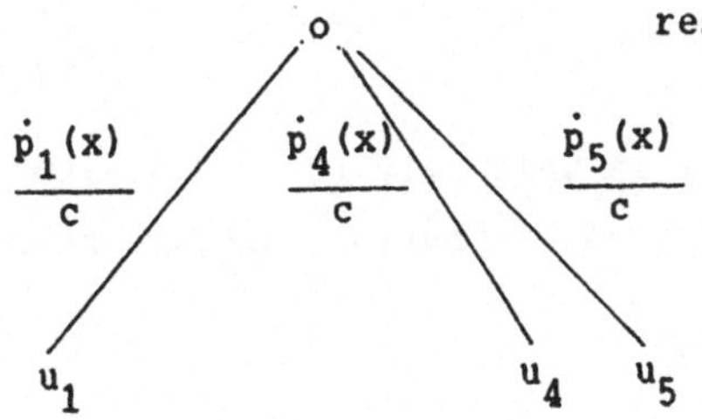

resp.

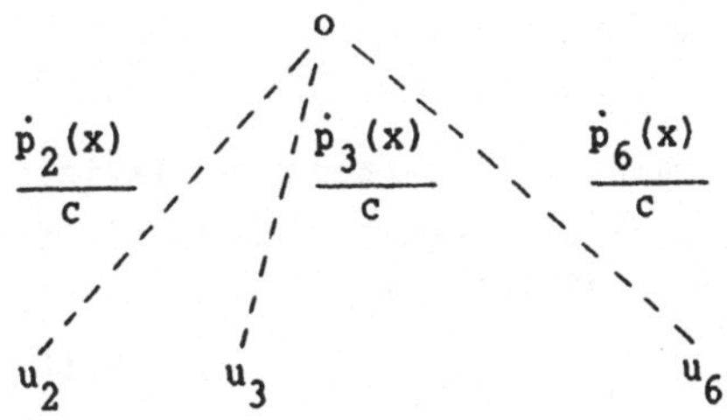

where $c = \sum_{i=1}^{m} (p_i(x))^+ = \sum_{i=1}^{m} (p_i(x))^-$.

Sometimes random numbers are generated in a two - or multiple step procedure. Let, for instance

$$\mu_x = \sum_{j=1}^{m} p_j(x) \cdot \nu_{x,j}$$

We sample μ_x by a two-stage procedure:

Let Z take the value j with probability p_j. If Z equals j then a $\nu_{x,j}$ distributed r.v. is sampled. If $x \longrightarrow \nu_{x,j}$ is differentiable with derivative $(c_j, \dot{\nu}_{x,j}, \ddot{\nu}_{x,j})$ and $x \longrightarrow p_j(x)$ is differentiable, then so is $x \longrightarrow \mu_x$ and the derivative is

$$\left[d, \frac{1}{d}\sum_{j=1}^{m} \dot{p}_j^{+}(x)\nu_{x,j} + \frac{1}{d}\sum_{j=1}^{m} p_j(x)\dot{\nu}_{x,j}, \frac{1}{d}\sum_{j=1}^{m} \dot{p}_j^{-}(x)\nu_{x,j} + \frac{1}{d}\sum_{j=1}^{m} p_j(x)\ddot{\nu}_{x,j}\right]$$

where $d = \sum_{j=1}^{m} \dot{p}_j^{+} + \sum_{j=1}^{m} p_j(x) \cdot c_j$. The sampling procedure is the following:

Let $c = \sum_{j=1}^{m} p_j^{+}(x)$ and $\dot{Z}$ resp. $\ddot{Z}$ the derivatives of Z.

With probability $\frac{c}{d}$ we sample $\dot{Z}$ and if $\dot{Z} = j$
we sample finally V_j.
(and analoguously for $\ddot{Z}$)

with probability $1-\frac{c}{d}$ we sample Z an if Z = j
we sample finally $\dot{V}_j$
(and analoguously $\ddot{V}_j$)

For a non-discrete probability μ_x there are several methods of sampling. We discuss here the transformation method:

Suppose that

$$Y_x := K(x, \xi) \tag{10}$$

is a μ_x-distributed random variable, where ξ is uniformly [0,1] distributed. A typical transformation method is the inverse-distribution function method, where K is defined as

$$K(x, \xi) = F_x^{-1}(\xi)$$

F_x being the d.f. of μ_x. Our goal is to find random variables $\dot{Y}_x$ resp. $\ddot{Y}_x$ which are distributed according to

$$\dot{Y}_x \sim \dot{\mu}_x \quad ; \quad \ddot{Y}_x \sim \ddot{\mu}_x$$

Theorem 5. Suppose that $(x,z) \longrightarrow K(x,z)$ is twice differentiable. Let, for $z \in [0,1]$

$$L(x,z) := \frac{\frac{\partial}{\partial z \partial x} K(x,z)}{\frac{\partial}{\partial z} K(x,z)} + \frac{\frac{\partial^2}{\partial z^2} K(x,z \frac{\partial}{\partial x} K(x,z)}{\left[\frac{\partial}{\partial z} K(x,z)\right]^2}$$

Let $L(x,z) = L^+(x,z) - L^-(x,z)$ be the decomposition of L into its positive resp. negative part. If η_1 resp. η_2 have Lebesgue densities

$$\text{const. } L^+(x,z) \quad \text{resp. const. } L^-(x,z)$$

on [0,1], then

$$\dot{Y}_x \sim K(x,\eta_1)$$

$$\ddot{Y}_x \sim K(x,\eta_2)$$

is a possible sampling realization for $\dot{\mu}_x$ resp. $\ddot{\mu}_x$.

Proof. By the transformation rule of measures we know that the density of $K(x,\xi)$ is

$$f_x(y) = \frac{1}{\frac{\partial}{\partial z} K(x,K^{-1}(x,y)}$$

Let μ_x' be the signed measure which is the derivative of $x \longrightarrow \mu_x$. Of course

$$\frac{d\mu_x'}{d\mu_x}(y) = \frac{\frac{\partial}{\partial x} f_x(y)}{f_x(y)} . \tag{11}$$

Let λ be the Lebesgue measure on [0,1] and let the signed measure ν_x' have the density

$$\frac{d\nu_x'}{dx}(z) = \frac{d\mu_x'}{d\lambda}(K(x,z)).$$

Everything is proved, if we show that

$$\frac{d\nu_x'}{dx}(z) = - \frac{\frac{\partial}{\partial z \partial x} K(x,z)}{\frac{\partial}{\partial z} K(x,z)} + \frac{\frac{\partial^2}{\partial z^2} K(x,z) \frac{\partial}{\partial x} K(x,z)}{\left[\frac{\partial}{\partial z} K(x,z)\right]^2}$$

which follows from simple calculus. □

Example 1. Let $K(x,\xi) = -\frac{1}{x}\ln(1-\xi)$; i.e. μ_x is an exponential distribution. Then $L(x,z) = \frac{1}{x}(\ln 1-z)+1)$. The positive resp. negative normed parts of L are

$$L^+(z) = e\cdot(\ln z+1)\cdot 1_{\{z\geq 1/e\}}$$

$$L^-(z) = -e\cdot(\ln z+1)\cdot 1_{\{z<1/e\}}$$

and $c = \frac{1}{e\cdot x}$.

If η_1 has density $L^-(z)$ and η_2 has density $L^-(z)$ then

$$\dot{Y}_x = K(x,\eta_1) \qquad \dot{Y}_x = K(x,\eta_2)$$

is a possible model for the derivatives. η_x may be sampled by a rejection method, since it has bounded density. $\sqrt{\eta_2}$ has the density $-2ez(2\ln z+1)1_{\{z\leq 1/e\}}$ which again is bounded and may be sampled by rejection. Note that the distributions of η_1 and η_2 both do not depend on x!

The very same method can be applied, if the distribution of μ_x is sampled by using a couple of uniform random variables, i.e.

$$Y_x = K(x,\xi_1,\ldots,\xi_k)$$

Suppose that for fixed $z_2,\ldots,z_k$

$$(x,z_1) \longrightarrow K(x,z_1,z_2,\ldots,z_k)$$

satisfies the assumptions of theorem 1, then we may view μ_x as a decomposition of the form

$$\mu_x = \int \nu_x(\cdot|z_2,\ldots,z_k)\ dz_2,\ldots,dz_k.$$

where ν_x is the image measure of $K(x,\xi_1,z_2,\ldots,z_k)$. Since the "mixture measure" $dz_2,\ldots,dz_k$ does not depend on x, we may differentiate μ_x by mixing over the derivatives of ν_x i.e.

$$(c,\dot{\mu}_x,\ddot{\mu}_x) = (c,\frac{1}{c}\int\dot{\nu}_x(\cdot(|z_2,\ldots,z_k)dz,\ldots,z_k,\ \frac{1}{c}\int\ddot{\nu}_x(\cdot|z_2,\ldots,z_k)dz_2,\ldots,z_k)$$

where c has to be chosen in such a way to make $\dot{\mu}_x$ and $\ddot{\mu}_x$ probability measures.

Example 2. Take a normally distributed random variable $Y_x \sim N(x,1)$, sampled by the Box-Muller method:

$$Y_x = \sqrt{-2\ln(1-\xi_1)}\ \cos(\pi\xi_2)+x$$

Here $K(x,z_1,z_2) = \sqrt{-2\ln(1-z_1)}\cos(2\pi z_2)+x$. $(x,z_1) \longrightarrow K(x,z_1,z_2)$ satisfies the assumptions of theorem 1 and we may calculate $L(x,z_1,z_2)$ for fixed z_2:

$$L(x,z_1,z_2) := \frac{1}{\sqrt{-2\ln(1-z)}} - \sqrt{-2\ln(1-z)}$$

We may decompose L in the two parts

$$L_1 = \frac{1}{\sqrt{-2\ln(1-z)}} \quad ; \qquad L_2 = \sqrt{-2\ln(1-z)}$$

We do this for the sake of simplicity - being quite aware that this is not the decomposition of L into positive and negative parts.

As $\int_0^1 \frac{1}{\sqrt{-2\ln(1-z)}}\,dz = \int \sqrt{-2\ln(1-z)}\,dz = \frac{1}{\sqrt{2}}\,\Gamma(\tfrac{1}{2}) = \frac{1}{\sqrt{2}}\sqrt{\pi}$

We have to sample random variables $\dot{Y}_x$ resp. $\ddot{Y}_x$ with densities

$$\frac{1}{\sqrt{\pi}}\,\frac{1}{\sqrt{-\ln(z)}} \text{ resp. } \frac{1}{\sqrt{\pi}}\sqrt{-\ln(z)}\,.$$

These densities are unbounded on [0,1] and the calculation of the distribution function requires the knowledge of the incomplete Γ-function.

So, we proceed a different way: The densities of $\sqrt{\dot{Y}_x^2}$ resp. $\sqrt{\ddot{Y}_x^2}$ are

$$\frac{1}{\sqrt{\pi}}\,\frac{z}{\sqrt{-\ln(z)}} \text{ resp. } \frac{1}{\sqrt{\pi}}\,z\sqrt{-\ln(z)}$$

which are both bounded by $\frac{1}{\sqrt{\pi}}\,e^{-1/2}$.

The final algorithm is:

1. Sample U_1, ξ_2, ξ_3 from uniform [0,1]
2. Accept

$$\dot{Y}_x = \sqrt{-2\ln(\xi_1^2)}\,\cos(\pi\xi_2)+x$$
$$\text{if } \xi_3 \cdot \le e^{1/2}\cdot\frac{z}{\sqrt{-2\ln z}}$$

   Accept

$$\ddot{Y}_x = \sqrt{-2\ln(\xi_1^2)}\,\cos(\pi\xi_2)+x$$

$$\text{if } \xi_3 \le e^{1/2}\cdot z\sqrt{-2\ln z}$$

3. The constant is $c = \sqrt{\pi}/2$.

Of course, one could also consider the function

$$(x, z_2) \longrightarrow K(x, z_1, z_2) \text{ for an application}$$

of theorem 5. Here

$$L(\nu, z_1, z_2) = \frac{\cos(\pi \cdot z_2)}{\sin^2(\pi \cdot z_2)}$$

which is not integrable in [0,1]! There are two poles: $z_2 = 0$ for the positive part and $z_2 = 1$ for the negative part. We are therefore led to try to take the point mass at 0 for the positive part and the point mass at 1 for the negative part.

Thus the algorithm is

1. Sample ξ_1 from the uniform distribution
2. Take

$$\dot{Y}_x = \sqrt{-2\ln(\xi_1)} + x$$

$$\ddot{Y}_x = -\sqrt{-2\ln(\xi_1)} + x$$

$$c = (2\pi)^{-1/2}$$

Obviously, this leads to a correct distribution.

REFERENCES

[1] Bickel, P.J. and Freedman, D.A. (1981). *Some asymptotic theory for the bootstrap*. The Annals of Statistics, Vol. 9, No. 6, 1196-1217.

[2] Devroye, L. (1986). *Non-Uniform Random Variate Generation*. Springer Verlag, New York, Berlin, Heidelberg.

[3] Dunford, N. and Schwartz, J.T. (1957). *Linear operators, Part I*. Interscience publishers, New York.

[4] Ermoliev, Yu. (1976). *Methods of stochastic programming*. Monographs in Optimization and Operations Research, Nauka, Moskwa.

[5] Glynn, P.W. (1987). *Construction of process-differentiable representations for parametric families of distributions*. University of Wisconsin-Madison, Mathematics Research Center.

[6] Ho, Y.C. and Cao X. (1983). *Perturbation Analysis and Optimization of Queueing Networks*. J. Optim. Theory Applic. Vol. 40, No. 4, 559-582.

[7] Kersting, G.D. (1978). *Die Geschwindigkeit der Glivenko-Cantelli Konvergenz, gemessen in der Prohorov-Metrik*. Math. Zeitschrift 163, No. 1, 65-102.

[8] Major, P. (1978). *On the invariance principle for sums of independent, identically distributed random variables*. Journ. of Multivariate Analysis 8, 487-501.

[9] Rubenstein, R.Y. (1981). *Simulation and the Monte Carlo Method*. John Wiley, New York.

[10] Suri, R. (1987). *Infinitesimal perturbation analysis for general discrete-event systems*. To appear in Journal of the ACM.

THE SEPARATION OF JETS AND SOME ASYMPTOTIC PROPERTIES OF RANDOM SEQUENCES

I.M. Sonin
Central Economic Mathematical Institute, USSR Academy of Sciences,
ul. Krasikov, 32 Moscow 117418, USSR

1 The present article is concerned with three seemingly unrelated problems which are actually profoundly interconnected. We shall refer to these as Problems 1,2,3.

Problem 1 deals with a mathematical model of a physical problem concerning the asymptotic behaviour of the nonhomogeneous solution in a system of vessels (discrete coloured streams) as time tends to infinity. Theorem 1 proved in [1] – a theorem on separation of jets – states that every coloured stream with a bounded number of vessels at each moment, can be decomposed into jets (a jet is any sequence of subsets of vessels) such that stabilization of volume and concentration takes place in every jet and the overflow between different jets is finite on an infinite time interval. From the physical point of view a coloured stream is an example of an irreversible process and in our opinion Theorem 1 is only a sketch of a general theorem about the asymptotic behaviour of such processes. From the probabilistic point of view Problem 1 is a problem on the asymptotic behaviour of a nonhomogeneous Markov chain. Using Theorem 1 we get Theorem 2 which describes such behaviour under the single condition that the number of elements in the state spaces of the Markov chain is bounded. Problem 2 deals with $P(\liminf (X_n \in D_n))$ for classes of random sequences with the same two-dimensional distributions $P(X_n, X_{n+1})$, $n \in \mathbb{N}$. Theorem 3 gives an interesting inequality for such probabilities. Problem 3 and the corresponding Theorem 4 are concerned with the existence of a nonrandom sequence (barrier) such that the expected number of intersections between this sequence and the trajectories of a martingale-type random sequence is finite on infinite time interval.

2 A sequence $(M_n, r_n^{ij}) \equiv (M, r)$ is called a (discrete) *stream* if its elements satisfy the following conditions:

$$M_n \subseteq \mathbb{N} = \{1, 2, \cdots\},\ r_n^{ij} \geq 0,\ i \in M_n,\ j \in M_{n+1},\ n \in \mathbb{N}\ ,$$
$$\sum_{i \in M_n} r_n^{ij} = \sum_{k \in M_{n+2}} r_{n+1}^{jk} < \infty,\ j \in M_{n+1},\ n \in \mathbb{N}\ . \tag{1}$$

The elements of a stream (M, r) have the following interpretation: M_n is a set of vessels in moment n, r_n^{ij} is the amount of liquid (solution) flowing from vessel $i \in M_n$

to vessel $j \in M_{n+1}$ in moment n.

Denote by $m_n^i = \sum_{j \in M_{n+1}} r_n^{ij}$ the amount of solution in vessel i in moment n. We assume that $\sum_{i \in M_1} m_1^i < \infty$ and hence from (1) without loss of generality

$$\sum_{i \in M_n} m_n^i = 1, \; n \in \mathbf{N} \; . \tag{2}$$

A sequence $(M_n, r_n^{ij}, O_n, \alpha_n^i) \equiv (M, r, O, \alpha)$ is called a (discrete) *coloured stream* (CS) if (M, r) is a stream, $O = (O_n)$ is a sequence of subsets, $O_n \subseteq M_n$, $n \in \mathbf{N}$, the numbers α_n^i satisfy the following conditions:

$$\begin{gathered} 0 \le \alpha_n^i \le 1, \; \alpha_n^i = 0, \; i \in O_n \; ; \\ \alpha_{n+1}^j = \sum_{i \in M_n} \alpha_n^i r_n^{ij} / m_{n+1}^j, \; j \in M_{n+1} \backslash O_{n+1} \; . \end{gathered} \tag{3}$$

The number α_n^i is interpreted as a concentration of the solution (temperature) in vessel i in a moment n. The sequence of sets (O_n) is called an "ocean" -- the concentration of the solution in vessels of this sequence is always zero and does not depend on the concentration of the solution flowing into these vessels. Every CS (M, r, O, α) is specified if a stream (M, r), an ocean (O_n) and initial values α_1^i for $i \in M_1 \backslash O_1$ are specified.

Problem 1 is to describe the asymptotic behaviour of the CS as time tends to infinity. For a fixed stream, any sequence $J = (J_n)$, $J_n \subseteq M_n$, $n \in \mathbf{N}$ is called a *jet*; a jet $J = (J_n)$ is called a *trap* if the total 'overflow' from jet J to other vessels is finite, i.e.

$$\sum_{n=1}^{\infty} \left[\sum_{i \in J_n, j \notin J_{n+1}} r_n^{ij} + \sum_{i \notin J_n, j \in J_{n+1}} r_n^{ij} \right] < \infty \; . \tag{4}$$

A tuple of jets $\gamma = (J^1, \ldots, J^c)$ is called a *partition* (of $M = \prod_{n=1}^{\infty} M_n$) if

a) $J_n^s \cap J_n^l = \emptyset$, $s \neq l$,

b) $\bigcup_{s=1}^{c} J_n^s = M_n$, $n \in \mathbf{N}$.

Let $|B|$ be the cardinality of the set B and $m_n(A) = \sum_{i \in M_n} m_n^i$, $A \subseteq M_n$. The following theorem was formulated and proved in [1].

THEOREM 1 *(Separation of jets). For any CS* (M, r, O, α) *such that* $|M_n \backslash O_n| \le N < \infty$, $n \in \mathbf{N}$ *there exist an integer* c, $1 \le c \le N+1$, *a tuple* $\tilde{m}^l, \tilde{\alpha}^l)$, $l = 1, \ldots, c$, $\tilde{m}^l > 0$, $0 \le \tilde{\alpha}^1 < \cdots < \tilde{\alpha}^c \le 1$, *a partition* $\gamma = (J^1, \ldots, J^c)$, $J^l = (J_n^l)$, $l = 1, \ldots, c$ *and a tuple of jets* $(\hat{J}^1, \ldots, \hat{J}^c)$ *such that* $\hat{J}_n^l \subseteq J_n^l$, $l = 1, \ldots, c$ *and*

a) *Stabilization of the volume in every jet* J^l *takes place,*

$$\lim_n m_n(J_n^l) = \tilde{m}^l, \; l = 1, \ldots, c \; ,$$

b) *Stabilization of the concentration in every jet* J^l *takes place, with the possible exception of some vessels whose volume tends to zero, i.e.*

$$\lim_n \alpha_n^i = \tilde{\alpha}^l, \; i \equiv i_n \in \hat{J}_n^l \subseteq J_n^l, \; \lim_n m_n(J_n^l \backslash \hat{J}_n^l) = 0, \; l = 1, \ldots, c \; ,$$

c) *Every jet J^l, $l = 1, \dots, c$ is a trap for the stream (M, r).*

REMARK 1 The existence of the partition γ with properties a) and b) is rather trivial. A nontrivial fact is the existence of γ with all three properties a), b), c). The main difficulties arise in the case of $\lim_n \inf_{i \in M_n} m_n^i = 0$.

REMARK 2 The partition γ in Theorem 1 was constructed in [1] recursively in an explicit form and it was proved also that the overflow between each jet and other jets can be estimated through the increments of the function $F(m_n, \alpha_n) \equiv \sum_{i \in M_n} m_n^i (\alpha_n^i)^2$, which is monotonically decreasing for every CS.

3 The probabilistic interpretation of Problem 1 is the following. Let (M, r) be a stream. Without loss of generality $m_n^i > 0$ for all i, n. Consider a sequence of matrices $P_n = [P_n^{ij}]$, where $P_n^{ij} = r_n^{ij}/m_n^i$, $i \in M_n$, $j \in M_{n+1}$, $n \in \mathbf{N}$. It is obvious that P_n are stochastic matrices and (P_n) with initial condition m_1^i, $i \in M_1$, uniquely specify a nonhomogeneous Markov chain (Z_n) with state spaces M_n, transition matrices P_n and initial distribution m_1^i. Conversely, if a Markov chain (Z_n) with values in discrete spaces M_n is given, then defining a sequence r_n^{ij} by formulae $r_n^{ij} = P(Z_n = i,\ Z_{n+1} = j)$, $i \in M_n$, $j \in M_{n+1}$ it is easy to check that (M, r) is a stream. Therefore, a stream and a nonhomogeneous Markov chain are two interpretations of the same model. Obviously the sequence $m_n^i = P(Z_n = i) = \sum_{j \in M_{n=1}} r_n^{ij}$ satisfies the following relations.

$$m_{n+1} = m_n P_n,\ n \in \mathbf{N}\ , \tag{5}$$

where $m_n = (m_n^1, \dots, m_n^{|M_n|})$ is a row-vector.

The relation (4) from the definition of a trap means that the expected number of the entrances into and exists from a jet (J_n) of a trajectory of the Markov chain (Z_n) is finite.

Any CS (M, r, O, α) also has a simple probabilistic interpretation. It is easy to show that a CS may also be defined in such a way that the initial condition α_1^i, $i \in M_1$ will take a form $\alpha_1^i = I_G(i)$, where $G \subseteq M_1 \backslash O_1$, $I_G(\cdot)$ is a characteristic function of the set G. Let (Z_n) be a Markov chain corresponding to a stream (M, r). Define a sequence of sets (D_n) by $D_1 = G$, $D_n = M_n \backslash O_n$ for $n > 1$ and introduce the posterior probabilities

$$\beta_n^i = P(Z_s \in D_s,\ s = 1, \dots, n | Z_n = i),\ i \in M_n,\ n \in \mathbf{N} \tag{6}$$

Using the Markov property of (Z_n) it is easy to show that β_n^i satisfy the same formulae as α_n^i in (3). It is evident also that $\beta_1^i = \alpha_1^i$ for $i \in M_1$.

Conversely, any nonhomogeneous Markov chain $Z = (Z_n)$ with values in some discrete spaces (M_n) and a sequence of sets $D = (D_n)$ $D_n \subseteq M_n$ (we shall say for the sake of brevity that (Z, D) is a *Markov pair*) specify a CS with a stream (M, r), $r_n^{ij} = P(Z_n = i,\ Z_{n+1} = j)$, $i \in M_n$, $j \in M_{n+1}$, with an ocean $O = (O_n)$, $O_n = M_n \backslash D_n$ and with α_n^i given by (6). So as in the case of a stream and a Markov chain, CS and a Markov pair provide two interpretations of the same mathematical model.

It is useful to note that, if we introduce the sequence of vectors (a_n), where $a_n = (a_n^i,\ i \in M_n)$, $a_n^i = \alpha_n^i m_n^i$ is equal to the amount of a solute in vessel i in a moment n and the substochastic matrices $\tilde{P}_n = [\tilde{P}_n^{ij}]$, with $\tilde{P}_n^{ij} = p_n^{ij}$ for $j \notin O_{n+1}$, $\tilde{p}_n^{ij} = 0$ for

$j \in O_{n+1}$, $n \in \mathbf{N}$, then, as in (5), we have

$$a_{n+1} = a_n \tilde{P}_n, \; n \in \mathbf{N} \quad . \tag{7}$$

The above relationship between CS and Markov pairs allows us to consider Theorem 1 as a theorem describing an asymptotic behaviour of posterior probabilities (6) for a given Markov pair (Z, D), $D = (D_n)$ in the case when $|D_n| \leq N < \infty$. Properties a) and b) of Theorem 1 take a form

a) $\quad \lim_n P(Z_n \in J_n^l) = \tilde{m}^l > 0, \; l = 1, \ldots, c \;,$ (8)

b) $\quad \lim_n P(Z_s \in D_s, \; 1 \leq s \leq n | Z_n = i) = \tilde{\alpha}^l, \; i \in \hat{J}_n^l \subseteq J_n^l \;,$ (9)

$\quad \lim_n P(Z_n \in (J_n^l \backslash \hat{J}_n^l)) = 0, \; l = 1, \ldots, c \;,$

From (9) it is easy to get

$$\tilde{\alpha}^l = \lim_n P(Z_s \in D_s, \; 1 \leq s \leq n | Z_n \in J_n^l) \quad . \tag{10}$$

4 We now present Theorem 2. We introduce some useful notions which play also an important role in the proof of the Theorem 3. Generalizing the previous definition of a trap for a Markov chain, let us say that a sequence of sets $J = (J_n)$ is a *trap* for a random sequence (r.s.) $X = (X_n)$ if as in (4)

$$\sum_{n=1}^{\infty} [P(X_n \in J_n, \; X_{n+1} \notin J_{n+1}) + P(X_n \notin J_n, \; X_{n+1} \in J_{n+1})] < \infty \quad . \tag{11}$$

If the r.s. X is fixed or there is no danger of confusion we shall omit the reference to X. If $J = (J_n)$ is a trap for $X = (X_n)$ let us call the limit $P(X_n \in J_n)$ (it always exists for any trap) as the volume of J for X. We say that a trap J is *indecomposable* if its volume is positive and J can not be represented as a sum of two traps with positive volumes, $J_n = J_n^1 + J_n^2$, $n \in \mathbf{N}$.

THEOREM 2 *Let (Z_n) be a (nonhomogeneous) Markov chain with values in M_n, $|M_n| \leq N < \infty$, $n \in N$. Then there exist a partition into indecomposable traps $\gamma = (J^1, \ldots, J^c)$ and a tuple of numbers $\tilde{m}^l$, $\tilde{m}_k^l(j)$, $j \in M_k$, $k \in \mathbf{N}$, $l = 1, \ldots, c$, such that*

a) $\lim_n P(Z_n \in J_n^l) = \tilde{m}^l > 0$, $\lim_n P(Z_n \in J_n^l | Z_k = j) = \tilde{m}_k^l(j)$,

b) $\lim_k \lim_n P(Z_k \in J_k^s, \; Z_n \in J_n^l) = \tilde{m}^l \delta^{sl}$,

c) $\lim_k \lim_n |P(Z_k = j, \; Z_n = i) \tilde{m}^l - P(Z_k = j) P(Z_n = i) \delta^{sl}| = 0$,

$j \equiv j_k \in J_k^s$, $i \equiv i_n \in J_n^l$, $s, l = 1, \ldots, c$. *Here δ^{sl} is the Kronecker symbol.*

REMARK 3 It is easy to prove that the existence of γ, specified with regards to traps with zero volume, and points a), b) and point c) for $s \neq l$ are valid for any r.s. with bounded number of values. A nontrivial fact, valid in general only for a Markov chain with a bounded number of values is point c) for $s = l$. In other words any Markov chain with bounded number of values has a mixing property inside every indecomposable trap. Theorem 1 is used only in the proof of this part of Theorem 2.

5 Let us dwell upon another interesting item related to Problem 1. We give first some basic definitions and notions regarding the so-called theory of majorization. Our starting point is the books of Marchall and Olkin [2] and of Dieter and Uhlman [3]. Let x, y be the vectors in R_+^N. The vector x is said to be majorized by the vector y ($y \succ x$) if for any continuous convex function g the inequality $\sum_i g(y_i) \geq \sum_i g(x_i)$ holds. The relation $\succ$ defines a partial pre-order and if $y \succ x$ and $x \succ y$ then x may be gotten from y by some permutation of coordinates. This relation also has some other interesting interpretations. In particular, $y \succ x$ if and only if there exists a doubly stochastic matrix P such that $x = yP$. According to an economic interpretation the coordinates of a vector $x = (x^1, \ldots, x^N)$ are the incomes of N economic agents and the transformation of y into x represents some "fair" redistribution of incomes. The background for geometric interpretation gives Birkhoff's theorem on the representation of a doubly stochastic matrix as a convex linear combination of permutation matrices. A function preserving the relation of majorization, i.e. such that $y \succ x$ iff $\varphi(y) \geq \varphi(x)$ is called Schur-convex or simply S-convex. As was mentioned above all the functions $\varphi(x) = \sum_i g(x_i)$, where g is convex are S-convex. If g is strictly convex then φ is strictly S-convex, i.e. $\varphi(y) > \varphi(x)$ if $y \succ x$ and $x \not\succ y$. The theory of majorization gives a unified approach to deriving a large number of different inequalities and results based on them in different fields of mathematics. The systematic consideration of these problems may be found in [2].

The generalization of the main notions of theory of majorization to abstract algebraic systems – W^*-algebras is described in a book [3]. Some sections of this book are devoted to physical interpretations of the majorization. If the vectors x, y are interpreted as the states of some physical systems and $y \succ x$ then x is called more mixed, more chaotic. This terminology is due to the fact that any sequence of vectors (x_n), where $x_{n+1} = x_n P_n$, P_n are doubly stochastic matrices, $n \in \mathbf{N}$, describes an irreversible process. The property of irreversibility fails to hold of course if P_n are only stochastic (not doubly stochastic matrices). But if a state of some physical system is described by a pair of vectors $z = (m, a)$, m, $a \in R_+^N$ and the transformations of a physical system ($z \succ\succ z'$) have a form $m_{n+1} = m_n P_n$, $a_{n+1} = a_n P_n$, P_n is an $N \times N$ stochastic matrix, $n \in \mathbf{N}$, then the property of irreversible holds. This fact follows immediately from the existence of functions preserving the relation $\succ\succ$ (see Section 1.9 of book [3]). An example of such a function is $\sum_i m^i g(a^i/m^i)$, where g is convex. Note that in our paper CS (without an ocean) is a sequence of states (m_n^i, a_n^i), $a_n^i = \alpha_n^i m_n^i$, $i \in M_n$, of the same type (see formulae (5) and (7)) and that the function $\sum_i m^i g(\cdot/\cdot)$ was used in [1] for $g(\lambda) = \lambda^2$ (see Remark 2).

Every sequence (x_n), $x_{n+1} = x_n P_n$, $x_n \in R_+^N$, where P_n are doubly stochastic, can be considered as a specific CS with $m_n^i = 1/N$, $\alpha_n^i = x_n^i / \sum_i x_1^i$, $i = 1, \ldots, N$, $n \in \mathbf{N}$. So Theorem 1 may be applied but in the case when $m_n^i =$ constant its proof is rather trivial (see Remark 1).

The idea of using the theory of majorization to the description of irreversible physical processes was elaborated in some papers following the pioneering work of Ruch [4]. But in these papers as in [2] and [3], two important problems were not touched: what is the limit state of an irreversible process and how is this limit achieved. Theorem 1 gives an answer to these questions in a more general situation than doubly stochastic transformations – that is in the case of CS. Moreover, in our opinion it is only a sketch of a general theorem about the asymptotic behaviour of an irreversible process. The heuristic formulation of this theorem is the following. The limit state exists and the system may be decomposed into subsystems such that there is mixing inside every subsystem, the limit state of the subsystem is "uniform" and the interaction between subsystems is in some sense "finite". The decomposition may depend on time. Of course this theorem would be valid under some conditions of the same type as in Theorem 1. We stress once more that the crucial point is the finiteness of the interaction.

6 Now we turn to Problem 2. The r.s. $X = (X_n)$ and $\tilde{X} = (\tilde{X}_n)$ with values in some discrete spaces (M_n) are called equivalent ($X \sim \tilde{X}$) if for all $A \subseteq M_n$, $B \subseteq M_{n+1}$, $n \in \mathbb{N}$

$$P(X_n \in A, X_{n+1} \in B) = P(\tilde{X}_n \in A, \tilde{X}_{n+1} \in B) \ . \tag{12}$$

It is obvious that every class of equivalent r.s. contains some nonhomogeneous Markov chain and conversely, every Markov chain defines some class of equivalent r.s.

Let $X = (X_n)$ be a r.s., $D = (D_n)$ be a sequence of sets. Denote by

$$i(X, D) = P(\liminf (X_n \in D_n)) = P(\bigcup_k \bigcap_{n \geq k} (X_n \in D_n))$$

the probability that the trajectories of X do not leave (D_n) after some (random) time, and by

$$s(X, D) = P(\limsup (X_n \in D_n)) = P(\bigcap_k \bigcup_{n \geq k} (X_n \in D_n))$$

the probability that X visits (D_n) infinitely often.

Problem 2 is to investigate $i(X, D)$, $(s(X, D))$ for some fixed class of equivalent r.s. The following theorem was formulated in [1] (the proof is not yet published).

THEOREM 3 *Let $Z = (Z_n)$ be some nonhomogeneous Markov chain with values in (M_n) and $D = (D_n)$ be some sequence of sets, $D_n \subseteq M_n$, $n \in \mathbb{N}$, such that $|D_n| \leq N < \infty$, $n \in \mathbb{N}$. Then for every r.s. $X \sim Z$ the following inequality holds*

$$i(Z, D) \leq i(X, D) \tag{13}$$

REMARK 4 Since $i(X, D) = 1 - s(X, M \backslash D)$, $M \backslash D = (M_n \backslash D_n)$ the inequality (13) is equivalent to the assertion that for any Markov chain Z and a sequence $G = (G_n)$ such that $|M_n \backslash G_n| \leq N < \infty, n \in \mathbb{N}$

$$s(Z, G) \geq s(X, G), X \sim Z \ . \tag{14}$$

REMARK 5 There is an example where $|D_n| \to \infty$ and the enequality (13) fails to hold but it is too complex to be presented here. The motivation for studying (13) was the following. The question of when Markov strategies ensure a payoff close to the value is one of the main questions arising in the theory of stochastic programming. Strange as it may seem, there is as yet no full answer to this question. In particular it is an open problem for a countable state space and a functional $E \limsup f(X_n^\pi)$, where (X_n^π) is a r.s. corresponding to a strategy π. In the fundamental work of Hill [6] a positive answer was given for a finite state space and a class of functionals $L(x_0, x_1 \cdots)$ called shift-and-permutation invariant. This class includes the classical finite fortune gambling problems of Dubins and Savage [5], $L = \limsup f(x_n)$, as well as a case $L = \liminf f(x_n)$ and their combinations, (in a recent paper of Hill and Pestien [7] the case of $L = \lim\inf$ was extended to a countable state space). Using the well-known fact that for any strategy π there corresponds a Markov strategy σ such that $X^\pi \sim X^\sigma$ and X^σ is a Markov chain, it is easy to get from Theorem 3 the result of Hill for a finite state space and $L = \lim\sup f(x_n)$. Note that the formulation of Problem 2 and the hypothesis about inequality (13) are due to E.A. Fainberg. Note also that the example showing that the inequality (13) fails to hold if $|D_n| \to \infty$, does not imply that theorem of Hill fails to hold for a case of countable state space. To prove Theorem 3 we prove also a lemma which may be called an analog of the Borel 0, 1 law for nonhomogeneous Markov chains. From this lemma and some other lemmas Theorem 4 below is an easy corollary from results of [1].

7 We first give some definitions. Let (a_n) and (b_n) be some nonrandom sequences. The sequence (a_n) intersects the sequence (b_n) in a moment k if $a_k \leq b_k$, $a_{k+1} > b_{k+1}$ or $a_k > b_k$, $a_{k+1} \leq b_{k+1}$. Let $X = (X_n)$ be a r.s., $d = (d_n)$ a nonrandom sequence. Denote by $V_T(X, d)$ the expected number of intersections of trajectories (X_n) with (d_n) on the time interval $(1, T)$. The nonrandom sequence $d = (d_n)$ is called a *barrier* for r.s. X if $V_\infty(X, d) < \infty$. In other words, d is a barrier for X if the jet (J_n), $J_n = (-\infty, d_n]$ is a trap for X. Problem 3 [1] is to describe the classes of random sequences for which barriers exist in some intervals (a, b).

Denote by $M^N(a, b)$ the class of r.s. $X = (X_n)$ taking no more than N values inside the interval (a, b) for all $n \in \mathbf{N}$, i.e. $X \in M^N(a, b)$ iff there exists a sequence of finite sets (G_n), such that $|G_n| \leq N < \infty$, $n \in \mathbf{N}$ and

$$P(X_n \notin G_n, X_n \in (a, b)) = 0, n \in \mathbf{N} .$$

(Outside the interval (a, b) the sets of values of X may be arbitrary). From the results of [1] (Theorem 3 and Lemma 4.3) we immediately get Theorem 4.

THEOREM 4 *Let (X_n) be a bounded sub(super)martingale in direct or reversed time and let $X \in M^N(a, b)$ for some a, b, N. Then there exists a barrier (d_n) for X, $d_n \in (a, b)$, $n \in \mathbf{N}$.*

The idea of using Theorem 4 is the following. If (M, r, O, α) is some CS and (Z_n) is the corresponding Markov chain, then a r.s. (Y_n) defined by $Y_n = \alpha_n^{Z_n} \equiv \sum_{i \in M_n} \alpha_n^i I_i(Z_n)$ is a submartingale (a martingale in the case of a CS without an ocean) in reversed time with respect to σ-algebras $\sigma(Z_n, Z_{n+1}, \cdots)$, $n \in \mathbf{N}$. If $|M_n| \leq N < \infty$, $n \in \mathbf{N}$ then (Y_n) satisfies the conditions of Theorem 4 for all (a, b).

Note that Theorem 4 does not follow from the well-known theorem of Doob which states that the expected number of intersections of every fixed interval by trajectories of bounded submartingale is finite. The example mentioned above shows also that Theorem 4 fails to hold if the condition $|G_n| \leq N < \infty$ is replaced by $|G_n| \to \infty$.

8 Let us mention some unsolved problems connected with these questions besides the mentioned theorem about general irreversible process. (1) To formulate an analog of Theorem 1 for the case $|M_n| = \infty$. Such an analog would give a possibility to consider the case of continuous space and time. (2) To estimate a value of "work" connected with given CS, i.e. $\sum_n \sum_{ij} r_n^{ij} |\alpha_{n+1}^j - \alpha_n^i|$ or in other words a value of $E \sum_n |Y_{n+1} - Y_n|$ for a martingale associated with given Markov pair (Z, D). (3) To generalize the inequality (13).

REFERENCES

[1] Sonin, I.M.: A theorem on separation of jets and some properties of random sequences. Stochastics 21, 231–250 (1987).

[2] Marshal, A.W. and I. Olkin: Inequalities: Theory of Majorization and its Application. Academic Press, New York, 1979.

[3] Alberti, P.M., and A. Uhlman: Stochasticity and Partial Ordering. Ser. Math. and its Appl., v 9, Berlin, D. Reidel Publishing Company, 1982.

[4] Ruch, E.: The principle of increasing mixing character. Theor. Chim. Acta 38, 167–175 (1975).

[5] Dubins, L.E. and L.I. Savage: Inequalities for Stochastic Processes (How to Gamble if You Must), New York, Dover Publications, 1976.

[6] Hill, T.: On the existence of good Markov strategies, Trans. Am. Math. Soc. 247, 157–176 (1979).

[7] Hill, T. and V. Pestien: The existence of good Markov strategies for decision processes with general payoffs. Stoch. Pr. Appl., 23, 412 (1986).

Lecture Notes in Control and Information Sciences

Edited by M. Thoma and A. Wyner

Vol. 62: Analysis and Optimization of Systems
Proceedings of the Sixth International Conference on Analysis and Optimization of Systems
Nice, June 19–22, 1984
Edited by A. Bensoussan, J. L. Lions
XIX, 591 pages. 1984.

Vol. 63: Analysis and Optimization of Systems
Proceedings of the Sixth International Conference on Analysis and Optimization of Systems
Nice, June 19–22, 1984
Edited by A. Bensoussan, J. L. Lions
XIX, 700 pages. 1984.

Vol. 64: Arunabha Bagchi
Stackelberg Differential Games in Economic Models
VIII, 203 pages, 1984

Vol. 65: Yaakov Yavin
Numerical Studies in Nonlinear Filtering
VIII, 273 pages, 1985.

Vol. 66: Systems and Optimization
Proceedings of the Twente Workshop
Enschede, The Netherlands, April 16–18, 1984
Edited by A. Bagchi, H. Th. Jongen
X, 206 pages, 1985.

Vol. 67: Real Time Control of Large Scale Systems
Proceedings of the First European Workshop
University of Patras, Greece, Juli 9–12, 1984
Edited by G. Schmidt, M. Singh, A. Titli, S. Tzafestas
XI, 650 pages, 1985.

Vol. 68: T. Kaczorek
Two-Dimensional Linear Systems
IX, 397 pages, 1985.

Vol. 69: Stochastic Differential Systems – Filtering and Control
Proceedings of the IFIP-WG 7/1 Working Conference
Marseille-Luminy, France, March 12-17, 1984
Edited by M. Metivier, E. Pardoux
X, 310 pages, 1985.

Vol. 70: Uncertainty and Control
Proceedings of a DFVLR International Colloquium
Bonn, Germany, March, 1985
Edited by J. Ackermann
IV, 236 pages, 1985.

Vol. 71: N. Baba
New Topics in Learning Automata Theory and Applications
VII, 231 pages, 1985.

Vol. 72: A. Isidori
Nonlinear Control Systems: An Introduction
VI, 297 pages, 1985.

Vol. 73: J. Zarzycki
Nonlinear Prediction Ladder-Filters for Higher-Order Stochastic Sequences
V, 132 pages, 1985.

Vol. 74: K. Ichikawa
Control System Design based on Exact Model Matching Techniques
VII, 129 pages, 1985.

Vol. 75: Distributed Parameter Systems
Proceedings of the 2nd International Conference, Vorau, Austria 1984
Edited by F. Kappel, K. Kunisch, W. Schappacher
VIII, 460 pages, 1985.

Vol. 76: Stochastic Programming
Edited by F. Archetti, G. Di Pillo, M. Lucertini
V, 285 pages, 1986.

Vol. 77: Detection of Abrupt Changes in Signals and Dynamical Systems
Edited by M. Basseville, A. Benveniste
X, 373 pages, 1986.

Vol. 78: Stochastic Differential Systems
Proceedings of the 3rd Bad Honnef Conference, June 3–7, 1985
Edited by N. Christopeit, K. Helmes, M. Kohlmann
V, 372 pages, 1986.

Vol. 79: Signal Processing for Control
Edited by K. Godfrey, P. Jones
XVIII, 413 pages, 1986.

Vol. 80: Artificial Intelligence and Man-Machine Systems
Edited by H. Winter
IV, 211 pages, 1986.

Lecture Notes in Control and Information Sciences

Edited by M. Thoma and A. Wyner

Lecture Notes in Control and Information Sciences

Edited by M. Thoma and A. Wyner

Vol. 97: I. Lasiecka/R. Triggiani (Eds.)
Control Problems for Systems
Described by Partial Differential Equations
and Applications
Proceedings of the IFIP-WG 7.2
Working Conference
Gainesville, Florida, February 3-6, 1986
VIII, 400 pages, 1987.

Vol. 98: A. Aloneftis
Stochastic Adaptive Control
Results and Simulation
XII, 120 pages, 1987.

Vol. 99: S. P. Bhattacharyya
Robust Stabilization Against
Structured Perturbations
IX, 172 pages, 1987.

Vol. 100: J. P. Zolésio (Editor)
Boundary Control and Boundary Variations
Proceedings of the IFIP WG 7.2 Conference
Nice, France, June 10-13, 1987
IV, 398 pages, 1988.

Vol. 101: P. E. Crouch,
A. J. van der Schaft
Variational and Hamiltonian
Control Systems
IV, 121 pages, 1987.

Vol. 102: F. Kappel, K. Kunisch,
W. Schappacher (Eds.)
Distributed Parameter Systems
Proceedings of the 3rd International Conference
Vorau, Styria, July 6–12, 1986
VII, 343 pages, 1987.

Vol. 103: P. Varaiya, A. B. Kurzhanski (Eds.)
Discrete Event Systems:
Models and Applications
IIASA Conference
Sopron, Hungary, August 3-7, 1987
IX, 282 pages, 1988.